DEHYDRATION OF FRUITS AND VEGETABLES

DEHYDRATION OF FRUITS AND VEGETABLES

M. Torrey

NOYES DATA CORPORATION

Park Ridge, New Jersey   London, England

1974

664.8402'72

TOR

FOREWORD

The detailed, descriptive information in this book is based on U.S. patents since the early 1960s relating to the dehydration of fruits and vegetables. Where it was necessary to round out the complete technological picture, some earlier but very relevant patents were included.

This book serves a double purpose in that it supplies detailed technical information and can be used as a guide to the U.S. patent literature in this field. By indicating all the information that is significant, and eliminating legal jargon and juristic phraseology, this book presents an advanced, technically oriented review of modern dehydration techniques applicable to fruits and vegetables.

The U.S. patent literature is the largest and most comprehensive collection of technical information in the world. There is more practical, commercial, timely process information assembled here than is available from any other source. The technical information obtained from a patent is extremely reliable and comprehensive; sufficient information must be included to avoid rejection for "insufficient disclosure."

The patent literature covers a substantial amount of information not available in the journal literature. The patent literature is a prime source of basic commercially useful information. This information is overlooked by those who rely primarily on the periodical journal literature. It is realized that there is a lag between a patent application on a new process development and the granting of a patent, but it is felt that this may roughly parallel or even anticipate the lag in putting that development into commercial practice.

Many of these patents are being utilized commercially. Whether used or not, they offer opportunities for technological transfer. Also, a major purpose of this book is to describe the number of technical possibilities available, which may open up profitable areas of research and development. The information contained in this book will allow you to establish a sound background before launching into research in this field.

Advanced composition and production methods developed by Noyes Data are employed to bring our new durably bound books to you in a minimum of time. Special techniques are used to close the gap between "manuscript" and "completed book." Industrial technology is progressing so rapidly that time-honored, conventional typesetting, binding and shipping methods are no longer suitable. We have bypassed the delays in the conventional book publishing cycle and provide the user with an effective and convenient means of reviewing up-to-date information in depth.

The Table of Contents is organized in such a way as to serve as a subject index. Other indexes by company, inventor and patent number help in providing easy access to the information contained in this book.

15 Reasons Why the U.S. Patent Office Literature Is Important to You —

1. The U.S. patent literature is the largest and most comprehensive collection of technical information in the world. There is more practical commercial process information assembled here than is available from any other source.

2. The technical information obtained from the patent literature is extremely comprehensive; sufficient information must be included to avoid rejection for "insufficient disclosure."

3. The patent literature is a prime source of basic commercially utilizable information. This information is overlooked by those who rely primarily on the periodical journal literature.

4. An important feature of the patent literature is that it can serve to avoid duplication of research and development.

5. Patents, unlike periodical literature, are bound by definition to contain new information, data and ideas.

6. It can serve as a source of new ideas in a different but related field, and may be outside the patent protection offered the original invention.

7. Since claims are narrowly defined, much valuable information is included that may be outside the legal protection afforded by the claims.

8. Patents discuss the difficulties associated with previous research, development or production techniques, and offer a specific method of overcoming problems. This gives clues to current process information that has not been published in periodicals or books.

9. Can aid in process design by providing a selection of alternate techniques. A powerful research and engineering tool.

10. Obtain licenses — many U.S. chemical patents have not been developed commercially.

11. Patents provide an excellent starting point for the next investigator.

12. Frequently, innovations derived from research are first disclosed in the patent literature, prior to coverage in the periodical literature.

13. Patents offer a most valuable method of keeping abreast of latest technologies, serving an individual's own "current awareness" program.

14. Copies of U.S. patents are easily obtained from the U.S. Patent Office at 50¢ a copy.

15. It is a creative source of ideas for those with imagination.

CONTENTS AND SUBJECT INDEX

INTRODUCTION

The dehydration of fruits and vegetables is most important in the production of the convenience foods which have become such a large part of American life. For example, about a million pounds of dehydrated onions are used each year in the United States for the manufacture of catsup and chili sauce. Dehydrated onions are also used in such varied products as soups, salad dressings, dog food, sausage and liverwurst.

Although the individual American consumer is using more and more dehydrated fruits and vegetables each year, the bulk of the dehydrated foods are used in the manufacture of other food products. The next largest amount is accounted for by institutional and military use.

The advantages of dehydrated products are fairly obvious: (1) their flavor stability at room temperature over long periods of time; (2) their protection from enzymatic and oxidative spoilage; (3) their light weight for shipping; (4) the fact that they do not need refrigeration; and (5) their availability at all times of the year.

An entire chapter of this book is devoted to the dehydration of potatoes, a business which grows larger each year. The world average annual per capita production of potatoes is about 200 pounds, and the use of dehydrated potatoes has grown steadily since the end of World War II. The most widely used dehydrated potato product is instant mashed potatoes, but dehydrated sliced potatoes are used by institutions for preparing hash browns, salads, etc. and dehydrated diced potatoes are used by manufacturers of canned beef stew and corned beef hash. Potato flour is used by the baking industry for such products as bread, rolls and doughnuts.

Other dried vegetables used in large quantities are onions, garlic, bell peppers, mushrooms, parsley, sweet potatoes, beans and peas. Of the 30,000 tons of garlic produced in the United States in 1971, for instance, 60% were dehydrated.

Individuals are using more and more dehydrated soup each year, and with the advent of envelopes of soup mix which make a single portion, the use for dehydrated vegetables has soared. In Europe, dehydrated soups have been available for a long time in a far greater variety.

A chapter has been included on dehydrated and concentrated fruit and vegetable juices. Completely dehydrated citrus juices have never become as popular here as the frozen, concentrated variety which is, of course, also partially dehydrated.

The production of dried fruits is an important business, particularly in California, where over $100 million worth of dried fruits is produced annually. In 1971, about 485,000 tons of dried fruits were produced in the United States, about one third of which, principally prunes, apples, figs and golden raisins, were processed in dehydrators, the rest being sun dried.

This book describes processes taken from over 200 patents, from 1960 to the present. The first chapter deals with general dehydration techniques, suitable for a wide variety of fruits and vegetables. The reader should keep in mind that a number of the processes which have been indexed under a specific product, in order to bring organization to the book, are usable for the dehydration of other individual fruits and vegetables as well.

GENERAL TECHNIQUES

HOT AIR DEHYDRATION

On Superposed Conveyors

In the most common commercial apparatus for the dehydration of fruits and vegetables, a batch of the material to be dried, usually in the form of dice or slices, is subjected to a stream of warm air, which is directed over or through a shallow bed of the material, usually supported on a woven wire tray. Due to the fact that air follows the path of least resistance it is usual for proportions of the pieces to be rather overdried and others underdried and, in consequence, there are frequently in the final product a number of pieces which can only be reconstituted with difficulty and decrease the quality.

R.A.S. Templeton; U.S. Patent 3,409,999; November 12, 1968 has developed an improved form of apparatus which enables a continuously moving bed of diced or sliced vegetable or fruit to be dried by means of a stream of warm air and which also provides means for disturbing the bed of material periodically to avoid difficulties due to irregular drying of the bed by changing the surfaces under treatment without mechanical impacts to avoid injury and by the arrangement of gravity drops and changes.

In the dehydration of fruits and vegetables the largest discrete pieces which are dried by the application of warmed air streams are normally about ⅜ inch cubes. Pieces above that size have a volume/surface area ratio too large to permit them to be dried economically. It follows that a successful dehydration apparatus must be capable of drying pieces of that size and to achieve good results means must be provided to ensure that the bed of material is maintained at substantially uniform thickness.

With this object in view there is provided here an apparatus for dehydration of fruit and vegetables, which comprises one or more endless conveyors arranged within a casing, through which a stream of warm air is passed, each conveyor being formed from a series of perforated plate-like members, pivoted about axes which extend perpendicularly to the direction of movement of the conveyor.

The plate-like members are adapted to be maintained in the same plane so that the upper and lower run of the conveyor acts as a tray for a bed of material and to pivot individually at a dropping point at the end of each conveyor run to drop the material supported thereon onto a conveyor run below. The width of each plate-like member in the direction of movement of the conveyor is related to its distance above the next conveyor run in such a manner that the material falling onto the next conveyor run forms a bed of substantially uniform thickness. It is found that the desirable width of the plate-like members to achieve this result is about 3 inches, when the vertical interval between adjacent conveyor runs is 1 to 2 feet, but in practice a width of the transversely pivoted plate-like members, which form the conveyor, of about 2 inches does produce a more even bed following the drop of products induced by gravity in the direction in which the conveyor travels.

The plate-like members are pivoted along their front edges on pivot members, which are connected between a pair of drive chains and ride on rails positioned close to the drive chains. The rails terminate close to the outlet end of each conveyor run and at such point the hinged plates tilt about their pivots to permit the material, supported on them, to drop by gravity onto the next conveyor run. It is found that with vegetable dice of the size quoted above, the width of 2 inches for the plate-like members is about the practical maximum to maintain a substantially uniform bed of material on the next conveyor run.

Since drying is effected by the passage of air through the run of the conveyor, each plate is chosen to have the minimum solid area consistent with having perforations small enough to prevent much material falling through onto the next run of the conveyor, and satisfactory strength. In practice it is found that stainless steel plates, having square apertures of $5/32$ to $3/16$ inch size at $5/16$ inch centers, are the most satisfactory for diced vegetables.

As will be readily understood the dehydration apparatus includes a plurality of conveyors arranged one above the other, the fresh material being fed in at the top and descending by gravity from each conveyor run to the conveyor run below it and being removed at the bottom of the apparatus. The conveyors are preferably housed within a casing and warmed air may be injected into it at one or more levels, the whole of the spent air conveniently being withdrawn at the top.

While the freshly cut material still has a damp surface at the beginning of the dehydration operation, it can be dried at a relatively rapid rate and therefore on the top of the conveyor a relatively shallow bed of material is subjected to a large flow of air at rather high temperature, say up to 300°F. The speed of the lower conveyor may be reduced as compared with the top conveyor because the surface dried material thereon can now be formed into a thicker bed by driving the conveyor more slowly.

At this stage the material can only tolerate a lower heat value than in the initial stage, 200°F. A third conveyor can extend the same advantages of thicker bed by retarded speed at again lower temperature, such as 150°F., and on the final run may be used to cool the product by using atmospheric air or refrigerated air. The patent includes drawings of the apparatus.

On Stationary Bed

A dehydration process operated without heat and without vacuum and which does not necessitate a large investment in apparatus is described by *W. Groth and P. Hussmann; U.S. Patent 3,490,355; January 20, 1970.* In the process, the starting product for the preparation of the dry powder is filled into a suitable reaction vessel. The starting product may be in the form of liquid or pasty materials, solutions, suspensions of foodstuffs or other products, especially a fruit juice or fruit juice paste, a vegetable slurry, mashed potatoes, blood, yeast, a plant extract, pectin, gelatin, glue, etc.

The reaction vessel has a lower net or grid shaped bottom and a top cover provided with gas outlet openings as well as rigid side walls. It has conveniently cylindrical shape. At a distance of conveniently 200 mm. from the lower tray, there is provided as an intermediate tray a microporous supporting substance which, for example, consists of a plastic material. The material to be dehydrated is placed on this support as a bed having a depth of from 1 to 1,000 mm. Then highly dried nitrogen or CO_2 in case of all products which are sensitive to oxidation or highly dried air if permitted by the product due to its resistance to oxidation is introduced through the bottom of the vessel and passed in upward direction through the porous support. By the forcing of this gas stream through the micro-porous support, the stream is divided into many small individual particles so that it represents a disperse phase. This divided gas is forced onto the bottom surface of the material being dried and arranged on the support and the pressure is increased until the finely dispersed gas is forced through the entire layer of the material to be dried and carried by the support. The degree of division and the velocity at which the gas is passed through the liquid layer are of importance to the efficiency of the process. In general, the velocity ranges between 0.2 and 2 meters/second when drying liquids. For purées, the velocities are somewhat lower.

The process may be operated batchwise or continuously. When operated continuously, the process may be carried out in a single reaction vessel, or a plurality of reaction vessels, especially 4 to 8 reaction vessels may be connected in parallel. Depending upon the product to be processed, they may be provided with a stirrer or not. They are arranged above the microporous support, and the material to be dehydrated is placed into the vessels to a layer depth of from 1 to 1,000 mm. The gaseous drying agent is then passed through the liquid layers in each of the vessels under a specific superatmospheric pressure of, for example, 50 to 1,200 mm. water column.

By means of the microporous supporting material, the gaseous medium is brought into an ultramicroporous dispersion and the gas is then present in a finely divided state similar to that of an emulsion. When the gas is forced through the layer of the material being dehydrated, the divided droplets are present in an order of magnitude which, for example, may be as low as 1 to 200 millimicrons. This permits permanent, very dynamic maintenance of a steady state throughout the process and leads to solidification of the product in pumiceous porous form as the concentration of the material being dehydrated increases, i.e., as the moisture content decreases.

In another advantageous mode of operation, the gas stream may be returned and repeatedly passed through the layer of the material to be dried. This merely requires sectioning of the gas supply perhaps in a manner which comprises passing the fresh, completely dried inert gas, at a first partial section of the lower surface of the material to be dried and at a given pressure, through the porous support and through the layer of the material to be dried; passing the gas emerging from the upper surface and loaded with moisture from the material being dried to the lower side of the second partial section of the material to be dried and forcing it again through the microporous support and through the layer of material to be dried.

In doing so, saturation with moisture increases. The inert gas is withdrawn from the upper surface of the second partial section and passed to the underside of the third partial section of the bed of material being dried where it is again forced through the proous support and in upward direction through the layer of material being dried, at the upper surface of which it is either vented or led to a fourth partial section as described above. Depending upon the material being treated and its quantity, units may be provided where the inert gas stream is recycled as many times as desired, it being possible as required to provide drying units for intermediate drying of the inert gas.

The reaction vessels are preferably made of chrome nickel steel or plastic lined iron. They may also be constructed of self-supporting plastic materials. In a unit equipped for continuous operation, from 4 to 8 reaction vessels may be advantageously operated in parallel. It is also possible, for example, to use an endless belt conveyor which may consist of the porous supporting material, slidingly moved on a support and subdivided into individual cells which are sealed against one another.

The dried inert gas is introduced into the first cell and, when emerging from the surface of the layer, returned to the underside of the second cell unit and so on to the last cell unit until it is completely saturated. It is possible with belt conveyors of this kind to operate with very thin layers of the material being dehydrated of a thickness from 1 to 100 mm. At the end of the belt, the dried material is desirably taken off and transferred into a larger reaction vessel which is filled therewith to a level of 200 to 1,000 mm. Here, super-drying with fresh and highly dehydrated inert gas or air may be effected whereupon the aroma and flavor components released from the special adsorption unit by selective desorption are added to the dried material.

Example: Preparation of Banana Purée – In this experiment, 672 kg./hr. of banana pulp containing 25% of dry substance are processed to form a dry powder containing 3% of residual moisture. The banana pulp is introduced into reaction vessels at a layer depth on the porous supporting tray of 330 mm. Flavor and aroma components had previously been removed from the banana pulp in a stripper column. The aroma and flavor components were accumulated and enriched in the adsorption bed.

The process was carried out with 60,000 cu. m./hr. of dehydrated nitrogen of 24°C. which was passed into the reaction vessels and then through the layers of the banana pulp to be dried. Dehydration of the moisture enriched nitrogen stream was effected in the central adsorption unit. The dehydration process was effected for 40 hours.

There was obtained per hour 172 kg. of banana powder which was transferred from the reaction vessels into the collecting vessel where the banana aroma and flavor components selectively desorbed from the adsorption bed was integrated in the dry banana powder. Thereafter, the powder was ground to the particle size desired and passed to the packaging unit. Upon reconstitution with water, banana pulp is obtained which is undistinguishable from the starting product. Dissolution takes place instantaneously in cold water.

Centrifugal Force Applied to Material Bed

In the dehydration of foods in particulate form, for example, potato dice, carrot dice, potato granules, etc., a known technique involves forming a bed of the particles on a screen or perforated metal plate and blowing hot air up through the bed. In conducting dehydration in this manner the rate of air flow through the bed may be held at a level such that the bed is static. More preferably, the air flow is high enough to cause the individual particles to move about or circulate within the bed.

Such movement of the particles generally provides a better rate of moisture evaporation and more uniform dehydration of individual particles. However, there is a limiting factor in that if the air velocity is too high the particles are blown out of the bed before they are properly dried. As a result, one cannot realize the full benefit to be gained from high velocity air flow.

An apparatus and method by which this problem may be solved is described by *D.F. Farkas and M.E. Lazar; U.S. Patent 3,500,552; March 17, 1970; assigned to the U.S. Secretary of Agriculture.* This apparatus makes it possible to hold the particles within the bed even though very high rates of air flow are applied. As a net result the process permits one to take full advantage of the benefits deriving from high air velocities by counterbalancing the levitating effect of the air with centrifugal force.

More particularly the procedure is as follows. A bed of particles of food is subjected to rotation, and, concomitantly, drying air is forced inwardly (i.e., toward the axis of rotation), through the bed. The rotation creates forces tending to move the particles outwardly (i.e., away from the axis of rotation), whereby it opposes the tendency of the air to move particles inwardly.

Accordingly, by suitable adjustment of the speed of rotation, one is enabled to employ air velocities higher than those which can be employed in conventional drying systems. As a result, many advantages are obtained, for example, faster rate of evaporation, and more uniform dehydration, not only of individual particles but also more uniform dehydration of portions of individual particles. Regarding rate of evaporation, a rate 7 times that attained with conventional systems can be achieved permitting the use of lower than conventional air temperatures and preserving vital attributes such as color, flavor and odor to a greater degree.

The apparatus, Figures 1.1a and 1.1b includes plenum chamber **1** communicating with hot air inlet **2**. In operation, air is forced by compressor **3** through heater **4** and into the plenum whereby this chamber is kept full of hot air under pressure higher than atmospheric. Mounted within the plenum is a rotatable centrifuge basket **5** which serves to support a bed of material under dehydration. Basket **5** includes a base member **6**, a throat **7** and a circular wall **8**, the last being of screening or perforated sheet metal. The openings in the wall are so selected that the particles of material are retained within the basket so that hot air from the plenum can pass through.

Basket **5** is journalled in bearings **9** and **10**, these including conventional gas seals to prevent escape of the hot air under pressure contained in the plenum. For rotation, the basket is keyed to shaft **11** which is driven by a variable speed motor (not illustrated), or the like.

For feeding material to be dried into the system there is provided hopper **12**, communicating via tube **13** with feed pipe **14**. Compressed air introduced from inlet **15** impels the material into feed pipe **14** and into the interior of the basket. By suitable adjustment of valve **16**, the material may be introduced into the system at a desired rate. The feed pipe extends into the basket with a slight twist so that material is not fed directly downward, but about 35° to 45° counter to the direction of rotation. This disposition of the feed pipe is preferred so that the entering material quickly forms a part of the bed.

Dried material discharged from the basket passes to chamber **17**, provided with an outlet duct **18**. For separation of the dried material from the accompanying flow of air, the stream exiting from duct **18** is conducted to a conventional cyclone separator, represented by block **19**. A sight glass **20** is provided in the top of the chamber. This permits the observer to see the bed of material within the basket so that he can make proper adjustments of speed of rotation, rate of hot air flow into the plenum etc., as may be required.

Generally, it is preferred to conduct dehydrations on a continuous basis (as opposed to batch operation) and the apparatus of the process is particularly useful in such applications. In this type of operation, material to be dried is fed at a predetermined rate through the feed pipe into the basket while the latter is rotated and hot air forced into the plenum. The speed of rotation and rate of air introduction are so correlated that the material forms a bed about the inside of the perforated wall and the particles in the bed are fluidized or at least tumble about so that there is a circulation of the particles within the bed. Also, the aforesaid variables are so correlated that as individual particles become dry and therefore become less dense, they are entrained in the stream of air leaving the bed and exiting into the chamber.

Reference is now made to Figure 1.1c which illustrates the bed, the flow of material, etc. during continuous dehydration. Reference numeral **22** designates the bed of particles, held in place about the inner face of wall **8** by the centrifugal force generated by rotation of the basket. The centrifugal force is represented by arrow **23**. Hot air from the plenum is concomitantly forced through the bed counter to the direction of the centrifugal force.

This flow of hot air, represented by arrow **24**, counteracts the compacting effect of the centrifugal force and causes individual particles in the bed to move about randomly or circulate within the bed, as indicated by arrows **25**. At the same time, there is a movement of particles from right to left caused by the continuous addition of fresh material adjacent to base plate **6**. Thus, in all, the individual particles not only move about in random circulatory paths within the bed, they also move toward throat **7**. As the particles approach the throat end of bed **22** they are in a dehydrated condition and have a low density so that they are entrained by the exiting air flow and pass to discharge chamber **17**.

The process is further demonstrated by the following illustrative example. The dryer used in these runs was as described above, and wherein basket **5** had the following dimensions: diameter 6 inches, depth 4 inches (excluding the throat), thus providing a drying surface (wall **8**) of 75 square inches.

FIGURE 1.1: CENTRIFUGAL FORCE APPLIED TO MATERIAL BED

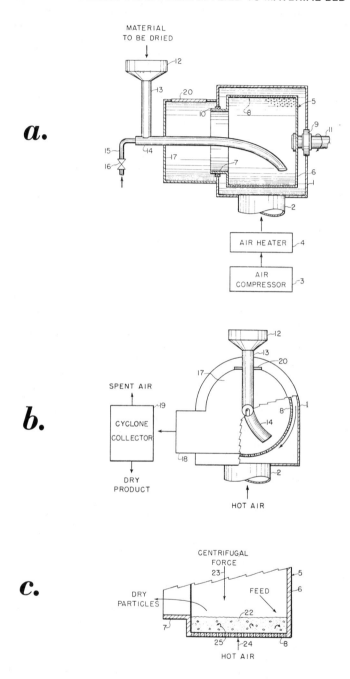

Source: D.F. Farkas and M.E. Lazar; U.S. Patent 3,500,552; March 17, 1970

Example: Continuous Dehydration of Diced Carrots — The starting material was ³/₈ inch diced carrots, having a moisture content of 90%. It was fed into the system at the rate of 13.2 lbs./hr. Basket **5** was rotated at 200 rpm and air at 205°F. was supplied into plenum chamber **1** at a rate of 216 cfm (measured at standard conditions). The pressure in plenum chamber **1** was 5 inches of H_2O above atmospheric. The product, having a moisture content of 79.5% was produced at the rate of 6.6 lbs./hr. (The piece residence time in the dryer was about 5 min.)

The product in this partially dehydrated condition, representing a weight loss of 50% by evaporation of water, is eminently suitable for preparing dehydrofrozen carrots. It was observed that the diced form of the starting material was retained in the product, and taste tests indicated that there was no damage to color or flavor.

It was calculated that the drying efficiency was 12 lbs. of water evaporated per hour per square foot of drying surface. This is 2 to 3 times the efficiency obtained in conventional belt dryers. Moreover, the residence time in the dryer of the process was only ¼ to ⅕ that required for a conventional belt dryer, indicating a fourfold to fivefold increase in evaporative capacity.

DRUM DRYING

Double Drum Drying

A dehydrating mechanism involving the principle of rotary partial dehydration is described by *F.G. Lamb; U.S. Patent 2,850,809; September 9, 1958.* The mechanism is characterized by the use of a double drum unit having concentric, double, perforate, peripheral walls, the product being contained between the walls of the two drums and, in the process of drying, being subjected to comparatively dry air heated to a predetermined, elevated temperature. The involved method tends to utilize, in a most efficient manner, all of the available drying medium largely because of the accurate channeling and control of such medium so as to effectively contact it with all of the food particles charged to the unit.

In this method only 35 to 60% of the moisture (depending upon the type of product being handled) is removed. The purpose of partial dehydrating is to produce a product that can be fully reconstituted by the addition of water, accomplishing hydration of the product to the extent that the moisture content thereof is increased to its full and original value. In complete dehydration of the product, reconstitution thereof is not possible. By a "reconstituted product" is meant one which, when hydrated, subsequent to partial dehydration, regains its original structure, taste, and appearance. The mechanism for the partial dehydration is fully illustrated and described in the patent.

Improvements in the method of application of heated gas to the double drum dehydrator and its utility are given by *F.G. Lamb; U.S. Patent 3,137,546; June 16, 1964; assigned to Lamb-Weston, Inc..* Primarily, it has been found that the material within the double walled structure of the rotary kiln can be more uniformly dried to the desired degree if first, the dehydrating medium is admitted into the interior of the inner perforate drum for passage through the product contained between the concentric walls.

Secondly, it has been found that far superior performance is additionally achieved, insofar as uniformity of drying, if a supplemental air supply be utilized which is admitted into that space between the two perforate drums where no product, in bulk, is permitted to remain because of the tumbling effect there obtained through rotational movement.

This latter concept is a fundamental one, in that such supplemental or classifying air supply, during the dehydration procedure, causes the loose particles such as peas, diced carrots, etc., to become air-borne. These will ordinarily be the lightest particles of the material undergoing treatment, in other words, the particles from which the most moisture has been removed. Being lighter, the classifying jet of air forces them, in air-borne fashion, to the rear or discharge end of the unit. Conversely, the heavier and less dehydrated particles of the material remain more nearly adjacent the inner drum, in closer and more prolonged contact with the hot air supply, and hence are subjected to more rapid and thorough drying, until they, too, become sufficiently dehydrated for discharge to the rear of the unit.

Hence, although the basic principle of the double drum dehydrator unit is utilized, the considerable improvement which is involved here over the previous patent, resides in the fundamental concept of admitting the main air supply to the interior of the inner drum and supplementing this main air supply, likewise through the interior of the inner drum, with an additional, classifying air supply that is jetted into that portion between the drums where, during rotation thereof, those particles reaching the desired amount of dehydration become largely air-borne or fluidized and consequently directed to the discharge end of the unit.

The process is illustrated with respect to a specific product, fresh peas, and it has been found that the pulp temperature of the peas, with a feed-in temperature of dehydrating medium of about 175°F., can be maintained at 80°F. with this system, a sufficiently low temperature to preclude pulp damage but which will effectuate continuous and thorough dehydration to the point desired without deleteriously affecting the pulp fibers of the vegetable. In the patent are complete drawings and descriptions of the process and apparatus used.

Stretching the Puree Film

In carrying out such a drum drying operation a number of difficulties are encountered which tend to impair the rehydration characteristics of the dried product. The partially dehydrated film of product is plastic at the point of removal from the drum by such means as a doctor blade tangentially engaging the smooth surface of the drum. Because of this there is a buildup of product on the doctor blades such that as the product is pushed away from the drum it assumes a wavy irregular surface of somewhat crepe-like appearance due to the deceleration of the film as it leaves the dryer. Buildup of this thickness of product is adverse to rapid rehydration of the doctored product.

G.J. Lorant, M.L. Rollins and D.C. Guterman; U.S. Patent 3,009,815; November 21, 1961; assigned to General Foods Corporation have developed a process for drying a fruit puree or similar product having a relatively high proportion of sugar and/or pectinous substances which tend to form continuous stretchable films rather than a powder upon removal in a concentrated form from a drying surface.

The process comprises applying a puree having finely divided food solids at a level usually in the order of 9 to 30% (but sometimes higher) to a smooth surface whereon the puree is heated, concentrated and converted into a continuous film having a syrupy viscous elastic condition while in the heated state; in this condition the finely divided food solids are suspended in a molten liquefied sugar and/or pectin-like sacchariferous molten carrier which supplies to the film a tensile strength varying from product to product, but in any event sufficient to permit the product to retain a continuous cohesive nature after it is stripped from the heating surface and while it is stretched.

Stripping is preferably facilitated by such means as a doctor blade. Usually, the moisture content of the stripped film will be less than 12% and higher than 2% by weight of the film. The film is continuously stripped in its molten condition.

Upon stripping, the film is subjected to tension substantially in excess of that required to prevent agglomeration at the doctor blade and below that tension at which the film will fail completely which failure would become evidenced by a continuous breaking or fracture resulting in interruption of the continuity of the film. The tension should be sufficient to stretch and thereby elongate the film whereby food solids are thinly and discretely dispersed in the still molten sacchariferous carrier and fibrous and other insoluble food solids contained in the film appear randomly arranged and unoriented to the unaided eye. Films of peach and pear purees, and to a lesser degree apple and tomato purees, will have a greater tensile strength than pudding films due to the higher proportion of pectinous materials in the fruits. Consequently, such fruit purees will generally call for a higher amount of applied tension than puddings in this connection, the temperature of the stripped film will contribute to its stretchability and although a rapid cooling of the film being stretched is desired, it should be sufficiently molten, at least in the initial stages of elongation, to create the stretched condition, however, the product temperature generally will not exceed 212°F. while it is being stretched.

Usually, this stretched condition will be evidenced by a random distribution of discontinuous openings, surface failures and void spaces throughout the stretched film and will also be evidenced in many natural fruit purees by minute blisters or puffs barely visible to the unaided eye, giving rise in many cases to a plurality of pustules some of which are broken and some of which are substantially intact.

Under a high powered microscope the stretched cooled product can be seen to comprise a translucent base layer or phase which is continuous and holds dispersed solids or agglomerates with the surface of the product also being translucent, the aforesaid discontinuous openings, surface failures, and void spaces in the product occurring in the continuous phase indicating a loss of elasticity at spaced points of the sacchariferous carrier for the food solids.

The thickness of the film of product after stretching will vary from product to product in accordance with the stretchability of the product and the degree of stretching required to render it instantly rehydratable. Usually the stripped film will be reduced to a thickness less than 0.050 inch and ranging anywhere from 0.005 inch upward. Thus, the tension employed for removal of the film from the heating surface is well in excess of that normally required to prevent agglomeration of the product and rapid removal of the film from the drying surface.

Stretching can be carried out by any positive controllable means for accelerating the rate of travel of the stripped film relative to the rate of film travel at the point of stripping. Because of the desired film characteristic such stretching means should be capable of creating such a rate increase uniformly without inducing excessive tension while assuring the creation of sufficient tension to establish and maintain a high degree of dispersion of food solids throughout the molten carrier.

By virtue of the condition of the film created by stretching, there is an ideal distribution of the water-soluble and water-insoluble food solids therein with the film displaying attractive colors and offering fresh flavors upon reconstitution. Most of the water-soluble food solids appear to be in the continuous phase, with most of the water-insoluble solids being thinly and discretely dispersed therein as a discontinuous phase. A majority of the surface of the film is comprised of the continuous phase. By this arrangement and condition of food solids, though the water-soluble solids will go into solution quickly in cold water, this does not occur at the expense of the water-insoluble solids which are ideally dispersed and have the greatest opportunity to rehydrate without clumping.

Flakes produced by breaking the film will have been broken into various sizes depending upon the rehydration characteristics of the particular puree being dried. For most fruit purees and puddings the film will preferably be flaked into a particle size where 100% of the product passes a 10 mesh U.S. Sieve Series screen (0.030 inch wire), although even larger size flakes may be produced with acceptable rehydration properties. Illustrations of apparatus for stretching the puree and samples which have been treated in various ways are included in the patent, which also contains a fairly exhaustive discussion of the conditions required by the process.

Reduction of Time Food Is in Contact with Drum

A commercially acceptable dry food product must be capable of being instantly reconstituted with the addition of water leaving no residual lumps. This requires a product in the form of fine particles or flakes that have little or no tendency to cake. The prime cause of caking is excessive water content. A successful process for the preparation of dried instant foods is therefore one which produces a very dry food in small particle form. These may be kept in a hermetically sealed package until used.

The drying of fruit and similar high sugar content foods of delicate flavor to obtain a suitably dry, brittle product without impairment of flavor has proved to be a difficult task. The drying of these foods is, in effect, a paradox. On the one hand, efficient water removal requires high heat and/or lengthy contact between the product and the heat. On the other hand, nonimpairment of flavor requires relatively low heat and as short a contact time with the heat as possible.

The drying of cereals and similar low sugar content foods has been conventionally practiced with internally heated drum driers. Conventional driers of this type comprise a pair of closely spaced drums. The cereal or other food in a liquid form is supplied to a puddle contained in the space between the upper adjacent surfaces of the two drums. The drums at their adjoining surfaces rotate downwardly, and each removes a thin film of the puddle contents on its outer surface.

Drying of the thin film continues throughout the interval that the sheet is in contact with the heated drum surface, and in the instance of a cereal product, the film sheet is fully dried by the time it contacts a doctor blade or knife placed near the top of each drum. The doctor blade peels the dried cereal film from the drum surface, and it is then passed to a flaker or other disintegrator.

In this design the puddle is held in direct contact with the upper adjacent surfaces of the two drums and has a prohibitive disadvantage against its use in the processing of a heat sensitive material like fruit due to its tendency to expose some of the puddle's content to heat for relatively long periods of time while awaiting application to the drums for drying. This results in an undesirably excessive heating of the food with concomitant impairment of the flavor.

A process and apparatus for drying fruit and other heat sensitive foods which overcomes this disadvantage has been developed by *D. Eolkin, E.R. Allard, L.H. Anderson and J.R. Lovasz; U.S. Patent 3,147,173; September 1, 1964; assigned to Gerber Products Company.* Broadly stated, the elements of the apparatus include a rotatable drum drier provided with a heated outer surface. In combination with the drum drier, the remaining elements comprise means for holding food in liquid form out of physical contact with the heated surface of the drum drier. This means for holding the food is adapted for supplying an adjustable quantity of the food to the heated surface of the drum drier as the drier is rotated.

The food is thereby applied to the drum drier in the form of a film. While this film is being dried, there is provided means for continuously removing the moistened atmosphere produced above and below the heated surface of the drum drier. When the food has been dried to a type of plastic sheet, means associated with the drum drier for removing the food from the heated surface, are employed to remove the food.

Means for transporting the removed food sheet away from its point of removal are provided. This transportation means carries the food sheet to terminal drying means adapted for removing moisture left in the food by the drum drier. A drawing of the apparatus and a complete description of the process are given in the patent.

Addition of Thickening Agent

Additional refinements to the process described in the previous patent for dehydration of fruits and other high sugar content foods are described by *D. Eolkin; U.S. Patent 3,197,312; July 27, 1965; assigned to Gerber Products Company.* One refinement for the process is practiced by suitably adapting a conventional drum drying apparatus. Optionally, an inert nonoxidizing atmosphere is employed therewith. Temperatures at various points in the process line are judiciously selected to improve the moisture elimination and assure proper product disintegration. In order to further improve the reconstitutable high sugar content food a thickening agent is employed in combination with a puree or slurry of the food. This has the effect of improving the film-forming characteristics of the product and partially nullifies the adverse effect of high sugar content in the food.

The thickening agent employed in the preferred form is one selected from the group of hydrophilic or lyophilic food colloids. Examples of hydrophilic colloids are starch, agar, gum arabic, gum acacia, carrageen, pectin, dextran, gelatin, sodium gluten sulfate, etc. Proportions of 1 to 5% by weight of a suitable colloid, as compared to slurry weight, are employed in the slurry of fruit or vegetable solids and water. The actual percentage of the thickening agent depends upon the colloid selected and on the percentage of fruit solids in the puree. A typical example of a fruit puree is one including 15 to 25% total solids to water, which total solids include 1 to 5% of a suitable thickener and 50 to 90% fruit solids.

The incorporation of a thickening agent results in a better rehydration because it absorbs some of the water which should be absorbed by the dry product. It is a fact that in dehydrating any food product, whether fruit, vegetable or otherwise, there is a loss of a portion of the gel characteristic of the plant or animal tissue, which portion is unrecoverable. Stated in another way, the process from natural product to rehydrated product is not without loss.

As a result, when a good product is rehydrated its consistency has a higher liquid content than that of the original product. The addition of a suitable amount of a colloidal thickener acts to absorb some of the water added during rehydration and permits the reconstituted food to more closely approximate the original from which it was derived.

Turning to the dehydrating part of the process, it is equally suited to being practiced with or without the use of an inert gas atmosphere at any of its stages. However, where desired an inert gas such as nitrogen or a process gas including nitrogen and carbon dioxide or in some cases superheated steam, the latter of which acts like an inert gas, may be employed. By using inert gas at certain places, it is possible to assist in preventing oxidation and control of scorching of the product. A satisfactory product is obtained, however, in the absence of the inert gas.

The remainder of the processing line from the drum drier to packaging is enclosed in a relatively cold inert gas atmosphere to improve the quality and output of the food processing line. More particularly, in the preferred form a supply of refrigerated gas is played on the film sheet product as it leaves the drum driers. This chills the sheet to reduce its plasticity and improve it for the flaking steps. The flaking and packaging steps are also maintained in the relatively cool inert gas atmosphere to prevent the ingress of oxygen and to maintain low plasticity.

In accordance with a common method of processing a high sugar content food, the puree or slurry, including approximately 10 to 23% (50 to 90% of the 15 to 25% solids) of fruit solids suspended in water, is mixed in a tank with the proper amount of thickening agent. The slurry is fed from the mixing tank into a gellation tank in which it is heated with agitation to a temperature in the range of $160°$ to $220°F$. so as to fully gel and activate the thickening agent preparatory to metering the slurry to drum driers.

Specifically, the slurry after the thickener has gelled is fed into a holding tank from whence it is metered into the drum drier. The slurry is metered onto the drums of the drier in an atmosphere of hot inert gas. The film of the food product formed on the drums is peeled off by doctor blades and fed through take-away screws where the sheet is cooled by relatively cool nitrogen to temperatures between about $40°$ to $70°F$. After the sheet has been cooled to improve its handling properties, it is directed through a flaker to a storage bin and ultimately to a packaging station, all while maintaining the cool ambient temperature.

External to the enclosed drier, chill screw, flaker, storage bin and packaging station is a gas supply and temperature control means to provide hot and cold nitrogen to the drum drier and chill screw, respectively. The entrained nitrogen may be evacuated both above and below the nip of the drums, and pumped back to the gas supply for moisture removal preparatory to heating and recycling in the system. In spite of the fact that the improved process described above has eliminated a number of serious problems, the problem of removing all moisture at the drum drier stage is still present.

The described process uses the same steps as the above sequence until the material is removed from the flaker or flaking mill. At this point it has been found advantageous to modify the process. The relatively dry, flaked or powdered product is moved through a relatively fine mesh screen, e.g., 20 to 40 mesh, and then deposited in a fluidized bed type of drier which includes a moisture control element and cooling bed. The product has further moisture removed from it in the terminal drier at a substantially lower temperature as compared to the drum drier. The product is then centrifugally separated and packaged for distribution.

This process has found application not only in the processing of high sugar content foods but with the usual low sugar content foods, such as cereals, soups and in some cases certain fruits as well. The addition of seasoning to the soups and certain vegetables is purposely added at the packaging terminus in order to conserve their volatile flavor content. This process permits an instantly reconstitutable food product to be formed which closely approximates the original product upon reconstitution by adding water and, at the same time, one which has a long shelf life.

Heated inert gas may also be employed to further reduce the moisture content of the food product in the moisture controller section of the terminal drier. In a typical situation where prunes are processed, hot nitrogen is supplied to the moisture controller at approximately 160°F. and the material moves through the controller at a rate sufficient to create a bed temperature of approximately 120°F. The actual bed temperature, i.e., the equilibrium temperature of the product, can be varied between 115° to 150°F. or so depending upon the particular food product being processed.

It can be appreciated that the relatively low temperatures maintained in the moisture controller permit further moisture removal without the deleterious effects upon the flavor, color or nutritive content of the food product. The cooling bed, which may also be a cool gas stream or equivalent thereof, reduces the temperature prior to depositing it in a centrifugal separator for separation prior to packaging. Inert gas optionally is maintained in the terminal drier as well as in the area surrounding the centrifugal separator and packaging stations to prevent oxidation from degrading the desired attributes of the product.

Example: A slurry of apricots containing about 17 to 20% solids was prepared and about 5% by weight (based on the finished reconstituted product) of the slurry of modified tapioca thickening agent was added thereto. The slurry with the thickening was heated with agitation at about 200°F. for about 10 seconds thereby causing the tapioca to gel.

The material was then applied onto two drum driers (12 inch diameter, 36 inches long) by forming relatively low puddles (about 3½ inches) in the upper nip of the drums while the drums were rotated. Steam at 230°F. in the drums caused the material applied on the surfaces thereof to be dried into a sheet. Residence time in the puddle and on the drums was about 2½ minutes. The dried sheet was thereafter stripped from the drums, chilled with cool nitrogen and flaked. The flaked material was terminally dried on a fluidized bed using nitrogen.

The sheet obtained from this procedure (where preheating and gelling was employed) yields a most satisfactory and desirable product which retains substanti-lly all of the natural flavoring and essence of the product and does not have a starchy aftertaste when reconstituted. This is due in part to the low puddle from which the slurry is fed to the drums, both of which help avoid burning or scorching of the product from excessive heating on the drums while the preheating step insures gelation.

Controlling Bulk Density

J.I. Wadsworth, A.S. Gallo, G.M. Ziegler, Jr., and J.J. Spadaro; U.S. Patent 3,494,050; February 10, 1970; and U.S. Patent 3,577,649; May 4, 1971; both assigned to U.S. Secretary of Agriculture have developed a method and apparatus for controlling the bulk density of high sugar content foodstuffs during drum drying by regulating the angle of attack of the doctor blade relative to the surface of a dryer drum. The "bulk density" of a dehydrated food product may be defined as weight per unit volume of material in the aggregate. This differs from the "true density" in that the volume measured is occupied by both the discrete particles and the void spaces between them.

A continuing problem in the manufacture of dehydrated food products is the variation in product bulk density during processing which can result in undesirable variation in package fill. For example, a processor is packaging his product on a volume basis, variation in bulk density can result in either an overweight package, which will reduce the processor's margin of profit, or an underweight package, which does not conform with label specifications.

If a processor is packaging his product on a weight basis, a high bulk density will result in a package appearing to be underfilled, which is undesirable from the sales appeal point of view, and a low bulk density will result in the volume of the product being greater than that of the package thereby preventing the processor from filling his package. For a successful operation, the product bulk density must be maintained within suitable limits. Whenever there is a variation in bulk density which exceeds these limits it is necessary to revert to costly blending operations or other additional processing steps to obtain a product with an acceptable bulk density.

The bulk density of processed food products varies not only with processing conditions but also with factors beyond the control of the processor, such as crop growing conditions, variety of the farm product being processed, soil conditions at the growing location, etc. This is especially a problem for drum drying foods containing a substantial amount of sugar. Even though the true density of the product remains fairly constant, large variations in bulk density result indirectly from differences in the composition of the raw materials being processed.

This is due to a ruffling effect which occurs as the dry but hot thermoplastic material is scraped from the drum surface by the doctor blade resulting in a sheet of material discharging from the dryer which is many times thicker than the film of material adhering to the drum surface. The thickness of the sheet varies with the composition of the material being dried, and this variation in sheet thickness produces variation in bulk density. For example, in the manufacture of dehydrated sweet potato flakes a thin film of sweet potato puree, which is applied to a steam heated drum dryer, is dehydrated to the desired moisture content in a partial revolution of the drum. The dried material is then scraped from the drum surface by a doctor blade and discharged from the dryer in the form of a continuous sheet.

As the sweet potato solids are scraped from the drum surface by the doctor blade a continuous ruffled sheet, which is much thicker than the film of sweet potato adhered to the drum surface, is formed. The ruffling effect is related to the sugar content of the sweet potato which will vary for different varieties, different growing areas, different climatic conditions, different harvesting times and different curing processes. A higher sugar content causes more ruffling which results in a thicker sheet.

Bulk density varies with sheet thickness because the flaking equipment is designed to break the dry sheet into flake-like particles, which have a nonvarying diameter size distribution. The only variation in particle size is in the flake thickness, therefore, the variations in bulk density are due to variations in sheet thickness which result from differences in the raw materials. The bulk density of dehydrated sweet potato flakes can range from 10 to 45 pounds per cubic foot, and this can be attributed to the variety, time of harvesting and the time that the sweet potato has been in storage. To illustrate the method employed in this process, reference is made to Figure 1.2. This drawing is a schematic presentation of the method of the process. This side elevation auxiliary view shows the periphery of the dryer drum **1** being touched by the doctor blade **2** at the point of contact **34**, which is also the edge of the doctor blade **12**. The fine line T is the edge of a surface tangent to the drum at line of contact **34**. Three angles are shown. These are referred to as α, γ and β.

The angle which affects the bulk density of drum dried high sugar content materials is the solid angle α, which is formed by the intersection of T, which is tangent to the surface of the drum at the line of contact **34** with surface B which is parallel to the beveled surface of the doctor blade. When the angle α is increased, the bulk density of drum dried high sugar content products is decreased, and when angle α is decreased, the bulk density is increased. It should be noted that variation of angles β and γ do not affect the bulk density when angle α is held constant.

An explanation for the variation of bulk density with angle α can be deduced from physical reasoning. The temperature of the material on the drum approaches the temperature of the drum surface by the time it reaches the doctor blade. Foodstuffs having a high sugar content tend to be thermoplastic and tacky at the operating temperatures of such machinery, even when the moisture content has been reduced to less than 4%. As the material is scraped from the drum surface by the doctor blade it ruffles to form a sheet discharging from the dryer which is much thicker than the film adhered to the drum surface. The bulk density, as stated previously, varies with the thickness of the sheet of food material.

The degree of ruffling (i.e., thickness) of the formed foodstuff sheet for a particular product depends upon the amount of impedance to the movement of the dry food material across the beveled surface of the doctor blade. For example, the degree of ruffling can be temporarily reduced (yielding a thinner sheet) by coating the doctor blade with a fluorocarbon spray thereby lowering the coefficient of friction between the material and the doctor blade which reduces the resistance to the movement of the sheet of foodstuff.

The variation of the angle α has a similar effect. Reducing α has the effect of increasing the impedance to the movement of the high sugar content material and increasing α lowers the impedance. For physical reasoning a rough analogy can be drawn with fluid flow through a pipe bend where an increasing deflection results in an increasing friction loss. In the table, the effects of change of doctor blade angle with respect to the surface of the drum (angle α) on the bulk density of three drum dried high sugar content food materials are shown.

Bulk Density (pounds per cubic foot)

Angle α (degrees)	Sweet Potato Flakes	Peach Flakes	Pumpkin Pie Mix
148	27.1	25.0	24.6
133	29.7	29.1	–
118	33.1	30.9	–
103	33.9	33.1	31.8
88	37.0	37.0	–

This process provides a method and an apparatus for the drum drying of high sugar content foodstuffs, such as sweet potatoes, peaches, pumpkin pie mix, pears, apples, apricots, plums, oranges, and the like in which bulk density in the preparation of the dehydrated foodstuffs is controllable. The ability to control the bulk density is significantly functional in that the bulk density can be adjusted without interrupting the continuous operation and immediate response is obtained. Bulk density is adjusted simply by the turning of knobs or wheels which in turn alter the angle which the doctor blade makes with tangent to the drum surface. Experience with the equipment has dictated certain limits that are

FIGURE 1.2: DOCTOR BLADE OF DRUM DRYER

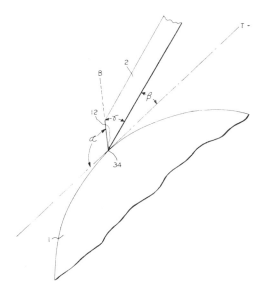

Source: J.I. Wadsworth, A.S. Gallo, G.M. Ziegler, Jr. and J.J. Spadaro; U.S. Patent
3,494,050; February 10, 1970.

applicable to the pertinent angles, the limits of angle α being about from 30° to 165°,
while angle γ would be limited to about from 15° to 75°. The initial settings of the doctor
blade so that angle α would be between 120° and 135°, and the use of a doctor blade which
would have an angle γ of about 15°, allow an ample range of control to both increase or
decrease angle α as the drying operation warrants. The patents provide drawings and detailed descriptions of a device which permits adjustment of the angle of attack of the doctor blade to the dryer drum.

HEAT TRANSFER UNDER REDUCED PRESSURE

Dispersed Oil

Food may be dehydrated by immersing it in a bath of circulating heated oil for a predetermined period, which bath is located in a subatmospheric environment. The subatmospheric environment sufficiently reduces the boiling point of moisture contained in the food
for the oil to evaporate and drive it off at a temperature below which the food is cooked
or otherwise altered. Immersion of the food in the oil bath, however, presents certain
problems. In the first place, only that portion of food located at the top of the bath is in
contact with the subatmospheric environment, whereas underlying food in the bath is subjected to progressively increased pressure as a result of the surrounding and overlying bath
whereby food located adjacent the bottom of the bath does not fully benefit from the
subatmospheric environment.

*W.R. Dorsey, R.L. Roberts and S.I. Strashun; U.S. Patent 3,194,670; July 13, 1965; as-
signed to Vacu-Dry Company* have developed a process in which food to be dried is

contacted with heated but dispersed edible oil in a subatmospheric environment, whereby outer surfaces of all the food are simultaneously subjected to the drying media and the subatmospheric moisture receiving environment. It is the object of this process to provide an efficient method and apparatus for dehydrating foods in a minimum period of time. An additional feature resides in terminal maximum evacuation of the drying chamber to render the heated oil from the food leaving only an insignificant and nondeleterious oil residue on the food.

Example 1: ⅜ inch thick slices of Pippin apples dehydrate from an initial moisture content of about 85% down to a finish moisture content of 5% (by weight) by approximately 10 minutes' exposure to dispersed coconut oil at 200°F. with a pressure of 1 inch of mercury absolute in the vacuum vessel.

Example 2: One inch cross-cut frozen green beans (Tendergreen variety) dehydrate from an initial moisture content of approximately 90% down to a finish moisture of 2% in about 40 minutes, employing coconut oil as the heat transfer medium at 155° to 175°F. and employing a pressure of 1 to 3 inches of mercury in the vacuum vessel.

Certain foods dehydrated according to the process do not reconstitute entirely satisfactorily in end use. This is particularly true with certain beans, meats and fish. By initially freezing these foods and then subjecting them to the process while in a frozen state it has been found these foods dry to an exceptional end product substantially better than when treated in an unfrozen state, and also they reconstitute exceptionally well.

Acetylated Monoglycerides as Heat Transfer Agents

An improved process for dehydration by heat transfer is described by *W.A. Heyman; U.S. Patent 3,408,210; October 29, 1968.* The food products are dehydrated at temperatures varying from about –6° to 93°C. (depending on the type of food), while under a high vacuum of about 100 to 25,000 microns while immersed in a liquid consisting essentially of the substantially fully acetylated monoglycerides of oleic, linoleic, palmitic, or stearic acids, either alone or in mixtures.

Foods which have been dehydrated by this process include fruits of all kinds such as bananas, peaches, pears, apricots, dates, pitted prunes, berries such as strawberries, raspberries, cranberries, blackberries, blueberries, grapes, cherries; poultry, fish, and vegetables such as edible green corn, peas, beans, spinach, tomatoes, etc.

The acetylated monoglycerides of oleic, linoleic or palmitic acids are available commercially. Such products can be made by reacting edible animal fats such as lard, or vegetable oils such as cottonseed or soybean oil, etc. (which are triglycerides of one or more of the fatty acids, oleic, linoleic, palmitic and stearic acids) with edible glycerol under such conditions as to provide the monoglycerides of the fatty acids and then acetylating the monoglycerides. The acetylated monoglycerides as defined, are water-white and highly fluid materials which do not solidify at various temperatures from about +8° to –5°C.

The acetylated monoglycerides, as defined, are soluble in 80% and higher, ethyl alcohol and can, when desired, be substantially removed from the dried food product by such alcohol. Acetylated monoglycerides as defined are entirely edible and nutritious food materials and commercial preparations thereof have been accepted by the U.S. Food and Drug Administration.

Example 1: Ripened bananas, the ripeness of which is indicated by a rich yellow skin on which there are a few black splotches, are selected, peeled and sliced to a thickness of ¼ to ½ inch. The slicing is preferably done in an atmosphere of inert gas and the slices are dropped immediately into a bath of acetylated monoglycerides consisting of a mixture of acetylated monoglycerides of oleic, linoleic and palmitic acids such as obtained from cottonseed oil. The container, usually a pan made of stainless steel or aluminum, is transferred to a conventional vacuum shelf drier provided with means for heating and cooling the shelves and with a transparent window. The interior of the vacuum shelf drier is exhausted to about 500 to 1,000 microns of mercury above absolute and heat is applied to the shelves, is transmitted to the pan and through the pan to the heat transfer agent and into the bananas. As long as there is substantial moisture in the bananas the temperature of the banana does not rise to the temperature of the heat transfer agent because heat is consumed in evaporating the moisture.

As the moisture is evaporated the heat is regulated and gradually lowered in order to adjust the amount of heat introduced in relation to the moisture remaining in the bananas. After the bubbles caused by the escaping vapors become greatly reduced in number and size indicating that the end of the operation is approaching, the heat is then raised to assure the absence of moisture on the interior of the slices. This heat can go as high as $65°C$. without causing a discoloration of the banana slices.

When the bubbling ceases at this temperature it is an indication that the banana slices are dried to a point of less than 3% moisture which is sufficiently dry to guarantee complete stability and preservation of the nutritional values of the banana. Slices are recovered by cooling to about $5°$ to $10°C$., breaking the vacuum by introducing nitrogen in the drier and draining the heat transfer agent therefrom. The slices have the original shape and appearance, but are somewhat shrunk from loss of water. On rehydration, the banana slices swell again to their original size and have the consistency of ripe banana slices. Their flavor is substantially equal to the original flavor.

Dried products prepared in this manner are crisp and are preferably packaged in moisture-proof air-tight containers to keep them in their crisp state. Such products have been thus preserved in hand sealed jars for at least about a year and still have their original characteristic banana taste and their physical and nutritive properties without rancid or off-color tastes.

Example 2: The process is conducted as in Example 1 except that whole ripe strawberries are substituted for the sliced bananas. The temperature of the acetylated monoglyceride in the vacuum drier is allowed to reach about $38°C$. and a vacuum of about 1,000 microns or less is applied. The dehydrated berries or fruits retain their volatile aroma and flavor as well as their enzymic properties, vitamins, minerals and eating qualities. They remain edible and can be rehydrated or eaten just as they are.

Moisture Released into Evacuated Space

M.P. Lankford; U.S. Patent 3,718,485; February 27, 1973; assigned to Vacu-Dry Company describes another method for drying comestibles by contacting them with a heated edible liquid heat transfer medium in a subatmospheric environment. The process, in general, includes development of an evacuated drying region maintained at high vacuum levels in the order of about 0.20 to 10 mm. of mercury absolute pressure. A supply of an edible hydrophobic heat transfer medium which remains liquid at the selected operating temperatures and pressures is presented to a portion of or a zone within the evacuated drying region.

The zone of medium can be either a bath as is disclosed in the following Forkner patent, or a shower such as that described previously by Dorsey, et al. The remainder of the drying region is a vapor space in which essentially no liquid heat transfer medium is present. The particular comestible to be processed is supplied either continuously or in batches to the drying region.

During dehydration each particle is alternately moved into contact with the heated liquid transfer medium to heat and vaporize its contained moisture and then removed from the medium substantially into the vapor space so that a major portion of its vaporized contained moisture is released directly into the vapor space. At such time as the desired moisture level is reached the liquid heat transfer medium is withdrawn from the evacuated region and the comestible centrifuged at gravities in the order of nine times the force of gravity to remove essentially all medium from the food particle surfaces.

The particles are then cooled, the vacuum released and the dried comestible discharged from the drying region. The entire process can be performed continuously or on the batch basis which is specifically described herein. The comestible may be subjected to a number of preparation steps before drying. The particulate comestible is then sealed in a vacuum chamber and vacuum is drawn on the chamber. Liquid heat transfer medium is heated and supplied to the chamber to contact the comestible under the manipulative conditions described above.

The medium preferably is recycled and continuously or intermittently purified to remove contaminants resulting from contact with the comestible or otherwise. Moisture from the chamber is removed either as condensate through a trap or as vapor through the vacuum system or both. At a selected moisture level medium is withdrawn from the chamber and the comestible spun or centrifuged to remove any remaining medium from the food surfaces while still under vacuum. The comestible is then cooled, the vacuum released, and the dry product unloaded from the vacuum chamber.

The liquid heat transfer medium has been handled in the system and apparatus described schematically in Figure 1.3 which is useful for processing on a batch basis. It includes a vacuum chamber 20 and means, not illustrated, for drawing vacuum within the chamber at 21. The vacuum means can be a positive displacement pump, oil diffusion pump or steam jet ejectors all of which are commercially available to develop the vacuum levels contemplated herein.

Within the vacuum chamber is basin 22 which confines the liquid heat transfer medium so that, if desired, it may be applied to the comestible as a bath. Within the basin is a generally cylindrical foraminous basket 23 rotated or rocked upon drive shaft 24 by a variable speed motor drive assembly 25 mounted exterior to and on the vacuum chamber wall.

The basket confines the food particles but is perforated substantially over its entire surface so that water vapor can freely move out of it and liquid heat transfer medium can move in. The interior of the basin through a plurality of perforations 26 communicates with the atmosphere of the vacuum chamber so that water vapor can freely pass from the basin to the vacuum system and the interior of basket is under vacuum. Liquid heat transfer medium through distributor 27 during the drying phase supplies a continuous flow of medium that falls by gravity over the particulate comestible 28 contained within the basket.

The heat transfer medium is withdrawn by pump 29 throughout outlet conduit 30 that drains the basin. The pump circulates medium through heater 31 and back into the vacuum chamber via inlet conduit 32 to the distributor. The heater adds sufficient heat to the medium to make up the heat lost through moisture vaporization in the basin. Depending upon the particular food being processed, a level of medium can be maintained in the basin to develop a zone of medium which is a bath 33.

Its contact with the food under process is accomplished by immersion of the latter in the bath. Alternatively, no liquid level is carried in the basin for many applications and in such cases contact of medium with the food being processed is entirely by shower from the distributor. In either alternative the movement of comestible particles also prevents their clumping as frequently occurs in prior art methods.

The medium circulation system also includes purifying the medium. Purification includes continuously or intermittently mixing all or a portion of the circulating medium at mixer 34 with an emulsifiable flushing liquid or solvent for the impurities in the medium; heating the mixed phases to assist extraction from the medium by the flushing liquid; and then separating, as at 35, the cleansed medium from the flushing liquid or solvent which now carries the impurities of the untreated medium and delivers them to further separation means or to the waste disposal drain as illustrated. During exposure of the processed food to the liquid heat transfer medium, the foraminous basket rotates at a low speed in the

FIGURE 1.3: DEHYDRATION BY HEAT TRANSFER IN A VACUUM

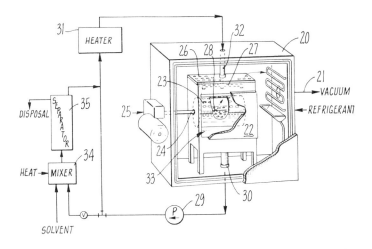

Source: M.P. Lankford; U.S. Patent 3,718,485; February 27, 1973

order of 10 to 50 rpm. In those instances where a liquid level is maintained in the basin this rotation of the basket periodically immerses the food particles under process in the bath and subsequently carries each particle substantially out of the bath into the vapor space lying above it. There most of the contained moisture vaporized by the heated medium is released directly into the vapor space through, at most, a thin residual film of medium carried on the food particle surfaces.

Since each particle periodically moves substantially out of the bath or at least to its surface, it never is exposed for any length of time to a head of liquid medium overlying it with the consequent undesirable hydrostatic pressure that otherwise partially nullifies the effect of the vacuum in the chamber, inhibits moisture release from the particles and/or drives medium into the voids remaining in the food particles after such moisture release.

In instances where heat transfer is entirely through a shower of medium, rotation of the basket again at low speeds in the order of 10 to 50 rpm tumbles the food being processed in the basket and uniformly exposes each particle to the descending shower. This eliminates the umbrella effect of the top layers of comestible in a static system which tends to shield lower layers from the medium shower and results in nonuniform exposure to the medium. This process is applicable to substantially all particulate foods and especially to a variety of cellular comestibles including vegetables and fruits. The following example is typical.

Example: Fresh or frozen green beans, such as California Bush varieties are cut to suit need and dipped in about a 1% magnesium or sodium bisulfite solution for color preservation. The beans have an approximate 90% moisture content. The beans are placed into a foraminous basket and vacuum on the chamber is drawn down to a level of about 10 millimeters of mercury absolute pressure. At that level heated medium, for example, the mixed cotton and soybean oil sold under the trade name Durkex 500, is started through the basin at temperatures in the range of 170° to 190°F. A liquid level in the basin 22 is allowed to build up to about the center line of the basket and the chamber pressure is drawn down to a level in the order of one-half millimeter of mercury absolute pressure.

The basket rotates slowly at about 20 rpm. The beans are processed under these conditions for about 45 to 60 minutes at which time their moisture content reduces to about 3%. The medium then is drained from the basin and the basket is rotated rapidly for 1 to 2 minutes under the same vacuum conditions at about 360 rpm or a speed sufficient to develop centrifugal forces at about nine times the force of gravity at the basket periphery. Next the beans are cooled and the vacuum broken with nitrogen. They have a residual moisture content of 2½ to 6% on a dry weight basis and residual medium in the range of 8 to 16%. The beans retain their original size, shape and color without shriveling of the soft tissue surrounding the seed cell.

HOT OIL IN PARTIAL VACUUM

Poor quality of dehydrated foods generally results from excess heating for prolonged periods of time, which detrimentally affects flavor and other heat sensitive properties, and from oxidation, which deteriorates color, flavor and nutritive value. Also it results from procedures which cause extensive shrinkage in volume and changes in physical form. In an effort to meet the demand for high quality dehydrated particulate products, vacuum evaporators have been employed which employ processing temperatures of the order of from about 140° to 230°F., under partial vacuums of the order of from 22 to 29 inches mercury column.

With such equipment, certain moist food products can be dehydrated without serious impairment of heat sensitive or oxidizable constituents, having reference particularly to food products which do not have a high water content. However, with conventional vacuum dehydration equipment, heat transfer to the material undergoing dehydration is at a relatively slow rate.

The masses or particles undergoing dehydration are supported in metal trays, and in many instances considerable physical impairment of the articles occurs due to overheating and burning of the surfaces in contact with the tray, to sticking to the tray due to exuding juices, and to changes in physical form and size. Because of the above and other difficulties, it is generally considered that conventional vacuum dehydration is impractical for fresh particulate fruits and vegetables of high water content.

Freeze drying, which has reference to vacuum drying while the product is frozen, is effective with some materials to produce good dehydrated products. Particular disadvantages of this method are the relatively high costs of equipment required for a given capacity, and high cost per unit weight of material dehydrated.

It is known that moist food particles can be dehydrated by immersion in hot oil. When carried out at atmospheric pressure the extended time and temperature factors involved cause undesirable changes in flavor, palatability, color, form and volume. If the oil temperature is increased to shorten the drying time, then serious burning occurs. With a lower oil temperature there is a tendency for the material to cook without substantial evaporation.

J.H. Forkner; U.S. Patent 3,261,694; July 19, 1966; assigned to The Pillsbury Company describes a process in which moist food material in particles of suitable size and at a low temperature level, is subjected to treatment under partial vacuum to complete a dehydration cycle, the treatment including dispersion of the particles in a mixed phase medium or foam consisting of water vapor and hot oil.

To start the cycle the material can be introduced into a quantity of hot oil under an applied partial vacuum. The temperature of the oil at the time of the initial contact is relatively high compared to the vaporization point of water at the applied partial vacuum, as for example, from 240° to 600°F., the range of from 325° to 440°F. being preferred. In a typical instance the material is frozen and is at a temperature level below 32°F. Because of the great temperature differential between the material and the hot oil, a rapid heat exchange takes place whereby the outer surface layer of the product is flash heated to the vaporization point of the hydrous juices present. Immediately the initial phase of the dehydration cycle proceeds at a rapid rate with almost explosive violence. The oil temperature drops immediately and rapidly and the evolving vapor creates a high rate of vapor flow to the evacuating means employed. The surfaces of the material and the outer layers through which dehydration progresses, are protected by the rapidly evolving vapor against burning by direct contact with the hot oil.

Thereafter (assuming that a low moisture content is desired) dehydration is continued at a lower temperature level to complete the cycle. The major part (e.g., 75 to 95%) of the moisture present in the material is removed in the short time (e.g., ½ to 4 minutes, depending largely on particle size, moisture content and initial temperature of the particles, ratio of particle weight to weight of oil, and oil temperature) of the first rapid evaporation phase, and the remaining moisture (except for residual) is removed at the lower temperature level. In general the overall time period of treatment in the hot oil, under applied partial vacuum, is relatively short, and may be, in typical instances, of the order of from 7 to 40 minutes. When the moisture content of the material has been reduced to the value desired, a separation of the free oil and the dehydrated material is made and then the material is centrifuged, after which the vacuum is broken.

In U.S. Patent 3,239,946 there is disclosed apparatus of the batch type suitable for carrying out the foregoing method. The apparatus consists of a tank of substantial height having provision for introduction and withdrawal of oil, and for the removal of the product at the end of the dehydrating cycle. While apparatus of the batch type for carrying out the foregoing method is workable and commercially practical, the capacity of batch equipment of a given size is limited. Also processing costs cannot be reduced beyond certain limits, due to such factors as labor cost and power consumption, which are inherent with batch equipment.

J.H. Forkner; U.S. Patent 3,314,160; April 18, 1967; and U.S. Patent 3,335,015; Aug. 8, 1967; both assigned to The Pillsbury Company are substantially identical and the latter will be used to describe apparatus for a continuous process for the dehydration of moist foods with hot oil under partial vacuum.

Figure 1.4a is a schematic view in side elevation illustrating the equipment used and Figure 1.4b is a flow diagram illustrating the method when carried out by use of the apparatus shown in Figure 1.4a. The apparatus shown in Figure 1.4a consists of a plurality of tanks or vessels **10**, **11** and **12**, which are disposed at successively lower levels, whereby material may move from one tank to the next.

A valve **13** is interposed between the lower end of tank **10** and the upper end of tank **11**, and may be of the butterfly type as illustrated. When closed, this valve interrupts communication between the tanks. When open, it permits material (i.e., oil and material being dehydrated) to pass from one tank to the next. A similar valve **14** is shown interposed between the tank **11** and tank **12**.

Means are provided for introducing a predetermined charge of moist frozen food material into the tank **10**. The charging means can consist of a hopper **16** having a normally sealed cover **17**, which can be removed for introducing the charge. Pipe **18** connects the hopper with suitable evacuating means for maintaining a partial vacuum, as for example, a vacuum corresponding to 25 to 29 inches mercury column.

A valve **19**, which can be of the butterfly type, is interposed between the lower end of hopper **16** and the housing **20**, the latter being of sufficient size to hold all of the charge. The lower portion of this housing is shown provided with a rotating feed screw **21**, which serves to move the charge through the annular housing portion **22**, into the upper portion of the tank **10**. A pipe **26** connects this upper portion of the tank to suitable vapor condensing and evacuating means. This equipment may consist of a condenser **27** of the waterspray type, which is sealed by a barometric leg **28**.

The vaccum line **29** from the condenser may connect with a second condenser of the tube type, which in turn is connected to a suitable evacuating pump.   Line **30** serves to deliver any carry-over of liquid into a trap, which is not shown.   Between the upper and lower ends of tank **11**, there is a barrier screen which is indicated at **31**. This screen is mounted to rotate 90° about its axis.   In the horizontal position shown, oil may circulate through the openings of the screen, but the openings are of such a size that the material being dehydrated cannot pass therethrough.   When the screen is turned 90° from the position shown, that is to a vertical position, movement of material is not restricted.

Line **33** is shown connected with the lower portion of tank **10**, and serves for introducing or removing oil as desired.   One branch **34** of line **33**, together with tank **35** and line **36**, forms a bypass whereby oil can be withdrawn from tank **10** and introduced into tank **11** at a level below the screen **31**.   Line **37** leading from tank **35**, connects with the evacuating means.

FIGURE 1.4: METHOD OF DEHYDRATING MOIST MATERIALS

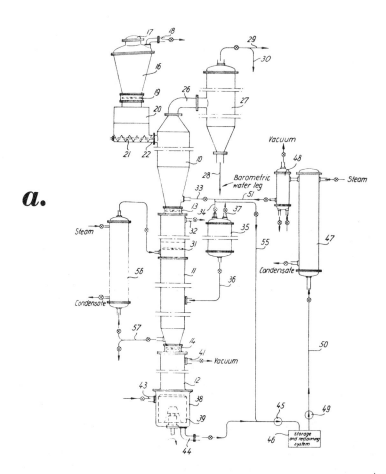

(continued)

FIGURE 1.4: (continued)

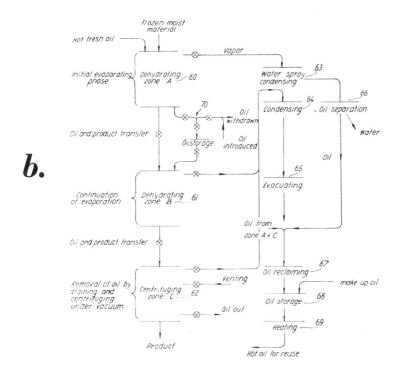

b.

Source: J.H. Forkner; U.S. Patent 3,335,015; August 8, 1967

The upper part of tank **11** is in continuous communication with the evacuating means, through pipe **32**. This means preferably is independent of the system connected to pipe **26**, and can consist of a suitable condenser and vacuum pump. The lower portion **38** of the tank **12** is a housing for the centrifuge basket **39**. Line **41** connects the upper part of tank **12** to evacuating means. Here again it is desirable to employ means independent of that connected to pipes **26** and **32**, such as a suitable evacuating pump of the jet type that may be capable of applying a vacuum somewhat higher than that normally applied to tanks **10** and **11**.

Line **43** is employed for breaking the vacuum, and may connect to a source of gas which is relatively anhydrous, such as dehumidified air, or anhydrous inert gas like nitrogen, or the like. Pipe **44** connects with the lower part of the housing **38**, and serves to remove oil at the end of the dehydrating cycle. Liquid pump **45** is shown for removing oil through the pipe and delivering it to a storage and oil reclaiming system **46**. The reclaiming system may include one or more centrifuges for removing solids, together with one or more clarifying filters.

Heat exchange means is provided for heating oil to a desired temperature level. The exchange means in this instance is in the form of two units **47** and **48** which are serially connected. Pump **49** and line **50** deliver clarified oil from the reclaiming system **46** to the first heat exchange unit **47**. Line **51** leading from the second exchange unit **48**, serves to introduce hot oil onto the tank **10** by way of pipe **33**.

One may withdraw oil from tank **10** and deliver it directly to the storage and reclaiming system **46**. For this purpose pipe line **55** connects between line **33** and the inlet side of pump **45**. Line **55** (with line **33** closed) can also be used when one may desire to recirculate oil through the heat exchange units **47** and **48**. A separate heat exchange unit **56** is shown connected to tank **11**, and may be used to maintain a desired oil temperature. Line **57** leading from the lower portion of tank **11** and connected to heat exchanger **56**, may be employed to draw off a part or all of the free oil, in instances where it is desired to partially or entirely replace the oil at this stage of the method.

Operation of the apparatus is as follows. A charge of frozen moist food material is introduced into the hopper **16**, after which the cover **17** is reapplied, and the frozen material evacuated by connecting line **18** with evacuating means. After such evacuation which serves to deaerate the material, valve **19** is opened to deliver the material into the housing **20**. A measured quantity of hot oil is introduced into the tank **10** from the storage and reclaiming system, by way of pump **49**, line **50**, heat exchangers **47** and **48**, and lines **51** and **33**.

The interior of the tank **10** is evacuated by virtue of the connection through **29** to evacuating means. The evacuating means functions continuously, and this is likewise true of the spray condenser **27**. After a predetermined quantity of the hot oil has been introduced into tank **10**, further introduction is discontinued, and the feed screw **21** is placed in operation to deliver the charge of frozen material into the tank and into the oil body. This starts the first evaporating phase, which is characterized by creation of a foam-like medium comprising a mixture of vapor and hot oil, the medium having a relatively large volume.

The volume may, for example, be from 3 to 10 times or more the original volume of the oil. After a short interval the foam subsides, and the charge is now approaching the time of transfer to the second tank **11**. In the first part of the initial phase, the material tends to be buoyant but is dispersed in the foam-like medium because of violent agitation. As the oil subsides, the material migrates to the upper part of the oil body. At the end of the initial phase in tank **10**, a substantial part of the oil is withdrawn from the lower part of the oil body, through pipe **33**, and a part of this oil is stored in bypass tank **35**, and another part withdrawn by way of line **55** to the storage and reclaiming system.

With the screen **31** in vertical or full open position valve **13** is opened to permit the partially dehydrated product, together with the remaining oil, to pass downwardly into the tank **11**, which at that time is likewise evacuated, because it is connected to evacuating means through line **32**. Valve **13** is now closed, and screen **31** returned to its horizontal position.

A quantity of hot clarified oil from heat exchange unit **48** is now supplied through line **51** to the tank **35** where it blends with the oil previously introduced in this tank, to produce an oil body of the desired temperature level. This oil is now introduced into the tank **11** through pipeline **36**. Introduction of this oil body serves to raise the oil in the tank **11** to a level above the screen **31**.

It may be explained that in the tank **11**, the material undergoing dehydration is buoyant. Without restraining means such as the screen **31**, the buoyant material would float as a thick mat upon the surface of the oil, thus elevating some of the particles above the oil line and preventing effective transfer of heat from the oil to all of the individual particles. The screen, when utilized in the manner described above, holds the buoyant particles below the oil level, where they are effectively contacted with the hot oil for good heat transfer.

Within the tank **11** evaporation of the material continues, but at a lower temperature level. Circulation of oil through heat exchanger **56** can be established to maintain the temperature level desired. Also, during or near the end of this part of the cycle, all or a part of the oil can be withdrawn and another oil substituted. At the end of the dehydrating cycle, butterfly valve **14,** is opened, thus delivering the dehydrated material and oil into the lowermost tank **12** which is at this time evacuated by virtue of its connection to an evacuating means through pipe **41**. After transfer of the charge to tank **12**, valve **14** is closed, and pump **45** placed in operation, with line **44** open, to withdraw oil from the lower part of the tank housing **38**. After the major part of the free oil has thus been removed from tank **12**, including oil that drains from the material, the centrifuge basket **39** is rotated whereby additional oil is removed by centrifugal force. Such oil likewise is removed through the pipe **44**.

After the centrifuging operation has been completed, the pipe connection **41** to the evacuating means is closed, and the vacuum within the tank **12** is broken by opening the pipe line **43**. As previously explained, this may serve to introduce substantially anhydrous gas into the tank **12**. After the vacuum is broken the housing part **38** of the tank can be separated from the upper tank part, and the product removed from the centrifuge basket.

The partial vacuums maintained in tank **10** may, for example, be at a value corresponding to 24 to 28 inches mercury column immediately before the charge is introduced. Shortly after commencing introduction of the charge the vacuum may fall to a value of the order of 22 to 26 inches. Toward the end of the initial phase of dehydration in tank **10**, the vacuum may rise to 26 or 28 inches. In tank **11** the vacuum at the time of transfer may be the same as in tank **10**, and for the period of treatment in this tank, the vacuum can be maintained within the range of 26 to 29 inches.

The partial vacuum in tank **12** should be the same as in tank **11** at the time of transfer. Thereafter it may be raised or lowered during the final drainage of oil and centrifuging. To prevent such evaporative cooling as may solidify remaining oil, the vacuum may be reduced to about 25 to 27 inches as the oil drains away from the product, and during the main part of centrifuging. Thereafter a higher value, e.g., 29 to 29½ inches, can be applied for final evaporation cooling and solidification of the oil. Application of such higher vacuum immediately before solidification also serves to somewhat expand the product and this increases its bulk volume.

The foregoing description traces the operations for one charge of material, from the time it is introduced into the hopper, until it is removed from the centrifuge basket at the end. In practice charges of frozen material are fed successively to the hopper and during the total aggregate time required for the processing of one charge, three charges are processed. In other words, immediately after a charge in tank **10** is delivered to the tank **11**, and valve **13** closed, another measured quantity of hot oil is supplied to tank **10** by way of lines **51** and **33**, and another charge of frozen material is introduced into the tank **10**, by operation of the feed screw **21**.

Likewise immediately after a charge is delivered to the tank **12**, through the valve **14**, from tank **11**, screen **31** is turned to its vertical position, and a new charge of material received from the tank **10**. Thus at any one time, each of the tanks is carrying out its part of a complete dehydration cycle, but on different charges. If in a typical instance the treatment in each of the tanks requires not more than 10 minutes, then over a period of 30 minutes of operation three charges are processed. Particular temperature and time factors, and particular operating conditions, can be selected and controlled in accordance with the material being dehydrated, and the overall results desired.

The flow diagram Figure 1.4b facilitates an understanding of the operating cycle. Step **60** represents the operations in tank **10**, where the initial rapid evaporation takes place. In most instances from 60 to 90% of the moisture content of the material is removed in this initial phase.

Step **61** represents the operations within tank **11**, where dehydration is continued under vacuum at a lower temperature level. Step **62** represents the operations in tank **12** and the tank housing **39**, where free oil is removed from the dehydrated material by pumping from housing **38** and by centrifuging, all with continued application of a partial vacuum. Following or near the end of centrifuging a somewhat higher vacuum can be applied for evaporative cooling. Step **63** represents the condensing of water vapor from the first step **60,** with further condensing at **64** and evacuation at **65.** Step **66** represents the removal of an oil fraction from the condensing water, with the return of such oil to the oil reclaiming system.

It will be evident from the foregoing that there is provided a method and apparatus which functions in a semicontinuous manner, for the dehydration of various moist food materials. The equipment has a relatively high capacity for its size. The cost of installation, and the labor requirement to maintain its operation are substantially less than with apparatus of the batch type. Substantially all of the major control operations of the apparatus can be cycled automatically, thus making for uniformity with respect to treatment of successive charges, and reducing the labor factor to a minimum.

The apparatus can be used for dehydration of a wide variety of food materials, including fruits and berries, vegetables, cereals, condiments, meats, fowl and seafood. The larger fruits like peaches, can be peeled, pitted and sliced, or cubed into pieces not bigger than $5/8$ inch thick. Smaller items, such as cherries, blueberries and grapes can be cleaned and frozen without reduction in size. It is desirable that the particles be of such size that they do not weigh in excess of about 10 grams.

Fresh vegetables can be cleaned and prepared by use of conventional procedures, such as are employed in the frozen food and canning industries. Blanching, with or without sulfiting, can be applied before freezing to minimize enzymatic activity. Multiple perforating or scarifying can be applied before or after freezing, to such materials as peas, Chinese peapods and the like. Here again where the items are of substantial size they are reduced to particles (e.g., slices or cubes) of a size suitable for processing.

A wide variety of fruits may be treated by this method, including fresh apples, peaches, apricots, pineapple, cherries, bananas, dates, strawberries, blueberries, and the like. Vegetables which are applicable include peas, carrots, potatoes, celery, cabbage, bean sprouts, onions, peppers, sweet potatoes, cereals and the like.

FOAM-MAT PROCESSES

In the dehydration of fruit and vegetable juices a principal problem lies in the difficulty of obtaining products which will reconstitute readily. The mere subjection of juices to conventional dehydrating conditions such as exposing them to hot air or to the heated surface of a drum dryer will yield a dense, leathery product which has no practical value as it is virtually impossible to reconstitute.

It has been shown that fruit and vegetable juices can be successfully dried by exposing a layer of concentrated juice to vacuum under temperature conditions at which the juice remains in a puffed or expanded condition. Although this process yields an excellent product, it requires expensive equipment because the drying mechanism must be in a vacuum-tight system and the maintenance of the vacuum by steam ejection or the like during dehydration involves a considerable expense.

The foam-mat dehydration process was developed by A. Morgan, Jr., L. Ginette and others at the U.S. Department of Agriculture. It is a simple and inexpensive process which is accomplished under atmospheric pressure instead of under vacuum. The essential steps in the process are:

    (a)   the fruit or vegetable juice is converted into a stable foam by incorporating therewith a substantial volume of air or other gas plus additives;

    (b)   the foam in the form of this relatively thin layer is exposed to a current of hot air until it is dehydrated; and

    (c)   the porous mass is crushed to form easily rehydratable flakes or powder.

Apparatus for the Basic Process

A.I. Morgan, Jr. and L.F. Ginnette; U.S. Patent 2,955,046; October 4, 1960; assigned to U.S. Secretary of Agriculture describe the process with reference to Figure 1.5. A liquid juice concentrate, for example, tomato juice concentrate, is fed into aerator **1** which may take the form of a conventional device commonly used for aerating ice cream, salad dressings, or the like. Air and a foam stabilizer are likewise fed into the aerator to provide a foam of the proper volume and stability. The juice concentrate, now in the form of a foam, is fed into hopper **2** of dehydrator **4**.

The dehydrator includes a flexible, endless belt **5** made of rubber, natural or synthetic, which is tautly disposed about drum **6** and rollers **7** and **8**. The drum is driven by suitable mechanism to continuously traverse belt **5** in the direction shown. The hopper and driven feed roller **3** extend in width essentially the same distance as the width of the belt **5**. The feed roller, in cooperation with the hopper, deposits on the belt **5** a thin layer of the foam. By suitable adjustment of the position of the hopper and roller **3** above the belt **5** and control of the speed of the feed roller, the foam is deposited in a thin layer on the order of 0.01 to 0.2 inch. The layer of foam on the belt **5** is carried through dehydrating chamber **9** where it is dehydrated by contact with hot air.

FIGURE 1.5: FOAM-MAT DEHYDRATION

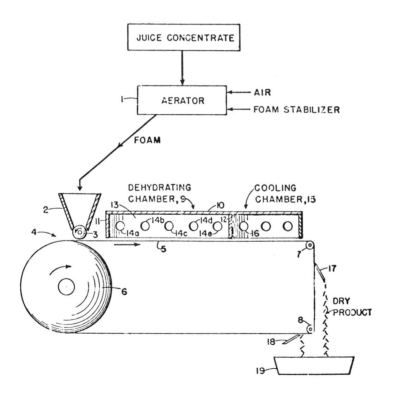

Source: A.I. Morgan, Jr. and L.F. Ginnette; U.S. Patent 2,955,046; October 4, 1960

The dehydrating chamber takes the form of a bottomless compartment, essentially as wide as the belt and is formed by top plate 10, end plates 11 and 12, and side plates 13, only one of the latter being shown in the drawing. Ports 14a, 14b, 14c, 14d, and 14e are provided for introduction of hot air. This hot air flows transversely across the layer of foam causing it to be dehydrated forming a solid, porous product. Ports or other suitable vents are provided in the opposite side wall 13 of the dehydrating chamber not shown in the drawing.

It is evident that the length of the dehydrating chamber, i.e., the distance from plate 11 to plate 12 is long enough to permit the layer of foam to be dehydrated in the time of its passage through the chamber. The chamber is not sealed from the atmosphere and the pressure therein is essentially normal atmospheric pressure. The temperature of the hot air entering the dehydrating chamber via ports 14a, 14b, etc. may range about from 120° to 220°F. During this dehydration, it is necessary that the volume of the foam be retained to yield a final product in porous, easily reconstitutable form.

After leaving the dehydrating chamber, the dehydrated product is next transported by the belt to cooling chamber 15 which is constructed essentially the same as the dehydrating chamber. Ports 16 are provided to contact the dehydrated product with a current of cool air. The cooling chamber is not sealed from the atmosphere and the pressure therein is essentially normal or atmospheric pressure.

The product as it leaves the dehydrating chamber is generally in a plastic state and would be relatively difficult to remove from the belt. By applying cooling, the product is put into a brittle state which facilitates its removal from the belt. Air having a low humidity is preferably employed in the cooling step to prevent moisture regain by the dry product. Generally, the volume and temperature of the air circulated through the cooling chamber are so regulated to reduce the product to a temperature of 100°F. or below, usually to about room temperature (70°F.).

The cool, dehydrated product is then carried by the belt about rollers 7 and 8. These rollers are deliberately of a small size so that the belt is subjected to a sudden change in direction. The belt being of flexible, rubbery material is able to repeatedly negotiate this abrupt turn without damage. However, the cooled, dehydrated product being now in a brittle condition is cracked into pieces and dislodged from the belt as it traverses rollers 7 and 8. The product now in the form of flakes or pellicles falls into receiver vessel 19. Doctor blades 17 and 18, spaced a small distance from the surface of the belt, assist in freeing the loosened particles of product.

The Basic Process

The basic process for which the apparatus was shown in the previous patent is fully described by *A.I. Morgan, Jr., J.M. Randall and R.P. Graham; U.S. Patent 2,967,109; Jan. 3, 1961; assigned to the U.S. Secretary of Agriculture.* In preparing the foam, a surface active agent is added to the liquid juice concentrate and air or other nontoxic gas such as nitrogen, carbon dioxide, nitrous oxide, helium, propane, n-butane, isobutane, dichlorodifluoromethane, trichloromonofluoromethane, trifluoromonochloromethane, etc. is incorporated. The proportion of surface active agent will vary depending on the properties of the juice concentrate, the properties of the agent in question, etc. In general, the proportion of surface active agent may vary about from 0.1 to 2.0%, by weight based on the weight of solids in the concentrate. It is naturally desirable to use the lowest proportion of surface active agent compatible with production of a stable foam.

A suitably stable foam is one which will retain its volume without any separation of gas from liquid for at least ½ hour, preferably at least 1 hour, when allowed to stand at room temperature. Incorporation of the gas into the concentrate containing added surface active agent may be accomplished by any of the conventional methods used, for example, in aerating ice cream, salad dressings, and the like. A simple and efficacious expedient is to subject the concentrate to a rotating wire whip which beats air into the material.

Another plan is to pump the concentrate through a conduit, a portion of which is of restricted cross section to form a venturi, the gas being introduced at the zone of high velocity and low pressure within the venturi and so thoroughly commingled with and dispersed into the concentrate. For best results, it is preferred that the gas bubbles in the foam be of uniform small size, i.e., about 100 microns or less in diameter. The proportion of gas incorporated into the concentrate is generally regulated so that the gasified concentrate (foam) has a volume at least 1.5 times that of the concentrate prior to introduction of the gas.

It is usually preferred that the foam have a volume about 2 to 3 times the volume of the concentrate to ensure formation of a highly porous dehydrated product. The foam volume may be increased above these levels to obtain even more highly porous products. Usually, however, it is desirable to limit the volume increase to about 5 times the original concentrate volume to avoid getting products having too low bulk density. That is, if excessive amounts of gas are added to the concentrate the dehydrated products although otherwise completely suitable from the standpoint of rehydration, taste, and color, will require too large a container to package a unit weight of product.

The concentrate may be cooled during introduction and dispersion of the gas; this generally promotes formation of a stable foam. If cooling is employed, any temperature below room temperature may be used provided the mass is not cooled enough to freeze it. Accordingly, temperatures not lower than about 35°F. are recommended. Having prepared a foam as described above, the foam is spread out in a relatively thin layer and subjected to dehydration at atmospheric pressure by contact with heated air. The thickness of the layer of foam may be varied. Generally, layers about ⅛ to ½ inch give satisfactory results. Suitable dehydration equipment is shown in the previous patent.

It is evident that during the dehydration the temperature of the product will rise and eventually equal that of the hot air stream. To avoid possibility of flavor damage by the product assuming too high a temperature, it is preferred to lower the air temperature in the final stages of the dehydration. Thus, for example, the air temperature in the final stage of dehydration may be at a maximum of 120° to 160°F. whereby the product temperature will not rise above these limits.

After contacting the layer of foam with hot air there is produced a solid dehydrated product having essentially the same volume as the foam and in a porous, spongy form. The product will generally have a moisture content of about 5%, or less. Generally, it is preferred to cool the dehydrated product before removing it from the tray, belt, or other equipment on which it was dehydrated.

When the dehydrated product is cooled to about 70° to 100°F. it is especially brittle and easy to remove from the surface on which it is located. The product breaks up on contact with spatulas or scrapers into a mass of flakes or particles. In such form the product is ready for use or packaging.

In preparing the foam for dehydration it is necessary to start with a juice in liquid concentrate form. Juices in their normal state are too thin and watery to form stable foams. The process may be applied to liquid juice concentrates prepared from any fruit or vegetable, for example, orange, grapefruit, lemon, lime, apple, pear, apricot, strawberry, raspberry, pineapple, grape, prune, plum, peach, cherry, tomato, celery, carrot, spinach, lettuce, cabbage, watercress, etc.

The juices may be prepared by subjecting the edible portions of the fruit or vegetable materials to such operations as reaming, pressing, macerating, crushing, comminuting, or extracting with water. The juice may be clear or contain suspended pulp. Methods of forming fruit and vegetable juices into liquid concentrates are well-known in the art. A typical method involves evaporating the juice under vacuum at a temperature of 50° to 150°F. to avoid heat damage to the product. For use in the process, the concentrate should have a solids content of at least 20% by weight. There is no upper limit in the solids content as long as the concentrate is liquid.

Example 1: Preparation of Foam — The starting material was a tomato juice concentrate containing 30% solids and of a pasty consistency. Into 100 parts of this concentrate was incorporated 0.24 part of sucrose dipalmitate (0.8% based on solids content of concentrate). The surfactant-containing concentrate was then whipped with a power operated egg beater, rotated at 500 to 700 rpm for 5 minutes. A foam having a density of 0.4 g./ml. was produced. A sample of this foam on standing at room temperature for two hours showed no change in height.

Example 2: Samples of the foam produced as described in Example 1 were spread on trays in layers ⅛ to ¼ inch thick. The trays carrying the foam layers were then placed in a cabinet type drier where they were subjected to hot air streams at 160° to 180°F. until dehydration was complete (final moisture content about 5%). The velocity of the air streams in each case was about 100 to 200 ft./min. In the following table are given the foam thickness, air temperature, and time for dehydration for each run.

Run	Foam Thickness (inches)	Air Temperature (°F.)	Drying Time (minutes)
A	¼	160	90
B	⅛	180	45
C	³⁄₁₆	170	60
D	¼	180	60

After dehydration was complete, the products were cooled to room temperature in a dry atmosphere to prevent reabsorption of moisture. It was observed that all the products had a porous texture and the reduction in thickness of the layers during dehydration amounted to less than 10%. By applying a spatula to the trays the products parted from the trays readily and broke up into flakes.

These flakes exhibited good rehydration properties. Thus by adding a suitable quantity of water and hand stirring with a spoon, reconstituted liquids free from lumps or grittiness were produced in 30 seconds or less. The proportions of dried product and water could be varied to obtain a reconstituted juice, a reconstituted concentrate or paste as desired. The reconstituted liquids showed no tendency of phase separation.

Hydrophilic Colloid Additive

The use of a nontoxic hydrophilic colloid which is added to the fruit or vegetable juice before the formation of the foam is described by *A.I. Morgan, Jr. and L.F. Ginnette; U.S. Patent 2,955,943; October 11, 1960; assigned to the U.S. Secretary of Agriculture.* The colloid is added to the liquid juice concentrate and air or other nontoxic gas such as nitrogen, carbon dioxide, nitrous oxide, helium, propane, dichlorodifluoromethane, trifluoro-monochloromethane, etc. is incorporated therein.

The chemical nature of the hydrophilic colloid is of no importance as long as it possesses the ability to stabilize foams. Various examples of suitable compounds are listed below. The proportion of hydrophilic colloid will vary depending on the properties of the juice concentrate, the properties of the colloid in question, etc. In general, the proportion of hydrophilic colloid may vary from about 0.1 to 4.0%, by weight based on the weight of solids in the concentrate. The foam is prepared for dehydration as described in the previous patent.

Typical examples of hydrophilic colloids which may be used in the process are listed below by way of illustration: soluble starch, sodium carboxymethylcellulose, methylcellulose, agar, gum tragacanth, gum arabic, gum acacia, carrageenan, sodium alginate, pectin, dextran, sodium carboxymethyl starch, sodium carboxymethylamylose, pentosans, albumin, gelatin, dried egg white, dried glucose-free egg white, and the like. Particularly preferred for use in accordance with the process is albumin and substances containing albumin such as dried egg white products.

Perforated Mat Process

An improvement on the original process is described by *L.F. Ginnette, R.P. Graham and A.I. Morgan, Jr.; U.S. Patent 2,981,629; April 25, 1961; assigned to the U.S. Secretary of Agriculture* which involves increasing the surface area of the foam by forming a mat of the foam to be dried and then impinging against the mat spaced jets of gas directed normal to the plane of the mat. Thereby the mat is perforated and its surface area is greatly increased. Consequently, when the perforated mat of foam is subjected to dehydration, elimination of moisture takes place more rapidly and with greater efficiency. The perforation is accomplished while the foam is on perforated support, as described below.

The foam to be dehydrated is spread onto a perforated surface, for example, a punched or drilled metal sheet. A blast of air or other gas is then directed through the perforations in the surface. This blast of air causes the portions of foam in and overlying the perforations to be moved away from the perforations toward imperforate sections of the surface. The net result is that the layer of foam is now perforated, the perforations in the sheet of foam corresponding with the perforations in the supporting surface. Because of the stiff nature of the foam, this new configuration is stable and is retained during subsequent treatment. The perforated foam is in prime condition for dehydration because its surface area has been multiplied many times.

Depending on such factors as the depth of the mat of foam and the structure of the supporting surface, particularly the proportion of free space therein, the surface area may be multiplied anywhere from 5 to 25 times, or more. Having prepared this perforated sheet or mat of foam, it is subjected to dehydration in any desired manner.

In a preferred modification, the perforated mat of foam is contacted with a current of hot air or other gas as in the known foam-mat method, except that in the present case, the tremendous area of the foam permits dehydration in a fraction of the time required with a solid sheet of foam, or, with equal time of dehydration a much greater amount of foam can be dehydrated.

Generally, mats about from 0.01 to 0.50 inch are used and give good results. Thicker layers can be used with advantage particularly where the relative volume of gas in the foam is high. In applying the foam onto the perforated sheet, the applicator means may be one that deposits the foam only onto the top surface of the sheet. Usually, however, it is preferred that the mat of foam extend into the perforations of the sheet.

This has the benefit that when the mat is cratered, the mounds about the individual craters are especially high whereby the surface of the cratered mat is especially uneven, hence offers a large surface area to the dehydrating medium. As the perforated surface, various structures can be used. A preferred structure is the ordinary perforated sheet metal of commerce which is provided with circular apertures in staggered rows.

Good results are obtained with such structures having holes from about $\frac{1}{16}$ to about $\frac{1}{2}$ inch in diameter spaced on centers to provide an open area of anywhere from 20 to 60% of the total area of the sheet. The maximum hole size is limited for any particular case so that the foam plug inside the hole does not flow out of the sheet before cratering. It is obviously desirable that the perforations, as is the case with commercial perforated sheet, be uniformly disposed over the area of the support. This contributes to providing the mat of foam with uniform spacing of perforations or craters so that diffusion of moisture will occur at essentially uniform rates from all areas of the mat.

For maximum increase in total surface area by perforation of the mat of foam, it is preferred that the support have as much open area as possible consistent with retention of sufficient strength in the sheet to make it usable. With supports having circular apertures, the maximum open area is generally around 60%. However, greater open areas with maintenance of structural strength can be provided with supports having square, rectangular, or triangular apertures.

The mat is then subjected to dehydration by any desirable means including vacuum dehydration, hot gas, etc. Examples will illustrate the process.

Example 1: One part of glyceryl monostearate is blended with nine parts of water at 160°F. The resulting emulsion is added to 300 parts of tomato paste (30% solids) at room temperature. This mixture is whipped to a stiff foam with nitrogen gas by passing it through a continuous mechanical foamer. Density of the foam is about 0.4 g./ml.

The foam is spread in a layer $\frac{1}{16}$ inch thick on one side and in the holes of a stainless steel perforated sheet. This sheet is provided with staggered rows of circular holes $\frac{3}{16}$ inch in diameter on $\frac{5}{16}$ inch centers; thickness of the sheet is about $\frac{1}{16}$ inch. The loaded sheet is then passed over an air manifold provided with a slot $\frac{1}{16}$ inch wide. Air is directed upwardly through this slot at a velocity of about 150 ft./sec. The layer of foam is thus perforated or cratered as described above. It is estimated that the surface area of the mat of foam is increased about five times by the cratering operation.

The sheet bearing the cratered mat of foam is then placed in a cabinet drier where it is subjected to a draft of air moving normal to the plane of the sheet. The air has a temperature of 160°F., relative humidity of 5%, and velocity of 5 ft./sec. After 8 minutes the sheet is removed from the drier. The product (containing 2% moisture) can be readily removed from the sheet by gentle scraping. None of the product has accumulated in the perforations of the sheet.

The product on spoon-stirring with cold water formed in a minute a reconstituted tomato paste indistinguishable from the original paste. In control experiments where imperforate layers of the tomato paste foam ($\frac{1}{16}$ inch thick) were exposed to the same drying conditions, drying times of 40 to 45 minutes were required.

Example 2: The process as described in Example 1 is repeated except that in this case 150 parts of orange juice concentrate (60% solids) is substituted for the tomato paste. A sample of the product added to cold water and stirred with a spoon for less than a minute formed a reconstituted orange juice indistinguishable in taste and texture from that prepared from the original concentrate.

Example 3: Tomato paste is foamed and formed into a perforated mat as described in Example 1. The sheets bearing the perforated or cratered foam are placed in a cabinet drier where they are subjected to a draft through the perforations. The air is initially at 180°F. and 5% relative humidity and is periodically changed over a 6 minute drying time to a final condition of 130°F. and 30% relative humidity.

This experiment duplicates conditions in a continuous through-flow drier employing concurrent air and product movement. The product containing 5% moisture, and still on the sheets, was removed from the moist air. It was then dried to 2% moisture in a stream of dry air at 120°F. A portion of the product on spoon-stirring with cold water for one minute formed a reconstituted tomato paste indistinguishable from the original paste.

Removal of Residual Moisture

In general, all of the known dehydration methods are effective to remove the bulk of the moisture from the food product. However, in all cases there remains in the product what may be termed a residual moisture content. The amount of this residual moisture will vary depending on the nature of the material being dried and the conditions of dehydration and in general it is about 5%, although it may be as low as about 2%. It is extremely difficult to remove this residual moisture content and, if it is attempted by conventional dehydration techniques, extremely long and costly processing times are involved.

A.I. Morgan, Jr., R.P. Graham and L.F. Ginnette; U.S. Patent 3,031,312; April 24, 1962; assigned to the U.S. Secretary of Agriculture have developed a process for removal of residual moisture which involves mixing the dehydrated product with a minor proportion of a volatile liquid and drying the resultant mixture. Addition of the volatile liquid causes a profound change so that when drying conditions are applied, the residual moisture (plus volatile liquid) is removed rapidly and effectively. The manner of drying is not critical, any conventional drying system whether under vacuum or at atmospheric pressure, can be used. Thus the volatile liquid has the critical effect of completely obviating the former difficulty of removing residual moisture.

The mechanism by which this drastic difference occurs is not entirely understood but it is believed that on addition of the liquid the water formerly bound to the solids in the product becomes associated with the liquid. In this condition the water has greatly increased mobility so that it is readily removed by applying evaporative conditions to the product containing the added liquid. It might also be said that the added liquid alters the properties of the solids-moisture system so that the moisture is provided with new, enhanced fugacity.

It will, of course, be appreciated that the particular volatile liquid for use in the process will be selected according to the use which is to be made of the final product. Thus where the product is intended for edible purposes, the liquid selected will be one which is edible or at least which may be ingested without harmful effects. Thus, for production of edible products it is preferred to use such liquids as ethanol.

Example: Orange juice was concentrated to 55% solids content. 100 parts of the concentrate was mixed with 0.4 part of solubilized soya protein and 0.1 part of low viscosity methylcellulose. The mixture was beaten in air to produce a foam having a density of 0.35 grams per milliliter. This foam was spread out as $\frac{1}{8}$ inch diameter extrusions on a belt moving through an air stream of 200 fpm velocity. The air temperature decreased from 180° to 130°F. during a 12 minute drying period. The crisp, dry extrusions containing 3% moisture were collected and crumbled.

10 parts of absolute ethanol was sprayed onto 100 parts of the above described dry, crumbled product, using enough mixing to form a homogeneous mealy mass. This mixture was then exposed to vacuum for 2 hours at 70°F. The resulting powder was found to contain 0.7% moisture and was free from ethanol odor. For comparative purposes, a sample of the dry, crumbled product, produced as described above, without addition of ethanol was exposed to the same vacuum for 2 hours at 70°F. The product in this case contained 2.6% moisture.

Addition of Thickening Agent for Improved Rehydration

An improved foam-mat process for dehydrating foods, such as fruits, with high sugar contents was developed by *D. Eolkin; U.S. Patent 3,066,030; November 27, 1962; assigned to Gerber Products Company.* The process comprises the steps of mixing a slurry of product solids and a thickening agent, foaming the slurry mixture with an inert gas, forming mats of the foam on a supporting medium, and dehydrating the foam mat formed by moving it through a dehydrating furnace having an atmosphere of inert gas in it.

Figure 1.6 illustrates the apparatus. The fruit, vegetable or other product puree is placed in a mixing tank 11 where the thickening agent, foam stabilizer and/or additional sugar may be added. Connected to the mixing tank is a source of a foam stabilizing or surface active agent 21, a supply of an appropriate thickening agent 22 and a source of sugar 23. Preselected percentages of the surface active agent, thickening agent and sugar may be added to the slurry in the mixing tank preparatory to placing the mixture in the gelation tank 12.

Heat is applied in the range of 160° to 200°F. to the mixture in the gelation tank in order to gel or fully activate the thickening agent in the mixture. From there the mixture is pumped into the flash cool unit 13 which cooperates with a condenser 26 and a vacuum pump 27 to reduce the temperature of the puree mixture preparatory to aerating the mixture in the modified foam mat apparatus 15. The slurry mixture is pumped into the foam generator 31 which is supplied with relatively cool inert gas from the temperature control

FIGURE 1.6: FOAM-MAT PROCESS USING THICKENER

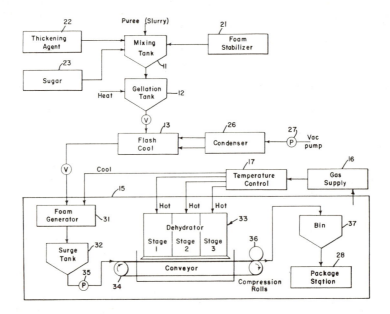

Source: D. Eolkin; U.S. Patent 3,066,030; November 27, 1962

means **17**. The inert gas is employed to aerate the slurry to foam the mixture to a volume in the neighborhood of 1½ to 5 times that of the unfoamed product. Any number of well-known methods may be employed to accomplish this. The slurry is cooled by the flash cool unit **13** and cool gas employed in order to improve the foaming characteristics of the slurry since heat has a tendency to adversely affect foam formation. This is especially true when air or some other oxygen containing gas is employed to foam the slurry.

In any event, the foamed slurry is moved from the foam generator into the surge tank **32** from whence it is carried by pump **35** and deposited on the belt of conveyor **34** as a relatively thin sheet. The moving conveyor carries the foam mat through the dehydrator or furnace.

In the dehydrator **33** three stages are employed in removing the moisture from the thin layers of foamed product. In the first stage, hot inert gas is supplied by way of temperature control means **17** to stage **1** at a temperature in the neighborhood of 300°F. As the foam becomes partially dehydrated, it passes through the second stage of the dehydrator which is supplied with inert gas in the 250°F. temperature range.

As the foam continues into the third stage of the dehydrator, the temperature of the inert gas is lowered to something on the order of 200°F. After the foam-mat has traversed all three stages, the moisture therein has been reduced to something less than 5% and the dehydrated foam-mat is passed through a pair of compression rolls **36** and deposited in bin **37** preparatory to flaking, packaging, etc., by package station **38**. The purpose of having a temperature gradient in the dehydrator is to take into account the propensity of the slurry product to scorch or burn as it is dependent on the amount of moisture in the foam-mat.

When the foam-mat first enters the dehydrator **33**, it is extremely moist and much higher temperatures may be used. As moisture is removed from the foam-mat, it is, to some extent, desirable to reduce the temperature in the furnace. Although nitrogen or a process gas, e.g., nitrogen and carbon dioxide, are supplied to all three stages of the dehydrator in the exemplary process, superheated steam, which acts as an inert gas, or a combination of air and nitrogen may be employed. For example, air might be used in the first stage where the moisture content is high and the product is not likely to scorch or burn.

Nitrogen might be used in the second and third stages where the tendency to scorch and adversely affect color and flavor is more pronounced. The use of air, however, would still have the disadvantage of oxidizing the product to some extent, thereby harming its nutritive content. It is also possible to vary the number of stages or eliminate them altogether.

After the dehydrated foam-mat leaves the dehydrator, it is important to maintain the inert gas atmosphere throughout the storage and packaging operations. It has been found that maintaining a relatively cool inert atmosphere for the storage, flaking, packaging, etc., steps provides an entirely satisfactory reconstitutable high sugar content product which has no tendency to cake or lump upon the addition of water. By relatively cool is meant somewhere in the neighborhood of 40°F., although it is possible to use lower temperatures if desired.

In discussing the dehydrator, temperature differentials in the order of 300°, 250° and 200°F. were noted. It should be apparent that other temperatures or gradients may be employed. As a matter of fact, it has been found that temperatures well above 300°F. may be employed without any adverse effects. This fact places the advantage of the inert gas dehydrating and processing in bold relief when compared to the use of air. In most prior art foam-mat processes in which hot air is used to dehydrate the foam-mat, the temperatures must not exceed 200°F. or so in order to avoid burning the fruit solids. With the use of an inert gas, the temperatures may be substantially raised and the overall process time substantially reduced.

Surface Sealing Treatment

A special characteristic of the foam-mat process is that it yields dehydrated products in an extremely porous state. In some instances certain problems have arisen which are attributable to the extremely porous character of the product. For example, when tomato juice concentrate is dehydrated by the foam-mat process, the product has a pink color rather than the typical red-orange tomato color. This color distinction is attributed to an optical phenomenon, that is, the light scattering effect of the myriad of minute voids in the dried particles.

Another point is that if the product is stored in contact with air it develops a brown color. This is attributed to an oxidation of the natural red coloring principle (lycopene) and is often noticeable in three days' storage in open containers at room temperature. The rapidity of the oxidation is again due to the porosity of the product, that is, tremendous surface area that it presents to the surrounding oxygen-containing atmosphere.

The problems explained above in connection with tomato products are of general occurrence with other materials dried by the known foam-mat process. Thus, the dried products exhibit an initial color which is paler than the raw material and when the products are stored in contact with air, oxidation reactions take place wherever the products contain oxidation-labile constituents, that is, constituents which are susceptible to oxidation by contact with oxygen.

The results of such reactions include color changes, development of unnatural odor and/or taste, destruction of vitamins, vitamin precursors, and other labile nutrient factors. In addition, many dry products prepared by the technique described above exhibit a very low bulk density. This may sometimes be undesirable as involving high packaging costs in marketing the product.

These disadvantages are avoided by a process developed by *R.P. Graham, L. F. Ginnette and A.I. Morgan, Jr.; U.S. Patent 3,093,488; June 11, 1963; assigned to the U.S. Secretary of Agriculture* which involves the following steps. The material to be dehydrated is gasified and formed into a stable foam. The resulting foam is then dehydrated, preferably by contacting it with hot air or other hot gas, at atmospheric pressure, to cause evaporation of moisture. The temperature of the gaseous dehydration medium is controlled so that the foam essentially retains its original volume during dehydration. When the material is thus converted into a porous dehydrated product it is subjected to a surface-sealing treatment.

Following this treatment, the material is again subjected to dehydration, if necessary, to remove any moisture added during the surface-sealing treatment. The resulting products are markedly different from those prepared by the usual foam-mat process. Thus, the products of the process exhibit the natural color of the starting material.

For example, a product prepared from tomato juice in accordance with the process exhibits a bright red color in contrast to the pale pink of prior foam-mat dried products. Moreover, the product of the process is stable in that it can be stored for long periods in contact with air without undergoing color, flavor, or nutritive changes. Furthermore, the bulk density of the products is materially increased so that packaging costs are less.

A convenient way to effect the surface-sealing is to place the dehydrated product on trays and expose it to heated moist air. Typically one may use air having a relative humidity of about from 60 to 100%. The temperature at which the step is carried out may be varied widely and generally more rapid surface-sealing is obtained with higher temperature. A convenient range of temperature is from about 100° to 212°F. In any case, the temperature should not be high enough to cause the particles to fuse or totally lose their porous character.

The time of treatment will vary by such factors as the properties of the product being treated, the moisture content of the moist air applied to the product, the temperature of treatment, and the degree of contact between the particles and the surrounding atmosphere of moist air. The proper time of treatment in any particular case may readily be gauged by observing the product under treatment. With pigmented products such as dehydrated tomato juice, orange juice, apricot puree, carrot puree, etc., an adequate surface-sealing effect is denoted by a marked change in color from a pale color to one distinctive of the commodity in question.

It is obvious that many alternatives are possible in applying the moist gas surface-sealing treatment. For example, as a matter of convenience, moist air is usually employed. However, the air in this case is merely a carrier or diluent and any other gas can be employed, as carbon dioxide, nitrogen, nitrous oxide, helium, etc. Naturally, when food products are being handled, a nontoxic gas is used as the carrier. The moisture can also be applied in the absence of a carrier gas as by treating the product in a sealed vessel whereby it may be contacted with pure water vapor at a selected pressure and temperature.

Although it is generally convenient to treat the product while it is supported on a flat surface, such as a perforated tray, one may put the product in a rotating cylinder of screening or perforated metal and contact it with moist gas. In this way the rotation of the cylinder will effect a tumbling of the particles so that faster and more uniform surface treatment will result.

It is evident that where the surface-sealing treatment involves an application of moisture to the product, it is necessary to then redry the product. This final dehydration can be accomplished by any of the methods previously described for the original dehydration. As an example, the moistened product is contacted with hot air, for instance at 100° to 300°F., until it is redried to the desired degree. Usually, the final product contains 6%, or less, of water. In a preferred method of conducting the surface-sealing treatment, compression is applied to the material. This compression is usually done by passing the material between a pair of rotating drums although it is evident that other systems can be applied such as rolling a drum over the material while it is supported on a flat surface. The drums, rollers, or other surfaces which contact the material are preferably heated as by internal circulation of steam or hot water. The amount of compression is regulated so that the product is made denser yet not compressed enough to eliminate all its porosity. In any particular case the proper degree of pressure can be gauged by observing the character of pilot batches of material which have been subjected to compression and selecting a degree of pressure which causes the surface of the material to be sealed, that is, fused or glazed, while the product, especially in inner portions thereof, is still in a porous condition.

To facilitate this pressing and to avoid rupture of cellular structure, etc., the material is tempered as by moistening and/or heating prior to application of the pressing operation. The degree of such tempering required in any particular case will depend on a variety of factors including the nature of the material, its moisture content, the temperature at which the pressing surfaces are maintained, the degree of densification to be achieved, and the like.

Example 1: (A) The starting material was a tomato juice concentrate containing 30% solids and of a pasty consistency. Into a lot of this paste was incorporated 1% of glycerol monostearate. The material was then whipped with a power operated beater until there was produced a foam having a density of 0.38 g./ml. The foam was extruded in the form of $\frac{1}{8}$ inch diameter spaghetti onto the surface of a Teflon-fiber glass belt.

The belt was passed through a cross-flow drier wherein the foam was contacted with air at 160°F. for 12 minutes and then with air at 130°F. for 3 minutes. The dehydrated foam product was cooled to room temperature and removed from the belt. The product had a moisture content of 3% and in color was pale pink.

(B) The dehydrated product prepared as described above was spread on a Teflon-glass fiber sheet at a loading of 50 grams per square foot and placed in a steam blanching chamber wherein the product was exposed to steam at 212°F. for 3 minutes. The product (moisture content 20%) was removed from the blanching chamber and redried in the cross-flow drier applying air at 130°F. for 5 minutes. The moisture content of the product was 3%; its color was a deep tomato red. The volume of the product was approximately 50% of that before applying the moistening step.

Example 2: A lot of dehydrated tomato concentrate prepared as in Example 1, part (A), was heated to 130°F. in an oven, then passed between two drums 12 inches in diameter, rotating at 1 rpm, heated to 200°F. and spaced 0.005 inch apart. The resulting product was observed to have a deep tomato red color.

Stabilization of Foam with Soy Protein and Cellulose Ethers

One difficulty which has been experienced with the foam-mat drying process is the lack of stability of the foam during the heating cycle. If the foam does not remain stable, cellular breakdown occurs causing serious impairment of the drying operation.

R.C. Gunther; U.S. Patent 3,119,698; January 28, 1964 and U.S. Patent 3,119,699; January 28, 1964; both assigned to Gunther Products, Inc. describe a process for producing heat stable foams by employing in the foam-mat drying process a water-soluble enzyme-modified soy protein in intimate association with a water-soluble cellulose alkyl ether or a water-soluble cellulose hydroxyalkyl ether, or a water-soluble cellulose in which the hydroxy groups of the cellulose are etherified with both alkyl and hydroxyalkyl groups.

Examples of suitable alkyl groups which may be present in the preparation of the water-soluble cellulose ethers are methyl, ethyl, hydroxyethyl, and/or hydroxypropyl groups. One specific example of a suitable water-soluble cellulose ether is the water-soluble cellulose methyl ether known as Methocel which is essentially the dimethyl ether of cellulose and which is made by etherifying purified cotton or wood cellulose by reacting an alkali cellulose with methyl chloride.

Specific cellulose ethers which are suitable are Methocel-60 HG, Methocel-65 HG, Methocel-70 HG, and Methocel-90 HG. The water-soluble cellulose ethers which are effective for the purpose of this process are all characterized by the fact that they form fluid solutions at room temperature, the viscosity of which increases when the solutions are heated and decreases again when the heated solutions are cool. The ability to thicken or stiffen on heating may be called thermal gelation and this in combination with the ability to become fluid again on cooling may be called reversible gelation.

This unique property allows the fluid mixes of products which it is desired to prepare in dry form to be whipped to the desired foam density with air or other gases and the foams produced to be stabilized on drying due to the thermal gelation property. The particular advantage of the reversible nature of the gelation property lies in the fact that these water-soluble cellulose ethers can be used to stabilize foams prepared from juices, such as prune, grape and the like, in which the reconstituted juice must have good clarity with no coagulated or gelled insolubles present.

For all practical purposes it may be stated that the concentration of the water-soluble cellulose ether may vary from 0.025 to 5.0% based on dry solids present in the system. For example, 200 parts by weight of prune juice with 25% solids would have 50 parts of dry solids, and 0.1% dimethyl ether of cellulose based on this would be 0.05 part of dimethyl ether of cellulose.

An enzyme-modified soy protein is another substance which may be employed in the process. Such soy proteins are made by steeping soy bean material in acidified water to remove the bulk of the solid nitrogen-free extract, subjecting the remaining material to hydrolysis with an enzyme, separating the solubles from the insolubles and concentrating the solubles. A suitable method for manufacturing an enzyme-modified soy protein for this purpose is described in U.S. Patent 2,489,173.

This substance is the essential ingredient of the substance designated D-100 whipping agent (Gunther Products, Inc.). The quantity of the enzyme-modified soy protein should be sufficient for whipping (i.e., to entrap air or other gas with the substance being treated). Generally, 0.1 to 5% by weight of the dry solids is sufficient and 0.4 to 1.5% by weight of the dry solids is preferred. The foamed products will usually have a density within the range of 0.15 to 0.60 g./cc.

Example 1: To 200 grams of concentrated, unsweetened orange juice (40% solids) was added 0.25 gram Methocel-65 HG (4,000 cp.) (10 grams of 2.5% solution) and 0.9 gram of D-100 vegetable protein whipping agent (3 grams of 30% solution). The mixture was stirred for a few minutes, then whipped on a Hobart Kitchen Aid mixer equipped with a wire whisk for 4 minutes. A firm, light foam was produced which weighed 0.23 gram per cubic centimeter.

This foam was spread to approximately ¼ to ⅜ inch thickness on a Teflon coated fiber glass mat and dried in an oven at 155°F. Drying time required was approximately 1 hour. The mat produced showed no signs of breakdown during drying. It could be readily crushed and the powder readily reconstituted in water to give an orange juice identical to the original juice.

Example 2: 200 grams prune juice (22% solids) was mixed with 0.375 grams Methocel-65 HG (4,000 cp.) (15 grams of 2.5% solution) and 0.3 gram D-100 whipping agent (1 gram of 30% solution). The mix was whipped with a wire whisk to produce a firm foam with a creamy consistency.

The foam had a density of 0.21 g./cc. A small sample dried at 170°F. and a larger sample dried on Teflon at 160°F. produced excellent products with no sign of breakdown or discoloration. The resulting powder could be reconstituted very readily in water to give a clear, excellent tasting juice.

Rehydration in Presence of Enzymes

It has been demonstrated that the flavor of dehydrated foods can be enhanced if they are rehydrated in the presence of certain enzyme preparations. It is postulated that the enzyme acts upon flavor precursors present in the dehydrated product, forming natural flavor. In a sense, the enzyme preparation may be considered as developing the latent flavor present in the dehydrated product. Such enzymes may therefore be termed as flavor-developing enzymes.

A.I. Morgan, Jr. and S. Schwimmer; U.S. Patent 3,170,803; February 23, 1965; assigned to the U.S. Secretary of Agriculture have developed a method in which two dehydrated ingredients are mixed prior to rehydration to produce foods which exhibit enhanced natural flavor. The two ingredients are listed as follows.

Ingredient A, which is used in major proportion, is a dehydrated food prepared by conventional methods. In the preparation of this material the natural produce has been subjected to substantial heating and as a consequence the dehydrated material is deficient in natural flavor.

Ingredient B which is used in minor proportion, is also a dehydrated food but of characteristics totally different from that of Ingredient A. Thus Ingredient B is a dehydrated food which has been prepared by a process where no enzyme-destructive heating is applied to the food during the course of its processing. Typically, Ingredient B is prepared from natural produce, for example, raw fruit, raw vegetables, raw meat, raw milk, or other food material containing active enzymes, by subjecting it to freeze drying.

This ingredient, not having been subjected to enzyme-inactivating heating programs, contains an essentially full complement of the enzymes appropriate to the produce in question. As a consequence, Ingredient B can be considered to contain the specific enzymes required to convert the latent flavor of Ingredient A into patent flavor. Thus, when the product of the process, containing Ingredients A and B in admixture, is rehydrated for use, the desired natural flavor will be developed in the reconstituted product.

The simplicity of the technique of the process is evident from the above description. Preparation of Ingredient A is simply an application of standard dehydrating methods and preparation of Ingredient B is simply an application of the lesser used but well-known freeze-drying method. Once Ingredients A and B are provided, they merely need be mixed together to provide the product.

In all cases only a minor proportion of Ingredient B is needed because the restoration of flavor is essentially a catalytic process. In general, Ingredient B is used in a proportion of about from 0.1 to 10 parts thereof per 100 parts of Ingredient A. In many cases, a proportion of about 1 part Ingredient B per 100 parts Ingredient A gives excellent results. Ordinarily, Ingredients A and B are both prepared from the same kind of food product. For example, both are prepared from tomato juice, orange juice, cabbage, or whatever food is selected. The preferred method of preparing Ingredient A is the foam-mat drying technique.

Example 1: Ingredient A in this case was commercial dehydrated cabbage. Ingredient B was prepared by chopping fresh cabbage, freezing it in a blast freezer to about –20°F. and subjecting it to vacuum until dry. The product was prepared by admixing 200 parts Ingredient A with 2 parts of Ingredient B. As a test, the product and a sample of Ingredient A by itself were each put into water (using 20 parts dehydrated material in 120 parts water), held at room temperature for one hour, then tasted. It was found that the product had a more natural cabbage flavor than did Ingredient A alone.

Example 2: Ingredient A was dehydrated grapefruit juice prepared by the foam-mat drying method. Ingredient B was prepared by peeling and slicing fresh grapefruit, freezing the grapefruit slices in a blast freezer, and subjecting the frozen material to vacuum until dry. The product was prepared by blending 1,000 parts of Ingredient A with 1 part of Ingredient B. To provide a comparative test, the product and a sample of Ingredient A by itself were reconstituted, using about 0.1 part of dehydrated material per part of water and holding the reconstituted juice at 37°C. for 3 hours. It was found that the juice prepared with the product had more natural grapefruit flavor than did the juice prepared with Ingredient A by itself.

Improvement of Drying Rate

R.J. Osborne, N.N. Potter, J.V. Fiore and T.K. Kelly; U.S. Patent 3,266,559; August 16, 1966; assigned to American Machine & Foundry Company have developed a process for providing a foam dehydration method which very significantly expedites the drying rate thereby lending improved commercial significance to foam material drying.

Heretofore it has been customary to dry foams in relatively thick layers, usually thicknesses in excess of ⅛ inch. Layers of this depth have required drying periods in the order of 15 to 30 minutes, and longer with many products. This relatively long drying period had detracted significantly from greater commercial utilization of this technique.

It was found that by careful control of a combination of conditions including foam layer thickness, underside heating of the support surface or belt, drying air temperature, velocity of air and direction of flow of air, a superior product is derived. By drying in accordance with the process, drying periods have been reduced to less than ¼ the time heretofore generally required and with some products less than ¹⁄₁₀ the period heretofore required for drying like amounts of the same materials.

A suitable food emulsion or suspension in a suitable concentration, generally about 20 to 60% solids, is introduced together with a foam stabilizing emulsion into a mixing vessel. The mixture is then fed into a foam generating apparatus where air or other suitable gas is incorporated, as by beating, into the liquid mix by an appropriate mechanism until a substantially rigid foam, in which the comminuted material to be dried is suspended, is produced. The foam prepared in this, or other suitable manner, is continuously fed onto an endless belt as by casting a layer in thicknesses which are generally kept below about 100 mils and preferably thicknesses between about 5 and 40 mils.

The foam fed from the foam generator is deposited by means of a suitable casting assembly onto an impervious metal belt of stainless steel. The foam layer is carried by the conveyor through the areas where drying occurs and moisture is reduced to about 1 to 10%. The belt is provided with a heating arrangement which may comprise one or more steam boxes to heat the underside of the belt. This under belt heating means preferably comprises a plurality of units in which each steam box unit may be separately regulated as to heat input although a single unit may extend the length of the belt which it is desired to heat.

Preferably, the steam box arrangement employed comprises steam impingement units into which steam is fed so as to impinge against, and condense steam on the underside of the belt. Optionally, the steam may be mixed with air and then the mixture is impinged against the belt underside. In the utilization of steam admixed with air, the air, introduced into the steam, is utilized to control the temperature of the steam in a given zone thereby permitting drying at lower temperatures which may vary, for example, over the range of from about 140° to about 212°F., using steam at normal atmospheric conditions. This concept offers an important advantage in the utilization of atmospheric steam while providing optimum control over a wide range of temperatures.

Drying of the product is further aided by the utilization of a relatively high velocity, hot air stream over the product. An important advantage in accelerating drying resides in conducting the drying material in the drying line through a series of zones into which a relatively high velocity drying gas or air is fed. The gas or air in these zones is introduced in a direction transverse to the movement of the foam material on the conveyor. It will be apparent that additional zones are contemplated depending on the apparatus used, the material in question to be dried, the temperature conditions desired, etc. The moisture laden air passing transversely over the product is exhausted below the belt surface. Preferably, the temperature at the upper surface of the belt is maintained between about 150° to 195°F. After a suitable period of drying the product containing relatively low moisture content in the range of about 1 to 4%, preferably 2 to 3%, is doctored and conveniently collected as by conveying through a conveyor arrangement to product packaging and storage facility. For certain products it may be desirable to utilize inert drying gas in lieu of a portion or all of the drying sequence. This may be effected, for example, by introducing an inert or desiccated gas or special treating gas into the last of the three chambers only or into two or more of the plurality of such gaseous treating units.

In general, the entire drying operation, from the time the material is cast on the belt until it is doctored off the conveyor support surface, is effected in less than 5 minutes. Preferably, drying is accomplished in less than 3 minutes. This time depends upon the product being dried and the selected temperature zones chosen both below and above the belt for optimum drying.

While it may be expected that as the film thickness is decreased, the loading per unit area would be decreased and the productivity of a drying unit would be reduced, it was found that not only is productivity not reduced, but indeed it is substantially increased (by a factor of at least two in going, for example, from an 80 to a 30 mil film). Productivity increase arises from the fact that the drying time to 3% moisture decreases much more rapidly than the film thickness. The following example illustrates the process.

Example: Commercial tomato paste, of 30% solids was blended with a 3% aqueous solution of polyglycerol stearate (Emcol-18) to give 0.75% of the foam stabilizer based on tomato paste solids. This mixture, at 50°F. was whipped in an upright mixer with wire whip planetary agitator to a foam density of 0.52. The resulting foam was cast onto the moving belt at a foam bed thickness of 25 mils. The foam deposited on the stainless steel belt was dried by passing through the steam box high velocity heated air drying unit of the kind previously described.

The belt entering the steam box zone was below 100°F., and emerged from the steam box zone at 190°F. This temperature was achieved by condensing steam on the bottom of the belt. High velocity, heated air at about 1,500 fpm was blown transverse to the direction of belt movement down on the material at the center of the belt and exhausted at the sides. The product was dried to 3% moisture in a residence time of 60 seconds. Under these conditions, a rate of 1.0 lb. of dried product per square foot of dryer area per hour was obtained. The dehydrated product was continuously doctored from the moving belt. The tomato powder was of high, uniform quality, which could be readily reconstituted in hot or cold water and possessed typical tomato taste, color and aroma.

Foam Stabilization with Glycerin Esters

J.V. Fiore; U.S. Patent 3,323,923; June 6, 1967; assigned to American Machine & Foundry Company describes foam stabilizers of the following classes: (1) ester derivatives of the monoglycerides of the higher fatty acids; (2) polyglycerin; and (3) esters of polyglycerin. Examples include: the acetyl, propionyl, butyryl or caproyl, etc., derivatives of glycerol monolaurate or monostearate or glycerol monooleate, etc.; the ester derivatives of oxalic, succinic, tartaric, citric, malic or oxalacetic, etc. acids and glycerol monomyristate or glycerol monopalmitate, etc.; and the sulfuric or phosphoric, etc., acid esters of glycerol monolaurate, glycerol monostearate, etc. An especially suitable compound in stabilizing food foams and in producing an excellent product of fine taste is the diacetyl tartaric acid ester of glycerol monostearate available commercially as Emcol D-66-1.

Examples of the polyglycerin compounds of the second group are glycerin polymers which are produced by the alkaline polymerization of glycerin under suitable conditions. A polyglycerin in mixtures of this kind is not a single compound but a mixture of compounds. Compounds of the third group useful as stabilizers are the esters of polyglycerin which include those derived from mono- or polycarboxylic acids as well as those from inorganic acids or mixtures of one or more of the aforementioned acids. Examples of this class of compounds include: the organic acid esters of polyglycerin such as the acetic, butyric, lauric, stearic, oleic, tartaric or citric acid esters, the inorganic acid esters of polyglycerin such as the sulfuric or phosphoric, etc., esters and the like. Especially preferred are the higher fatty acid ester derivatives of polyglycerin, in particular, polyglycerin stearate available commercially as Emcol 18.

In utilizing the compounds in the preparation of foam material to be dehydrated, a liquid concentrate of the material to be dehydrated is first diluted or concentrated, depending on the material, to a fairly viscous condition and converted into a stable foam by incorporating therewith a minor proportion of the foam stabilizing agent of the type described above together with a substantial volume of air or other gas. The foam so produced is then exposed to heat in the form of a relatively thin layer until it is dry.

Example: Polyglycerin prepared in accordance with Example 4 of U.S. Patent 2,023,388, was added to tomato paste 30% solids at a level of 1.0% based on tomato solids. The mixture was whipped in a vertical beater with a wire beater for 10 minutes. The resulting foam of 0.52 g./cc density was cast on a stainless steel belt in a layer thickness of 25 mils and dried using a belt temperature of 195°F. and air temperature over the belt of 175°F. The drying rate was 1.2 lbs./hr. per sq. ft. of drying surface. The product had good solubility and reconstitution properties and had good flavor, aroma and color.

Prevention of Foam Formation During Rehydration

The foam-mat dehydration process gives the best results when a foaming agent is used. However, the use of foaming agents causes the formation of extremely small microscopic bubbles which produce a large surface and permit rapid and more complete drying. Certain physical treatments of the liquid also permit the formation of fine bubble structures. Treatments to form these fine bubble structures have several advantages, including the removal of moisture to a more complete degree under less severe conditions of time and temperature.

This permits the retention of more flavor and aromatic character to certain juices, particularly fruit juices, and permits the preparation of powders having a high degree of storage stability. However, the powders thus produced often contain incorporated microscopic air bubbles. Upon reconstitution of these powders, these microscopic air bubbles cause the formation of air sols or colloidal dispersions of these air bubbles dispersed throughout the solution. These bubbles impart to the reconstituted juice an unnatural, white, cloudy or milky appearance.

After the reconstituted product stands a very short time, this air sol rises and collects on the surface as a dense layer of foam. The presence of this milky appearance and particularly the presence of the foam are detrimental to the physical appearance of many reconstituted products and adversely affect their commercial acceptance.

The process by *R.E. Berry, O.W. Bissett, C.J. Wagner, Jr. and M.K. Veldhuis; U.S. Patent 3,379,538; April 23, 1968; assigned to the U.S. Secretary of Agriculture* avoids the above disadvantages. It involves the following steps:

(1) the dried pieces are first prepared by one of the common foam-mat processes;
(2) the minute air bubbles contained therein are then removed by the application of a high vacuum. The length of time required to permit the diffusion of the gases from the dried pieces to a point of equilibrium is dependent not only upon the size of the pieces but also upon the extent of the vacuum applied, and, of course, may be readily determined by those skilled in the art; and
(3) after the absorbed gases have been removed by the vacuum treatment, the vacuum may be retained by sealing the container or the vacuum

may be released by means of a nontoxic water-soluble gas, such as carbon dioxide, and the tendency to foam upon reconstitution will still be inhibited.

It is an advantage of the process that the above treatments may be applied to products which have been dried by any of the prior art methods.

In general, in carrying out the process, a microscopic air bubble containing dehydrated food product, prepared by a foam-mat process, is deposited into an open container. The food product may be subjected to a warm-rolling treatment prior to being deposited into the container and may be in the form of a powder, flake or granule. When a powder is used, the particle size ranges from about 0.015 to 0.025 inch.

The open container with its contents is then subjected to a vacuum ranging from about 1 to 5 mm. Hg absolute pressure, and the vacuum maintained until a state of equilibrium is obtained, but not exceeding 72 hours. The container may then be directly sealed to maintain the equilibrium, or, as an alternative, the vacuum may be released with a water-soluble gas, such as carbon dioxide, sulfur dioxide, nitrous oxide, and ethylene oxide to give an absolute pressure ranging from about 38 cm. Hg to about 760 mm. Hg, and the container then sealed to maintain the equilibrium. Thereafter, the sealed container is re-opened and an aqueous fluid which may be a juice, concentrate or puree, is reconstituted from the resulting dehydrated product. The following examples illustrate the process.

Example 1: In the following example a concentrated orange juice is dried by the foam-mat process in which a 95% glyceryl monostearate is used as the foaming agent. A portion of the dried powder (product) is then reconstituted to give a liquid comprising the ingredients of freshly prepared juice from the tree ripened oranges. The resultant juice has a milky appearance and, after a period of time, a substantial layer of foam forms on the surface making this reconstituted juice commercially unacceptable.

Example 2: Another portion of the dried product of Example 1 is treated by the warm-rolling process of U.S. Patent 3,093,488 after which a portion of the dried product is reconstituted. The resultant fluid product also has a milky appearance but the formation of foam is less than in Example 1. This reconstituted product likewise is not commercially acceptable.

Example 3: In this example a portion of the dried product of Example 1 is finely ground and the portion that passed through a 35 mesh screen is subjected to a warm-rolling process as in Example 2. The reconstituted product has a slightly milky appearance and some foam forms after a period of time but the reconstituted product is superior to that of Example 2.

Example 4: Three portions of the dried product of Example 2 (treated by the warm-rolling process) are placed under vacuum (1 mm. Hg absolute pressure) for 24, 48 and 72 hours, respectively. Each of the three samples shows a noticeable improvement in the appearance of the reconstituted juice and the reconstituted juice from the 72 hours sample closely resembles the juice from a reconstituted concentrate. These reconstituted products are commercially acceptable and are clearly superior to a juice reconstituted from a powder which has been packed under nitrogen gas. There is a distinct improvement in appearance as the time under vacuum increases up to 72 hours.

Increasing Density of Foam-Mat Dehydrated Foams

The products of foam-mat dehydrated orange juice, tomato juice or similar foods are highly foamed concentrations in particulate size, readily reconstitutable or soluble by reintroduction of the aqueous or solvent vehicle. By and large, such products maintain all the characteristics and advantages of the natural product without being encumbered by the bulk and weight of water or solvent. Notwithstanding the success of such product and its value as food concentrate, the foamed product formed by this method, and, indeed, by any other method, has a singular drawback. The foam particles obtained as a concentrate have a low density and high bulk due to a high degree of entrapped air. This is, of course, desirable from a resolubility standpoint, but not desirable from a packaging or marketing standpoint. Such product requires relatively large packages (primarily glass jars, tin cans or flexible pouches are used) which are not only initially costly but costly to handle and transport.

T.K. Kelly and H.J. Light; U.S. Patent 3,573,938; April 6, 1971; assigned to AMF Inc. have developed a process for producing a foamed food concentrate denser than heretofore known and yet of the same or higher degree of solubility than heretofore obtained. Such a product would have the advantage of greater weight per volume and therefore packable in smaller, more economical units.

A suitable food emulsion or suspension is prepared in accordance with known techniques and deposited within a storage device and casting box from which it is continuously and uniformly fed on to a moving endless belt in a thickness preferably below 100 mils. The surface is preferably of stainless steel which may, if desired, be coated with an antisticking material such as silicone, Mylar, etc. The belt is suitably supported by roller drums, one of which is driven by conventional connection to a drive motor. The food layer is carried by the conveyor through areas which are preferably enclosed from the surrounding space and where drying occurs and the moisture content of the slurry rendered below 10%.

To accomplish this, the belt system is provided with one or more independently controlled steam box dryers. Drying is further enhanced by the employment of circulating air above the belt. Upon drying of the product, it may be cooled by application of liquid or air cooling media through the cooler or by an air blast directly on the surface of the food. The product is then removed as it passes over the forward roller and collected in a storage container prior to packaging. A doctor knife may be employed to facilitate removal of the product from the belt.

The particles have an open cellular structure of relatively large bulk volume and low density. Such product is highly soluble and reconstitutable but requires large packaging to accomodate its bulk. In accordance with the described process, the product is compacted while on the belt and in a hot condition. Accordingly, there is provided a roller mounted above the belt and located outside the area of the last heating zone. The roller may, if desired, be coated with an antisticking agent, silicone, Mylar, etc., or be covered with Mylar or Teflon roll cover and is preferably mounted on vertically adjustable bearings so that its weight and the amount of compaction on the product may be adjusted as desired or required.

Suitable guard means and guide means may also be employed to overcome tracking, rubbing or other problems attendant upon the rolling of a drum over an endless belt. A fixed bar or idler roll may be placed on the underside of the belt to permit application of roll pressure uniformly to the product on the top side of the belt.

As the product passes from the last of the heating zones, it has essentially been reduced in moisture content to the desired dry level. However, it is still hot. Compaction at this point results in producing a product which unexpectedly retains its foam open cellular structure but which takes on an overall flake configuration. The compacted product retains the foam characteristics of the earlier product but is denser, darker in appearance and more flaky in structure. The denser particle and its flaky structure result in advantageous packaging features since its greater weight/volume ratios enable it to be packed in smaller containers. Because of its bulk/density ratios more accurate measures may be obtained and individual units may now be produced.

DRYING AT ULTRAHIGH FREQUENCIES

Many drying processes are slow and wasteful of space because of the difficulty in supplying the heat of vaporization of the natural fluids or juices within the material. In ordinary oven drying, the drying progresses from the outer surfaces inward and the energy required to evaporate the moisture on the inside must be conducted through the outer portion. It has been found that when discrete pieces or chunks of materials to be dried, such as fruit, vegetables, or meat, are dried by conventional hot air processes, the final dried product has an outer shell which is hardened more than the rest of the product and the product as a whole is wrinkled, shrunken and tough. Also, materials and food products dried in this manner rehydrate, very slowly and incompletely.

With the advent of radio frequency heating apparatus, a means of generating heat uniformly throughout a nonconducting mass was provided, and its use resulted in the rapid and uniform drying of many materials. Investigations in radio frequency heating included the dehydration of pharmaceutical products, foods, and even the cooking of foods. These investigations indicated that in the preparing of food pieces bad electrical effects including burning resulted when the process was carried out at the conventional radio frequencies. Accordingly, developmental work was conducted to determine the optimum frequency relationships for transferring the heat energy to the food particles being processed.

G. Tooby; U.S. Patent 3,249,446; May 3, 1966; assigned to Hammtronics Systems, Inc. has developed an improved method and apparatus for drying or desiccating food pieces or chunks on a substantially large order allowing it to be employed in commercial applications. The process results in an improved palatable, dried product which retains its natural color and is of a low density; even an improved puffed or expanded product may be produced. The foods to be dried are preferably prepared in pieces or chunks of bite size and may be fruit, vegetables, meat and the like having a natural fluid or juice therein responsive to the application of heat; i.e., heat-sensitive foods.

The drying method comprises subjecting the food pieces to electromagnetic radiation in the microwave region to supply the heat of vaporization of the natural fluids or juices therein so as to evaporate them from within the food pieces. During the time the food pieces are exposed to the electromagnetic radiation, they are also subjected to a heated, dry fluid arranged so as to carry away the fluids or vapors expelled from the surface of the food pieces.

This process is controlled so that the food pieces are not only free of burns or any other deleterious electrical effects, but also the outer surface is prevented from drying faster than the remainder of the particle, thereby avoiding the formation of a hard outer shell. The temperature of the fluid for conveying the expelled vapors away from the food pieces can be controllably reduced after a preselected drying interval to maintain a maximum rate of drying consistent with optimum product quality. This temperature change is a function of the moisture content of the food pieces which, while exposed to electromagnetic radiation, can most readily be measured by the power absorbed by the food pieces.

If a puffed product is desired, the electromagnetic radiation is increased, evaporating the water within the piece at a higher rate than it can diffuse to the surface so as to expand the food piece and which expanded shape will be retained after the drying procedure is completed. While following the procedure to obtain a puffed product, and during the final phase of the drying period, a short, high burst of energy is applied to the food piece to assure puffing.

This process is carried out by means of drying apparatus comprising a drying cabinet having a metallic casing including electrically conducting perforated walls within the casing for defining an electromagnetic cavity therein. The cavity defined in this fashion is provided with a plurality of low loss dielectric trays or shelves supported in the cavity, and which shelves are adapted to receive the food pieces to be dried.

The trays are also perforated in the same fashion as the cavity walls. An ultrahigh frequency generator for generating electromagnetic radiation, preferably of 1,000 megacycles, is associated with the drying cabinet. The ultrahigh frequency generator is coupled to the cavity so as to permeate the cavity completely with the electromagnetic radiation.

Means are also provided for introducing a heated, dry fluid, such as the ambient air, into the drying cabinet to carry away expelled fluids and vapors. Means are also provided for exhausting this vapor laden fluid from the drying cabinet through the cavity, the casing being arranged and constructed to provide a passage for the conveying fluid therebetween and the perforations or holes in the walls of the cavity and in the trays provide the means of communication between this passage and the cavity.

The product resulting from this drying method is one in which each food piece is dried throughout to substantially the same low moisture content, producing a food product which may be stored for a long time without deterioration. The dried food product will retain its natural color and have a low density, and as well may have a puffed appearance. The patent contains drawings and a detailed description of the process.

OSMOTIC DEHYDRATION

In a typical example of a dehydration process utilizing osmosis, the food to be treated, for example, sliced apple, is mixed with sugar or sugar syrup and the mixture allowed to stand. Thereby moisture diffuses from the food pieces into the surrounding sugary medium with the result that dehydration of the food is achieved.

W.M. Camirand and R.R. Forrey; U.S. Patent 3,425,848; February 4, 1969; assigned to the U.S. Secretary of Agriculture has found that superior results are obtained when the food is coated with a water-permeable membrane prior to subjecting it to the osmotic treatment. Thus, the food to be dehydrated is first provided with a membrane coating. The coated food is then contacted with an osmotic medium, for example, sugar or sugar syrup, in order to achieve dehydration by osmosis, i.e., diffusion of moisture from the food into the surrounding osmotic medium.

A primary advantage of the process is that it yields a greater degree of osmotic dehydration than is obtained without the membrane coating but under otherwise identical conditions. In fact, the membrane coating enhances the dehydration to such an extent that some foods can be dehydrated to a self-preserving level by the use of osmotic dehydration alone, that is, supplemental treatment by air-drying or vacuum-drying is rendered unnecessary.

Another advantage is that the membrane minimizes loss of desired constituents, e.g., flavor and nutrient materials, through diffusion into the osmotic medium. Moreover, the membrane coating minimizes diffusion of the solute (used in the osmotic medium) into the food. Thus, although the membrane coating enhances diffusion of water from the food to the surrounding osmotic medium, it impedes diffusion of nutrients and flavor components from the food to the medium, and also impedes diffusion of solute from the medium into the food. Another advantageous item is that the membrane coating exerts a protective effect. For example, the coating protects the final product from air, dirt, bacterial contamination, etc., while in storage.

Various substances may be used to provide the water-permeable coating. It is obvious that the substance must be edible and be film-forming, i.e., it must be of high molecular weight so that it will form a continuous film.

Typical substances which may be used are as follows: pectin or pectin derivatives, i.e., ordinary (high methoxyl) pectin; low methoxyl pectins; pectic acid or pectates such as sodium pectate or other water-soluble salt of pectic acid; cellulose derivatives such as ethyl-cellulose, carboxymethylcellulose, and the like; starch and starch derivatives, e.g., carboxymethyl starch, carboxymethyl amylopectin, and the like; polysaccharide gums such as tragacanth, arabic, karaya, etc.; and proteins such as gelatin, casein, zein, gluten, soybean protein and the like. The food may be coated with the selected substance in any of the usual ways well-known in the art. Particularly good coatings for the purposes of this process are provided by the low methoxyl pectins. The coating is formed as follows: an aqueous solution of a water-soluble low methoxyl pectin is prepared. The concentration of the solution is not critical; usually it contains about 1 to 5% of the low methoxyl pectin. The food is dipped in this solution, then removed and drained. It is further preferred to harden the pectinate coating by dipping the coated food in a solution of a nontoxic, soluble, ionizable calcium salt such as calcium chloride, acetate or nitrate. When this is done a metathesis occurs.

The alkali ions (usually sodium, potassium or ammonium) of the pectinate are at least partly replaced by calcium with the result that the liquid coating is changed to a gel coating which is smooth and slippery; hence, the coated food pieces show no tendency to cohere to one another. After dipping in the calcium hardening bath, the coated food pieces are drained, or, they may be additionally rinsed in water to remove any residual hardening solution.

After the food pieces have been provided with the coating, they are contacted with a suitable medium to attain the desired osmotic effect, that is, diffusion of moisture from the food into the surrounding osmotic medium. As the osmotic medium, one can use any nontoxic water-soluble substance, or mixtures of such substances. Sugars are preferred substances and, typically, one may use sucrose, dextrose, lactose, maltose, invert sugar, or other individual sugar or mixture of sugars.

The osmotic medium is preferably an aqueous solution of the substance. This provides several advantages. For example, the solution can be easily conveyed with pipes and conventional pumps to the treating vessel. After use, the spent solution can be readily concentrated to put it into condition for treating another batch of food. Also with the solution, separation of the food after the osmosis step is easier and more efficient than where a solid substance is employed.

In order to drive the osmosis in the desired direction, it is obvious that the solution should be hypertonic with respect to the food, that is, the concentration of solute in the solution should be higher than the concentration of water-soluble solids in the food. Generally, it is preferred to use aqueous solutions of sugars, i.e., syrups, especially those which contain at least 60% of sugar. Particularly preferred are sucrose or invert sugar syrups of at least 65 to 75% concentration. Invert sugar has the advantage that it is more soluble than sucrose so that it may be formed into syrups of higher concentration. Ordinarily, the osmosis step is carried out for convenience at room temperature.

In order to attain a useful rate and degree of dehydration, it is necessary to contact the food pieces with an adequate amount of the osmotic medium. For example, when a dry solute is used in the osmosis, one should use at least one part thereof per part of food. With the solutions or syrups even larger proportions are preferred, i.e., at least three parts per part of food. Typical examples of proportions of syrups of different concentrations which provide good results are: 75% syrup, 3 parts per part of food; 70% syrup, 5 parts per part of food; and 65% syrup, 7 to 10 parts per part of food.

In conducting the osmosis step it is simply necessary to contact the coated food pieces with the selected medium, in dry or solution form. The process can be conducted statically, i.e., by mixing the food and medium and allowing the composite material to stand. In the alternative, the composite may be subjected to continuous or periodic stirring or tumbling to sweep the food surfaces with fresh portions of medium, whereby to avoid formation of local areas of decreased solute concentration.

A similar result can be achieved by recirculating the medium about the food pieces, through the use of a pump, or the like. Generally, the use of dynamic conditions, such as stirring, tumbling, or recirculation of medium, provide a faster rate of dehydration. The osmosis is continued until the desired degree of dehydration has been attained. Evaporation may be applied concomitantly with osmotic dehydration. This may be done, for example, by forcing a current of warm air through the mixture of the coated food pieces and osmotic medium whereby to achieve dehydration of the food not only by osmosis but also by evaporation. Another technique is to apply vacuum to the system to achieve evaporation by vacuum concomitantly with dehydration by osmosis. After completion of the osmosis step, the food is separated from the medium. The separation may be readily accomplished by allowing the composite material to drain on a screen or centrifugation may be applied to attain a more complete removal of medium from the surfaces of the food pieces. If it is desired to get even more complete removal of medium, the fruit pieces may be given a quick rinse with water.

The resulting osmotically-dehydrated food may be further treated in various ways as may be desired. For example, in cases where the osmosis is continued long enough to reduce the moisture content to a low level so that the product is self-preserving, it may be packaged in the same manner as other dehydrated foods and stored at ambient temperatures until required for use.

In the event that the osmosis is continued only long enough to get a partial dehydration, for example, to reduce the weight of the food by 40 to 60% through loss of moisture, the product may be frozen or the partially dehydrated foods may be further dehydrated by conventional techniques, for example, by drying in a current of warm air; by vacuum drying with or without application of heat; and by freeze-drying, etc.

The process is of wide applicability and can be used for dehydrating solid foods of all kinds, for example, fruits, vegetables, meats, herbs, etc. Generally, as a preliminary step the food material is cut into strips, slices, cubes, chips, or other pieces as conventional in dehydration practice.

DEHYDRATION BY PRESSURIZED NITROUS OXIDE

K.L. Miles; U.S. Patent 3,511,671; May 12, 1970 describes a process which consists in intimately contacting foods with nitrous oxide gas. Nitrous oxide gas has the property of displacing moisture from within foods and appears to associate with foods in some unexplained manner. Contact of the nitrous oxide gas with a food is continued until the moisture contained in the food has decreased to the desired level. Nitrous oxide gas may be used alone or in combination with other gases.

Since nitrous oxide gas is relatively expensive, it may be preferred to use a gas comprising nitrous oxide together with a gas such as air, nitrogen or other inert gas. However, use of other gases in combination with nitrous oxide requires longer contact of the dehydrating gas with the food than when nitrous oxide is used alone. On balance, it is usually preferred to use a combination of gases containing a majority of nitrous oxide.

Furthermore, it is preferable for economic reasons to combine nitrous oxide dehydration steps with standard dehydration steps such as, for example, vacuum drying. Thus, for example, the process may comprise in combination a preliminary drying step using vacuum drying and a final drying step using nitrous oxide gas or it may comprise four steps where the vacuum drying and nitrous oxide drying steps are alternated or a series of cycles where each cycle comprises a standard drying step and a nitrous oxide drying step.

The latter process is particularly useful where sensitive foodstuffs such as grapes are being treated. The cycling process introduces nitrous oxide into the food after each small quantity of moisture is removed by standard techniques thereby maintaining the cellular structure.

The preferred method of performing this process comprises a sequential combination of the following steps: a preliminary preparation of the foodstuff including washing and trimming in a standard fashion (Step 1); a preliminary drying step (Step 2); a second drying step in the presence of a nitrous oxide (Step 3); a third drying step (Step 4); a final treatment with nitrous oxide (Step 5); and packaging for market or shipping (Step 6). Various modifications and additions may be made at different steps in the above general outline.

The first step comprising the preliminary preparation of the foodstuff is a conventional step, that is, it is fairly standard in food processing. This step involves the substeps of washing, eliminating undesirable parts and cutting. Depending upon the particular foodstuff to be cleaned and perhaps on economic factors, washing may be carried out by using water, steam or a suitable organic solvent. Following the washing period, all undesirable parts such as stems, seeds and blemishes are removed either manually or by mechanical means. Next, the foodstuff is cut into any desired shape appropriate for marketing or shipping. After the latter substep, the foodstuff material is conveyed into a dehydration vessel preparatory to proceeding with the second step.

As previously noted, a preliminary drying step not involving the use of nitrous oxide, is preferred for economic reasons. Therefore, the preliminary drying step is preferably carried out by a conventional method such as, for example, vacuum drying, hot air drying or sun drying.

The maximum amount of moisture removed in this step depends to a major extent on the particular food being processed. It has been found that the maximum amount to be removed should fall within the range of from approximately 4 to 45% by weight of the total initial moisture content. It is the usual practice to remove as much moisture in the preliminary drying step as possible without reaching the overdrying point. It has been found that removal of too little moisture in this step reduces the effectiveness of the nitrous oxide dehydration step (Step 3). By way of example, for best results with fruits and leafy vegetables, the moisture removed falls between about 30 and 45% of the initial moisture content.

The third step involves further drying the foodstuff or other material in an atmosphere of nitrous oxide. If nitrous oxide is used in combination with another gas, it is preferable that the nitrous oxide comprise a major portion of the gas mixture for economic reasons, i.e., to reduce the time required to form the nitrous oxide-dehydrated foodstuff bond. This step is of major importance since it is believed that the nitrous oxide, in addition to removing some of the contained moisture and inhibiting enzymatic oxidation, is retained by and supports the natural cellular structure of the food. A dehydrated product is thus produced which, upon subsequent hydration, has much improved appearance, structure and flavor.

If the preliminary drying is performed under vacuum, virtually all the oxygen will have been removed from the dehydration chamber and the nitrous oxide employed in Step 3 can be introduced immediately upon closing off the vacuum line to the dehydration chamber. Should another drying method have been used in the preliminary drying step, there will be air present which should be removed before the introduction of nitrous oxide. This is best accomplished by drawing a vacuum.

After the air and/or other gases are substantially removed, pressurized nitrous oxide is allowed into the dehydration chamber. The nitrous oxide in the dehydration chamber is normally kept under a pressure of approximately 7 to 21 psig. For most foodstuffs a pressure of 10 to 12 psig is preferred. However, for citrus fruits and tomatoes it has been found that a pressure of 16 to 18 psig produces the best results.

The nitrous oxide is circulated through or past the food material until an additional 10 to 15% of the original moisture content has been removed. These limits should be closely approximated to produce a superior foodstuff product.

Step 4 is a third drying step and is carried out using, for example, a conventional vacuum dryer. The dehydration is continued until the desired moisture content is reached. This is usually between 6 and 8% of the final dehydrated food weight; however, moisture contents of less than 1% may be attained. Although a vacuum will remove additional moisture and other gases in the food structure, it does not appear to remove any of the combined nitrous oxide, that is, that nitrous oxide apparently forming a unique bond with the foodstuff. Thus, the nitrous oxide will continue to support the foodstuff structure as the dehydration is continued in Step 4. In some cases the process may be terminated at the end of Step 4. However, in order to maintain the natural color and flavor of the foodstuff and to practically eliminate enzymatic activity for long periods of time, further exposure of the food to nitrous oxide is made by Step 5. Low pressure nitrous oxide, at from about 1.5 to 6.0 psig, is passed through the foodstuff for several minutes. Although the nitrous oxide could remain in contact with the food indefinitely, it has been found that a 2 to 4 minute nitrous oxide gas flow or exposure at 2 to 3 psig produces optimum results.

In addition to producing a dehydrated food which has exceptionally long storage life, this step is also useful for producing a food consistency which is especially suitable for grinding, i.e., for making into a flour. Following Step 5, the foodstuff may be put into any form desired for selling purposes such as, for example, a powder, a meal or granules. The food-stuff may then be packaged in any of the conventional ways of packing.

As described, this process may be used to dehydrate fruit such as apples, pears and bananas; citrus fruit such as grapefruit, lemons and oranges; berries such as blackberries, cranberries and gooseberries; leafy vegetables such as asparagus, cabbage, celery and spinach; vine grown vegetables such as beans, peppers and tomatoes; root vegetables such as beets, carrots and potatoes; and miscellaneous products such as meat, fish, spices, wood, coffee and tobacco.

These previously described steps may be performed either as a batch or as a continuous process. In the batch process, the foodstuff is placed in a vacuum pressure vessel having associated with it requisite heaters and vacuum and nitrous oxide lines. Each of the de-hydrating steps is then performed in the sequence previously described. In the continuous process, the dehydrating vessel of the batch process is replaced by a series of separate chambers containing foodstuff and capable of being moved to various locations such that, at each location, the foodstuff is subjected to one of the steps in numerically ascending order.

Example: This example indicates how vegetables may be mixed and as a mixture may be treated by this process. Onions, carrots, celery and potatoes were prepared by cutting off the slender roots and the undesirable tops, washed in warm water in conjunction with brushing and then diced into approximately ½ inch squares. Green peas, separated from their pods were then added and the vegetables mixed to an even distribution for a stew mixture.

The mixture was then placed in a container which, in turn, was put into a vacuum dryer. A vacuum of about 28.5 inches of mercury was applied. Heat was applied to keep the vegetables with their moisture at about the boiling point. This was continued until approximately 35% of the moisture was removed. The vacuum system was blocked off and nitrous oxides was introduced until the pressure was about 12 psig. Nitrous oxide at this pressure was then passed over the vegetables until an additional 10% of the original moisture was removed.

At this time about 45% of the total moisture had been removed. The vegetables were then subjected to a vacuum and dried until only 6% of the original moisture remained. To maintain a high storage capability in the vegetables, nitrous oxide was again introduced and maintained at about 3 psig for about 3 minutes. The mixture of vegetables was then removed from the dryer and packaged in polyethylene bags and sent to storage.

Under these conditions, the stew mix remained usable for a long time. To use, the mixture was put in cool or warm water. As rehydration took place, the foodstuff resumed approximately its original shape and condition. It was then cooked in a conventional manner. The patent contains drawings and detailed description of an apparatus which can be used in this process.

DISPERSION IN HYDROPHOBIC FLUID

Various processes are known for partially or totally dehydrating heat sensitive materials, such as food products. However, they all have certain disadvantages, most of which appear in the quality of the final product. Generally, the properties of the material have been impaired due to long exposure to heat and/or by local overheating at certain points, particularly in products which have been totally dehydrated. As a result, the natural flavors are lost, off-flavors may be imparted by the overheating, or the products are difficult to rehydrate. Those processes which dehydrate the materials with the least loss of their desirable properties are too expensive to permit their use on an economical basis.

A process for dehydrating heat sensitive materials has been developed by *R. Calderon-Pedroza and I. Resano-Gonzalez; U.S. Patent 3,567,469; March 2, 1971; assigned to CPC International Inc.* It comprises dehydrating various materials by forming a stable dispersion of such materials in finely divided form and in a hydrophobic substance, and continuously and rapidly evaporating moisture from the dispersion in a wiped film type of evaporator.

In this type of evaporation, a thin layer of the material to be dried is spread by means of a wiping blade against a heated surface and under vacuum and the resultant heated layer is continuously and rapidly displaced to the outlet of the system by means of the same blade. Simultaneously the interface of the material in contact with the heated surface, as well as the opposite one, directly exposed to the vacuum, are renewed, also, by means of the same blade.

A stable dispersion of the material to be dehydrated and the hydrophobic substance should first be prepared. The term "stable" is intended to include dispersions which are stable only long enough to permit the same to be spread in a thin layer in the evaporator as well as those which are stable indefinitely.

In preparing the materials to be dispersed, it may be necessary to first remove some portion of the original material, for example, peels and seeds from fruits and vegetables. The material thus obtained is then chopped or ground directly to a particle size not exceeding about 100 microns in diameter. Depending upon the sensitivity of the material, chopping and grinding should take place in an inert atmosphere and with the addition of preservatives and, in some cases, under refrigeration as well.

If the material contains a high percentage of water, such as tomatoes, it may be treated by mechanical action to break the cells thereby releasing a certain amount of juice. In order to retain flavoring compounds in the juice, the mechanical operation may be carried out in the presence of the hydrophobic substance, or the latter may be added after such operation and mixed thoroughly with the material. Following this, the free juice may be eliminated by mechanical means from the rest of the pulp.

It may also be desirable in some cases to treat the ground pectinous material to remove enzymes already present from it. For example, enzymes may be removed from avocado pulp, or the like, by grinding it in the presence of the hydrophobic substance and, if convenient, a dilute solution of calcium chloride may be added as well to reduce the dispersibility in water of the pectinous material and thus allow for a sharper centrifugal removal of water phase, which contains extracted enzymes.

The hydrophobic material may be added prior, during or after grinding but, at any rate, the products must be blended properly to obtain a high degree of dispersion. Should the dispersion exhibit too low viscosity for proper spreading, a hydrocolloid may be added.

If the viscosity exceeds the value specified either more hydrophobic substance may be added or the material to be dried may be treated mechanically to allow enzymatic breakdown of the hydrocolloids, an operation which may be carried out in a shorter period of time by addition of hydrolytic or other enzymes. The dispersion consists of the proper mixture of components having the following characteristics. The diameter of the individual particles therein should not exceed about 100 microns and they should be evenly distributed in the mass. The viscosity thereof should not exceed about 200,000 cp. measured at the operating temperature of the inlet of the dehydrating equipment under vacuum. The viscosity may be measured indirectly through the net power consumption of the feeding pump used. The viscosity should be sufficiently high to spread in a thin layer on the heating surface of the dryer and will vary with the material to be dehydrated and the hydrophobic substance used.

The practical range of operating ratios of hydrophobic substance to the solids of the material to be dehydrated, on a dry substance basis, has a minimum of about 0.1 hydrophobic to 1 of solids, and an upper limit of about 6 hydrophobic to 1 of solids. Any material, but particularly heat sensitive materials, may be treated by the process. Among the various materials which may be included are: fruits, in general; vegetables, in general; pectinous and proteinaceous materials, e.g., gelatin; fish and seafood products; meat and meat by-products; pulps from fruits and vegetables, such as from avocado and tomato; onions; garlic; horseradish; fruit juices; etc.

Among the hydrophobic materials suitable for food use are vegetable oils, particularly those which have low polarity. As already mentioned, a wiped film type of evaporator is used for purposes of dehydration. This type of evaporator or dryer operates under vacuum and the material fed thereto is continuously dehydrated in the form of an agitated thin layer. In addition, this type of equipment provides a short holding time for drying and low volume capacity.

The dispersion to be dehydrated should be spread in a thin layer onto the heated surface of the evaporator to a thickness not exceeding about 3 mm. The residence time of the individual particles in the thin layer of the dispersion in the evaporator may vary from about 2 or 3 seconds to about 10 minutes. The temperature may vary in accordance with known principles, but, generally, the temperature used in the evaporation should be such that the temperature of the resultant dehydrated product at the outlet of the evaporator does not exceed about 65°C. for heat sensitive materials and about 100°C. for those products which have been previously cooked or require such a temperature level to develop adequate final flavor.

The resultant dehydrated products made in accordance with the process above described may be used as such or, if too much hydrophobic substance, e.g., vegetable oil, is present, it may be removed and reused in the process, or otherwise handled, such as by refining and then reusing.

The dehydrated product may require conditioning so that it may be easily rehydrated and regenerated into a product having appearance, consistency and organoleptic properties similar to those of the original product, or of different specifications as required. The adjuncts used should be able to either absorb water or easily disperse in it. Among the many suitable products, we have found that the pectins, alginates, carrageenins, gums, mucilages, modified starches, starch derivatives such as ethers or esters, dextrins and cellulose derivatives may be satisfactory when used in adequate amounts and in proper combinations. In some cases, it may be necessary to add emulsifiers and antioxidants as well.

One of the most important advantages of this process is that the products obtained are easily rehydratable because they are not denatured nor damaged in any appreciable way. The process has the further important advantage that it can be carried out in well-known equipment under simple and economical operating conditions.

Example: 1,000 grams of avocado pulp cooled to about 5°C. and having a moisture content of 69% were mixed with a dilute ascorbic acid solution and the mixture was fed into a disk mill, at the same time that a stream of CO_2 was injected into the system. The inert atmosphere produced by the CO_2 prevented the contact of oxygen with the avocado pulp. The amount of ascorbic acid added to the pulp was 0.1%, fresh pulp weight basis. After disintegrating, the resulting paste of avocado was mixed with 200 grams of sesame oil at about 5°C. and, then, the mixture was passed through a colloid mill, while the inert atmosphere of CO_2 was still maintained.

The homogenized mixture was then fed to a short holdup evaporator (wiped film type) at a rate of 1,000 g./hr. The residence time of the material in the evaporator was of 4 minutes and the maximum temperature reached by the product in this equipment was 42°C. After properly mixing all of the pulp and segregated oil coming out from the evaporator, the resulting product had a moisture content of 4%. A mixture of hydrocolloids and other materials was separately prepared with the following composition.

	Percent
Sodium chloride	10.5
Pectin, high viscosity	13.0
Methylcellulose, 4,000 cp.	20.0
Dextromaltose	19.0
Dried skim milk	37.5

Ten grams of this mixture were added to an amount of dehydrated material equivalent to 100 grams of the original pulp. The blend was bottled in airtight containers under nitrogen atmosphere. The blend was kept under these conditions, at room temperature for several weeks, at the end of which it did not show any substantial change in flavor and color.

The rehydration of this product was carried out by adding water to it, in the amount of approximately 2.5 times the weight of the dehydrated material, with the aid of a fork. A product with physical and organoleptic characteristics similar to the original avocado pulp was obtained. After 3 weeks, unlabeled samples of the regenerated avocado and fresh ground avocado pulp were given to 9 persons. Two of them noticed a slightly stronger taste to avocado flavor in the regenerated avocado samples with no other difference in taste.

DEHYDRATION BY SOLVENT DISTILLATION

A process is described by *N.F. Toussaint; U.S. Patent 3,628,967; December 21, 1971; assigned to Florasynth, Inc.* which comprises immersing the food product containing removable water in a hydrocarbon liquid which boils between about 60° and 140°C. at atmospheric pressure, distilling the hydrocarbon liquid together with water at a temperature between 20° and 75°C. until substantially all of the removable water has distilled off, removing excess hydrocarbon liquid from the food but leaving enough to wet it then extracting the hydrocarbon liquid-wetted food with a second organic water-immiscible liquid boiling between about –50° and 30°C. until all traces of the first solvent are removed.

Apparatus suitable for carrying out the process is illustrated in Figure 1.7. Figure 1.7a shows a cross-sectional elevation of an apparatus suitable for the first distillation step of the process and Figure 1.7b shows a cross-sectional elevation of an apparatus suitable for the second liquid treatment step of the process.

In Figure 1.7a, food and hydrocarbon liquid are charged to two-necked flask 11 through one neck 12 which is thereafter sealed with stopper 13. Heating element 14 brings the flask to the desired distillation temperature. Hydrocarbon vapors pass through neck 16 and 17 past thermometer 18 and into condenser 19 which has a cooling jacket 21 through which ice water is circulated from a pump (not shown) to inlet 22 and from outlet 23 back to the pump.

Hydrocarbon and water vapors condensed in the condenser drop into trap 24 in which separation takes place because of the higher density of the water. Water is drawn off through stopcock 26 into trap 27 and finally through stopcock 28 to discard. After the condensed hydrocarbon liquid fills the trap 24, excess liquid spills back into the flask 11 through tube 17 so that the distillation can continue until all of the removable water has been distilled over. When the hydrocarbon liquid used is one which has a normal boiling point higher than the desired distillation temperature, reduced pressure must be applied

FIGURE 1.7: DEHYDRATION BY SOLVENT DISTILLATION

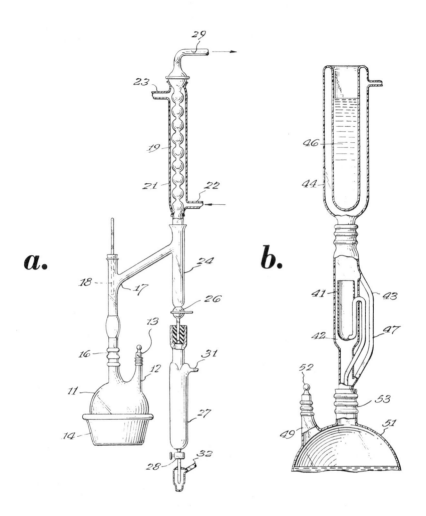

Source: N.F. Toussaint; U.S. Patent 3,628,967; December 21, 1971

to the system; and for this purpose, tube **29** leading from the condenser, and tubes **31** and **32** lead to a source of suction (not shown). After completion of the distillation of the removable water (as indicated by the fact that the fresh condensate contains no water) the hydrocarbon liquid is drawn from trap **24** to stop the return flow of condensate through tube **17** and the distillation of the hydrocarbon liquid is carried forward until all excess is removed from the food in flask **11**, although the food is still wet with the liquid.

At this stage, the wet food product is removed from the flask **11** of the apparatus and placed into extracting thimble **41** of a Soxhlet extractor **42** of the apparatus of Figure 1.7b for treatment with a highly volatile, water-immiscible second liquid, such as a liquid having a normal boiling point between about -50° and 30°C. The volatile liquid is charged through neck **49** into flask **51**, and stopper **52** is then reinserted to seal the flask.

In some cases, the volatile liquid will be charged in vapor form and will then condense and drop back into the flask as hereinafter described. Under ambient conditions, boiling will commence at once in most volatile liquids boiling within the above specified range. For some solvents boiling at the upper end of the above specified range, a slight heating may be required, and for those boiling at the lower end of the range cooling and/or superatmospheric pressure may be used for better control.

Vapors of the highly volatile liquid pass through neck **53** into Soxhlet extractor **41** through tube **43** and then into jacket **44** of condenser **46**. The inner chamber of the condenser is filled with a very cold liquid such as a Dry Ice-trichloroethylene mixture and the vapors in the condenser jacket condense therein dropping back into the Soxhlet extractor **42**. As the level of condensed liquid rises in the Soxhlet extractor, it immerses the food product within the filtering thimble and eventually rises to the level of the U-shaped bend at the top of tube **47**, causing the siphoning of condensed liquid back into the flask for repetition of the cycle as many times as desired.

The number of cycles of treatment with the highly volatile liquid to achieve complete removal of the hydrocarbon liquid from within the pores of the food product will, of course, vary with the nature of the food product, the nature of the hydrocarbon liquid and the nature of the highly volatile liquid. For some systems a single cycle may be sufficient, but it is generally preferred to use at least five cycles.

After treatment with the highly volatile liquid, it may be desirable, particularly with liquids having boiling points at the upper end of the acceptable range, to subject the food product to the final drying step of short duration, e.g., about 5 minutes, to remove all taste of the volatile solvent. Suitable hydrocarbon liquids include the paraffins and cycloparaffins having from 6 to 8 carbon atoms and aromatic compounds, such as benzene and toluene. The preferred hydrocarbon liquids are n-hexane, n-heptane, isoheptane and mixtures containing two or more of these materials.

As stated above, it is preferred to carry out the distillation step at reduced pressures when the hydrocarbon liquid is one which boils near the upper end of the acceptable range. A suitable pressure for distillation with n-heptane is about 200 mm. of mercury, at which pressure the liquid boils at 58.5°C., and a suitable pressure or distillation with n-octane is about 50 mm. of mercury, at which pressure the liquid boils at 50.6°C.

Suitable highly volatile liquids include low-boiling paraffins, and specifically propane, n-butane, isobutane, and mixtures thereof; and low-boiling fluoroparaffins and chlorofluoroparaffins such as dichlorodifluoromethane, 1,1-difluoroethane, 1-chloro-1,1-difluoroethane, symmetrical dichlorotetrafluoroethane and mixtures thereof. This process is generally applicable to food products which are naturally small in size, e.g., berries, or which may be cut into pieces. Typical food products include strawberries, raspberries, lemons, limes, oranges, tomatoes, apples, mushrooms, onions and ground beef.

Example: 50 grams of cut strawberries were charged to a round bottom (200 ml.) two-necked flask with 100 mm. of n-hexane and refluxed for 4 hours, the water separating in the trap. The strawberry pulp was then placed in a vacuum oven for 20 minutes at 40°C. to remove a portion of the liquid hexane. Some of the partially dried food was placed in a Soxhlet thimble and extracted with symmetrical dichlorotetrafluoroethane under a Dry Ice-trichloroethylene condenser at atmospheric pressure for about an hour (or for about 30 cycles of treatment). The product was placed in a vacuum oven for about 5 minutes to remove all taste of the dichlorofluoroethane. The product was a thoroughly dried product of good appearance and had the full taste of fresh strawberries without any trace of a foreign taste.

ADDITION OF STARCH

In a process developed by *H. Griffon; U.S. Patent 3,431,119; March 4, 1969* foods are dehydrated by mixing finely divided particles of the food with less than 50%, calculated on dry solids, of starch and sufficient water to form a fluid creamy mixture which is then spread into a thin layer about 2 mm. thick and dehydrated to 6 to 15% moisture content with a current of air at a temperature such that the temperature of the mixture during drying does not exceed 37°C.

The process consists in preparing by malaxation or crushing, with the addition of the quantity of water necessary, a fluid suspension-emulsion of fine particles of the substance to be dehydrated which contains amylaceous substances, e.g., starch, and subjecting the fluid mixture thus obtained in the form of a continuous thin sheet, to a current of air or of inert gas, the temperature of which is so low that the temperature of the treated substance does not substantially exceed 37°C., throughout the operation until the residual water content is reduced to the desired value, for example until the emulsion is dried to the solid state.

The manner in which the initial mixture of the substance to be dehydrated is first prepared may vary in accordance with the form of the physical chemical structure of this substance. Starch is employed in the crude state, i.e., in its natural form, before any heat treatment which is likely to convert it into starch paste and thus deprive it of its hygroscopic properties. The crushing and malaxating operations are carried out without the addition of foaming products and without injection of air, so as to obtain a paste rather than a foam which contains little if any occluded air and which is sufficiently fluid to be spread in thin layers.

The substance to be treated, when it is in the form of a homogeneous paste, either naturally or as a result of a prior preparation, is mixed with starch, which will generally be added in a proportion of the same order of magnitude as that of the substance to be treated, calculated on the dry material. For example, crude starch can be added to the substance to be dehydrated in a proportion such that the total proportion of amylaceous materials in the mixture is between 20 and 50% of the dry weight of the substance to be dehydrated. The mixture of starch and substance to be treated is malaxated until an intimate and homogeneous mixture of pasty consistency is obtained.

The quantity of water to be added during the crushing depends essentially upon the substance treated, e.g., being the proportion of water already contained in the substance in the natural state, and the viscosity of the paste obtained after crushing. This quantity will be so chosen in each instance as to favor the crushing and to obtain a good spreading in a thin layer on the support surfaces of the dehydrating apparatus. Thus, it is unnecessary to add water in the case of certain fruits, for example, tomatoes and strawberries, whereas this addition may be desirable in various proportions in the case of a number of products.

This mixture is thereafter treated by bringing it into the form of a continuous thin layer which is then washed by a current of air, or inert gas, at low temperature. For example, such a thin layer may be produced by spreading the paste on smooth plates, or better still on a smooth endless belt of water-repelling material, such as a synthetic resin, so as to produce a continuous dehydration.

The current of drying air is so directed as to play upon the free surface of the sheet of pasty mixture. The dehydration is consequently the result of a surface evaporation effect. The thinner the layer, the more rapid is the dehydration, at equal temperature, but too thin a layer results in the dehydration taking place too quickly. Experience has shown that good results are obtained with a thickness of the order of 2 mm. or even less.

The dehydration resulting from the application of the process is obtained owing to the hygroscopicity and the absorbent properties of starch especially for fatty substances. In a first phase, in the course of the preparation of the mixture, the starch swells and absorbs water of constitution, while promoting the dispersion of the substance intimately mixed therewith. In a second phase, which consists in the consecutive drying under the influence of the current of air or inert gas, the mixture gives up this water in the form of vapor until the water content of the substance is in equilibrium with the environment of the drying gas under the experimental conditions produced, this content generally being between 6 and 15%.

At the end of the dehydrating process, the layer of treated substance, whose thickness was initially 2 mm. is converted into a fine film of about 0.2 mm. in thickness. If, as had been stated above, the support of this layer consists of a water-repelling material, the substance becomes readily detached therefrom in the form of a thin wafer or film. The latter may be reduced to the form of flakes, scales or powder by any appropriate mechanical means, such as more or less intense crushing, optionally followed by screening.

The final product obtained is substantially balanced in regard to its residual water content with the hygrometric state of the atmosphere. It can generally be stored or distributed in various packings. However, in the case of some products it may be desirable to use hermetically sealed packages affording protection against humidity and light.

Starches which may be used include starch from tubers, e.g., potato starch, or from cereal grains, e.g., wheat, maize and rice. It is also possible to use seeds of Leguminosae, for example, peas and beans, or of any other origin, for example, bananas, which can be used in the form of flours.

Example 1: 100 parts of carrots admixed with 50 parts of water are so crushed as to produce a homogeneous pasty mixture of creamy texture. 40 parts of dehydrated maize starch are incorporated and the mixture is malaxated to produce a homogeneous smooth paste, which is spread in the drying apparatus and subjected to the action of a current of air heated at 35° to 37°C. until the residual water content is about 10%. The product is crushed to a fine powder.

Example 2: 100 grams of tomato are ground (without any water having to be added, since this fruit contains sufficient water) with 2 grams of maize starch, which corresponds on average to 20 to 30% of dry substance, and dried as in Example 1.

The dehydrated products obtained by the process take the form of flakes, scales or powder, of which the organoleptic properties, color, flavor and odor correspond to those of the original products. The volatile constituents (aromas) are preserved notably by reason of its absorbent properties. The dehydrated products may be directly employed in this form. This is the case, for example, with flakes and powders, which can be used in the preparation of soups. They may also be rehydrated in the cold by the addition of an appropriate quantity of water. They will then undergo a regeneration which restores to them the organoleptic properties of the initial substance before dehydration.

IMPROVING REHYDRATABILITY

By Explosion-Puffing Followed by Compression

J. Cording, Jr. and R.K. Eskew; U.S. Patent 3,038,813; June 12, 1962; assigned to the U.S. Secretary of Agriculture describe a process for dehydrating fruit and vegetable pieces to a moisture content between about 20 to 50% and then explosively puffing the pieces (see p. 240).

Although these products are far superior to conventionally hot air dried comestibles, and in some respects are superior to freeze dried material in that they are less costly to make and much less friable, they nevertheless are bulky compared with pieces of the same size dried conventionally in hot air. For example, 3/8 inch carrot dice when dried conventionally in hot air have a bulk density of about 26 lbs. per cubic foot. However, they require 45 minutes boiling to rehydrate and to be soft enough to eat. In contrast, pieces cut the same size and prepared according to the described process, although they require only 5 minutes boiling to reconstitute, have, before reconstitution, a bulk density of about

20 pounds per cubic foot and are hence more costly to package.

R.K. Eskew and J. Cording, Jr.; U.S. Patent 3,408,209; October 29, 1968; assigned to the U.S. Secretary of Agruculture have discovered means for greatly reducing the bulk of explosive puffed products without in any way impairing their rapid rehydratability, appearance, flavor, or nutritive value on reconstitution. In fact their dry bulk may even be reduced below that of conventionally air dried unpuffed material, e.g., to a bulk density of 30 lbs. per cubic foot or more.

This process comprises compressing the explosive puffed pieces of fruit or vegetable in their slightly moist plastic state after puffing but before final processing to stabilized form, as for example, by further drying. The compression of the explosive puffed pieces has the primary effect of temporarily collapsing the pores created by the puffing step.

The compression of conventionally hot air dried fruit or vegetable materials to eliminate interstitial voids and thus to reduce their bulk is admittedly a common practice. This process, however, shows that at the moisture content of between about 20 and 50%, typically that at which pieces of fruits or vegetables are discharged from the gun, they may be individually compressed, even to the extent of closing the porous canals created by puffing, and that they may thereafter be dried as, for example, in hot air to retain this compact form.

However, on immersion in boiling water the canals created by explosive puffing reappear, permitting rapid rehydration of the compressed pieces and expansion to their original size and shape. This rehydration or reconstitution takes place just as rapidly as though the pieces had not been compressed. The advantages of a compact product that possesses the other good attributes of a bulky product entail savings in packaging, storage and transportation. Such products are especially adapted to military use.

Compression can be done by any convenient means, for example, by passing the pieces in one or more stages between closely set rolls. These rolls may or may not be heated and the pieces may be either compressed while still warm from the gun or after cooling or they can be compressed in bulk not merely decreasing interstitial voids but compressing the individual pieces as herein described.

Example: Carrots of the Red Core Chantenay variety were peeled by immersing them in a lye solution (20% by weight sodium hydroxide) at 160°F. for 2½ minutes and then subjecting them to high-pressure water sprays to remove the lye-loosened skins. The caps and roots and blemishes were trimmed off by hand.

The trimmed whole carrots were blanched in steam at atmospheric pressure for 10 minutes to soften them slightly to prevent shattering in the cutting operation to follow. They were then cut into nominal ⅜ inch cubes. The product of the cutter was passed over a vibrating screen with ³⁄₁₆ inch wide openings.

Approximately 13% by weight of the product of the cutter passed through the openings and was processed separately. The 87% by weight larger than the ³⁄₁₆ inch minimum dimension were dipped 2 minutes in an aqueous solution containing 0.5% sodium bisulfite and 0.5% citric acid.

The pieces were blanched 4 minutes in steam at atmospheric pressure to inhibit enzyme activity and then dried in hot air at 200°F. dry bulb temperature to a moisture content of about 25%. The partially dried pieces were placed in a puffing gun to give them a porous structure permitting their rapid rehydration upon use after final drying.

A puffing gun is a cylindrical vessel, rotatable around its long axis, which can be tilted muzzle up for charging, level for heating and muzzle down for exploding or discharging. The gun is fitted with a hinged lid on the muzzle end which hermetically seals the gun during heating and which can be opened instantly for explosive discharge of the contents. The gun is heated externally, while rotating, with a gas flame. Superheated steam is introduced under pressure into the rotating gun at the end opposite the muzzle through a rotary joint. A charge of carrot pieces of 17½ lbs. at 28.7% moisture content prepared as described above was introduced into the gun, the lid was sealed and rotation was started. Heat was applied by the external gas burners until a temperature of 340° to 350°F. was reached (as measured by a sliding thermocouple) on the outer surface of the gun barrel. The temperature was maintained at 340° to 250°F. for 1½ minutes with the gas flame and then superheated steam at 35 psig and 500°F. was introduced.

Steam was caused to flow through the carrot pieces inside the gun barrel and out through a restricting orifice in the hinged lid until the pressure inside the gun reached 35 psig. Flow was further continued (with heat still being applied externally to maintain the outer gun wall at 340° to 350°F.) for ½ minute after the pressure inside the gun reached 35 psig. Then the gas heat was turned off, the gun tilted to the firing or discharge position of 22° below the horizontal, the lid was opened instantly and the superheated steam flow was stopped.

Upon opening of the lid the carrot pieces discharged explosively from the gun into a collecting system. The rapid reduction of pressure from 35 psig (above atmospheric) to atmospheric caused a portion of the water within the pieces to flash into vapor, creating a porous structure within the pieces. The product of the gun was divided into two fractions, A and B. Fraction A was placed in a tray of a through-circulation hot air drier, loaded to 9.25 lbs. per square foot, and air was passed through the material at 156°F. and about 270 fpm velocity until a stable moisture content of about 4% was obtained. The time required to reach the desired moisture content was about 90 minutes.

The carrot pieces in Fraction B were compressed between two rollers running at the same speed and in opposite directions, towards each other at the top nip, the clearance between them set at 1/32 inch. The pieces before compression varied in size; their minimum dimension was about 1/8 inch; their maximum dimension about 7/16 inch. Those pieces passing through the rollers with their minimum dimension parallel to the vertical axis of the rollers were thereby compressed to one-fourth of that dimension.

Those pieces passing through the rollers with their maximum dimension parallel to the vertical axis of the rollers were thereby compressed to one-fourteenth of that dimension. The compressed pieces thus prepared from Fraction B were dried to 4% moisture at exactly the same conditions of tray loading, air temperature and air velocity as was Fraction A.

The dried products of Fraction A and B were compared for apparent (bulk) density. That of Fraction A was about 19 lbs. per cubic foot that of Fraction B was 27 lbs. per cubic foot. The reciprocal of these values represents the packaging volume requirements for each product, i.e., 0.0525 cubic foot per pound of Fraction A against 0.0370 cubic foot per pound of Fraction B; Fraction B requiring only about 70% of the volume required for Fraction A.

The dried products were compared for rehydratability by placing samples of each in simmering water for 5 minutes. At the end of that time each sample had fully rehydrated, the weight of each rehydrated product being five times that of its dry weight. Both had equally regained the generally cubical shape of the carrot pieces entering the first drying step and both were equally soft, cooked, and edible.

Products of Fractions A and B were compared with conventionally hot air dried carrot pieces of the same size, i.e., carrot pieces dried in hot air from their fresh moisture content of about 4% without specific treatment to impart a porous structure. This product had an apparent (bulk) density of about 26 pounds per cubic foot, requiring a package volume of about 0.0385 cubic foot per pound. After simmering this product for 5 minutes, it was still tough and inedible, was not rehydrated, and its weight after simmering was only three times that of its dry weight. The example shows that Fraction B, having been compressed after puffing, retained the rapid rehydration qualities of Fraction A which was puffed but not compressed. At the same time the apparent density of Fraction B was increased greatly over that of Fraction A and equal to or slightly greater than that of conventionally dried carrot pieces of the same size.

By Irradiation at 0°C.

C.K. Wadsworth; U.S. Patent 3,484,253; December 16, 1969; assigned to U.S. Secretary of the Army describes a process for irradiating dehydrated vegetables and fruits which reduces rehydration time and tenderizes the irradiated items without substantially affecting flavor, odor, color or texture thereof. Dehydrated vegetables have been treated with high energy, ionizing radiation, such as gamma rays or electron beams, to reduce the rehydration time but it has been found that such irradiation imparts certain undesirable properties to the vegetables such as an undesirable change in or loss of color, loss or deterioration of flavor, development of off-odors, and excessive softening in texture.

It has been found, however, that the undesirable side effects of the high energy, ionizing radiation of dehydrated vegetables can be eliminated by irradiating the dehydrated vegetables while these vegetables are maintained at a temperature of 0°C. or lower. The radiation dosages received by the dehydrated items will range from about 0.5 to about 12 megarads and preferably from about 1 to 10 megarads. Rad is a unit of absorbed dose of ionizing radiation and is equal to an energy of 100 ergs per gram of irradiated material. Suitable types of high energy, ionizing radiation for use in this process include, for example, electron beams, gamma rays and X-rays.

The dehydrated vegetables and fruits which may be treated include, for example, cabbage, carrots, green and red bell peppers, celery, onions, leek, lima beans, okra, peas, corn, lentils, split peas, navy beans, pea beans, mushrooms, prunes, apricots, figs, peaches, pears, bananas, strawberries, papayas, raisins, dates, avocados, mangoes, etc. Dehydration is accomplished by any of the well-known standard commercial procedures and has as its purpose the removal of sufficient moisture from the food item so that it may be stored at ambient temperatures for extended periods of time without undergoing a loss in quality. The moisture level of the food items will be 20% or less by weight of the dehydrated item and preferably is less than 5% by weight.

Irradiation of the food items described in the following example was accomplished by exposing the food items to gamma radiation emanating from a 900,000 curie cobalt 60 source. The physical arrangement of the source consists of two spaced apart, parallel plaques, 42 inches high by 56 inches in length, containing the radioisotope. The samples which were packed in sealed tin cans were placed in stacked aluminum canisters and carried by a monorail conveyor between the plaques for a predetermined period of time. Ceric sulfate dosimeters were distributed throughout representative samples being irradiated to measure the dose absorbed.

Example 1: Cabbage cubes, ⅜ inch in size, were dehydrated to a moisture content of 5% by weight based on the weight of the dehydrated cabbage and hermetically sealed in a number of tin-clad steel cans. Half of the total number of cans of dehydrated cabbage cubes so prepared were chilled by immersion in liquid nitrogen and while at a temperature of approximately -180° ± 5°C. were exposed to gamma radiation as hereinbefore described at various dosage levels. The other half of the total sample of cans were irradiated at ambient temperatures within the radiation chamber which ranged from about 30° to more than 72°C.

The low temperature irradiated cabbage was compared with the ambient temperature irradiated cabbage. The evaluation was made in this example by a panel of experienced food testers. Rehydration of the control was accomplished by combining the cabbage with an excess of cold water (15°C.) and holding at room temperature for 45 minutes, the time required to rehydrate the control. Dehydrated cabbage which was irradiated to a dosage level of 2 megarads at -180° ± 5°C. rehydrated in less than one-half the time required by the nonirradiated control and gave a product which was essentially comparable in color, odor, flavor, texture and appearance to the control.

On the other hand, the sample irradiated to the same dose level at ambient temperatures was significantly poorer in flavor and texture. The differences between the ambient and low temperature irradiated samples becomes greater with increasing dosages which cause color, odor and appearance of the ambient samples to fall far below those of the low temperature treated samples.

By Dehydration, Grinding, Heating and Rolling

R.L. Roberts and R.E. Faulkner; U.S. Patent 3,174,869; March 23, 1965; assigned to Vacu-Dry Company describe a method for producing a dried fruit or vegetable in a flaked form which can be rehydrated quickly. Referring to the diagram of Figure 1.8a, box **10** represents the steps of washing, peeling, coring, slicing and otherwise preparing a fresh fruit. Box **11** represents the step of drying the treated fruit to a moisture content of less than 5% by weight. Many methods for this drying step are well-known such as, for example, kiln drying, vacuum drying, or conveyor belt methods. Boxes **12, 13** and **14** show additional steps commonly used in drying apples.

FIGURE 1.8: DRIED FRUIT AND VEGETABLE FLAKES

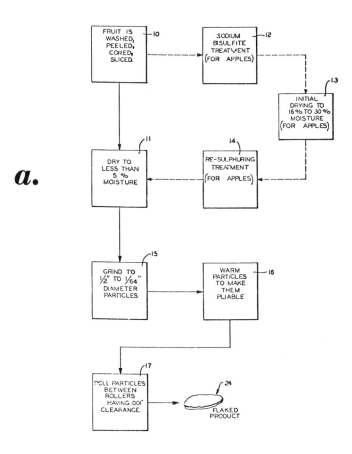

Flaked Fruit (continued)

FIGURE 1.8:   (continued)

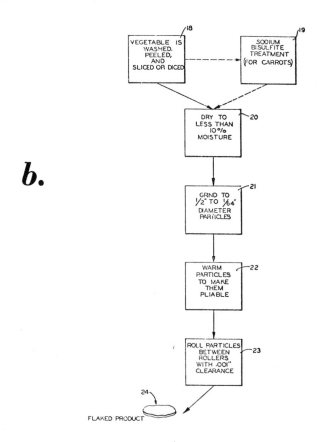

b.

Flaked Vegetable

Source: R.L. Roberts and R.E. Faulkner; U.S. Patent 3,174,869; March 23, 1965

Apples are usually treated with sodium bisulfite, as at box **12**, to prevent oxidation and browning of the newly exposed surfaces after the apples are peeled and sliced. Apples are dried in two stages and box **13** represents the initial drying stage usually to a moisture content of 16 to 30%. It is common to resulfur the apples at the end of the initial drying stage to restore bisulfite removed during the initial drying. Box **14** indicates this resulfuring step. After resulfuring, the apples are then dried to less than 5% moisture, as previously described.

Referring to the diagram of Figure 1.8b, box **13** represents the steps of washing, peeling, slicing, dicing and otherwise treating the fresh vegetables by methods well-known to the art. Box **20** represents the step of drying the treated vegetables to a moisture content of less than 10% by weight. Box **19** shows an additional treatment step commonly used in drying carrots by exposing them to sodium bisulfite to prevent oxidation and browning of the newly exposed surfaces after the carrots have been scraped and sliced.

The low moisture fruit or vegetable then is ground preferably to particle diameters ranging between ½ inch maximum and ¹⁄₆₄ inch minimum. However, larger particle sizes such as sliced apples for pies approximately ¼ inch thick and having 1½ square inches of top surface area may also be processed by this method. Grinding may be done by well-known methods, for example, a cutting mill such as a common corn cutter may be used with satisfactory results. Boxes **15** and **21** represent the grinding step.

After grinding the particles are warmed until they are pliable. Boxes **16** and **22** indicate this warming step in which radiant type heating has been found to be satisfactory. It is important that the particles are rendered pliable prior to performing the subsequent rolling step, otherwise the particles shatter and disintegrate when rolled.

After the particles are warmed, they are passed between rollers spaced 0.001 inch apart as indicated in boxes **17, 23**. This produces a flaked product having a thickness of approximately 0.01 inch because the product is resilient and partially tends to assume its initial thickness. If a thicker final product is desired, the spacing between rollers is increased so that the product is within the range of 0.01 to 0.06 inch thick. Rollers of hardened steel, for example, have been successfully used, each roller measuring 15 inches in diameter and 36 inches long. The rollers rotate at about 300 rpm.

The final product emitted from the rollers is thinly flaked. The flakes measure approximately 0.01 to 0.06 inch in thickness and vary between ¾ and ¼ inch in diameter. The characteristic flake is generally circular or somewhat elliptical in shape and has generally continuous and unbroken top and bottom surfaces. The flakes have a compressed but unruptured cellular structure which during rehydration rapidly resumes its original size and shape.

It has been observed that generally spherical particles screened through 8 mesh screen average about ¹⁄₁₀ inch in average diameter and usually flake out to a size slightly under ¼ inch. Generally cubical particles have also been processed having average dimensions of ⅛ x ⅛ x ⅛ inch and flaking out to about ¼ x ¼ x ¹⁄₁₀₀ inch. Particles measuring approximately ⅛ x ⅛ x ⅜ inch usually flake out to about ½ x ¾ x ¹⁄₁₀₀ inch.

The flaked product rehydrates much more rapidly than similar products. Moreover, the rehydration rate of the flake described herein is easily controlled merely by controlling the thickness of the final product. The thinness of the flake cross-sections together with their compressed but unruptured cellular structure contribute to a rapid rate of rehydration.

DRY FRUIT AND VEGETABLE JUICES

DRUM DRYING

Fruit juice or fruit concentrate is not readily dried by conventional techniques. This is due to the relatively large amount of natural sugar present in the fruit juice. Thus, when fruit juice or concentrate is dried on a drum dryer, the sugar present prevents the formation of a sheet which can be easily removed by the dryer's doctor blade or scraper. Instead of being able to remove a continuous dried sheet from the dryer as with other types of foods, the dried fruit juice yields a gummy mass after heating which collects at the doctor blade and disrupts the drying operation.

In the drying of high sugar content food other than fruit juice, the same problem has been encountered. To overcome the problem with other foods, starch thickeners have been added to the food slurry to be dried. The starch thickened mass of high sugar content food, after being heated to gelatinize the starch, can then be dried on a drum dryer for example, relatively free from the foregoing problem. It has not been feasible to use this same approach with fruit juices however, since the starch thickener, after reconstituting the dried juice, imparted a thick consistency to the juice wholly unlike and uncharacteristic of fresh juices.

L.H. Anderson; U.S. Patent 3,117,878; January 14, 1964; assigned to Gerber Products Co. describes a process which solves the problem of dehydrating juices and permits their reconstitution to their characteristic liquid form by combining an amylolytic enzyme with the dried, starch thickened juice. When water is added to the combination starch thickened juice and enzyme product, the starch is hydrolyzed by the enzyme to a sufficient degree to impart the desired consistency to the reconstituted product.

One suitable way of carrying out the process is to form a slurry with the selected fruit juice or concentrate with a starch thickener. Any of the common starches may be used for this purpose and include corn, wheat and rice; tapioca, potato and arrowroot; or sago. One suitable operable range for the quantity of starch to be added is from 35 to 45% by weight of the fruit juice. The mixture of starch and fruit juice is heated with agitation to gelatinize the starch. The mixture is then dried to a moisture content of 1 to 5%, although the precise moisture level after drying is subject to considerable variation. Under the above conditions, it is possible to drum dry fruit juices to form a continuous sheet that is conveniently handled for subsequent operations.

The dried product is then combined with an amylolytic enzyme. Generally, the enzyme may be added to the dried, thickened juice in an amount by weight of 0.5 to 0.8% and an

66

acceptable product will be obtained. The dried product mixed with the enzyme is then ready for use by the consumer. The consumer merely need add a sufficient quantity of water to replace that which has been removed in the processing and a juice closely approximating that of the preprocessed fresh variety is obtained.

Any amylolytic enzyme which will liquefy the starch may be used. One example is Rhozyme H-39.

DRYING TOWER FOR HEAT-SENSITIVE LIQUIDS

Product in Form of Spherical, Porous Granules

P. Hussmann; U.S. Patent 3,190,343; June 22, 1965; assigned to Birs Beteiligungs- und Verwaltungsgesellschaft AG, Switzerland has developed a process for dehydrating aqueous dispersions, suspensions or solutions of solids in which the material to be dried is dispersed in the upper region of a high tower in the form of a dense umbrella of droplets descending in the tower through a countercurrently flowing stream of relatively cool or cold drying air, the air being introduced in the lower region of the tower in a highly dehumidified state, for instance, with a water content of 0.35 gram/cubic meter, at a temperature not exceeding 60°C. In view of the low temperature of the drying gas, the residence time of the material in the tower must be relatively long and the volume of drying gas must be accordingly large.

In all embodiments of the process, use is made of a drying tower wherein the throughput of the material to be dried, its droplet size and distribution, the drying gas velocity and temperature are so correlated that the material falling through the dehumidified and upwardly streaming drying gas reaches the tower bottom in the form of porous and fully spherical granules, which remain on the bottom without collapsing or adhering to each other. This granular structure of the dried material makes it possible to introduce the gaseous drying agent into the tower through the bottom, even if it carries a layer of dried powder up to 20 cm. thick, and without imparting turbulence to the powder layer. The powder remains lying on the tower bottom until it has reached the desired degree of dehydration and is then conveyed on the shortest possible path to the vacuum packing station (which does not form part of the process). Alternatively, the powder may be removed from the bottom of the tower as soon as it may be mechanically or pneumatically conveyed to a sieve-like support on which it is additionally dried to the desired degree of dehydration without turbulence.

Since this drying method uses relatively cool air and so requires drying towers of large volume and correspondingly large volumes of drying gas, the process also provides for operation of the drying tower with as little heat requirement as possible and with a maximum recovery of the required heat. For this purpose, the drying tower is connected with a special air conditioning plant. Drying towers useful for the practice of this process must have a height in the range of 50 to 200 meters, tower heights of about 70 meters having been found most useful for most purposes.

Figures 2.1a and 2.1b illustrate a structural embodiment of the plant for conditioning the drying air in the indicated manner, Figure 2.1a being a vertical section of the tower and Figure 2.1b showing a horizontal section thereof. To facilitate a schematic showing of the conduits, the elements **2a**, **2b** and **3a**, **3b** are shown in Figure 2.1a one behind the other, while they are illustrated adjacent one another in Figure 2.1b. In practice, they are arranged only at one of the two locations. Elements **6** and **7** also are positioned at the same location.

The drying air enters the drying tower at inlet opening **9** and passes upwardly to outlet opening **11**. The liquid material distributing means (not shown) is mounted in the tower below the drying gas outlet opening. The tower, which may have circular or rectangular walls, is surrounded by a jacket formed by an inner and outer tower wall defining a space

FIGURE 2.1: DRYING TOWER FOR HEAT-SENSITIVE LIQUIDS

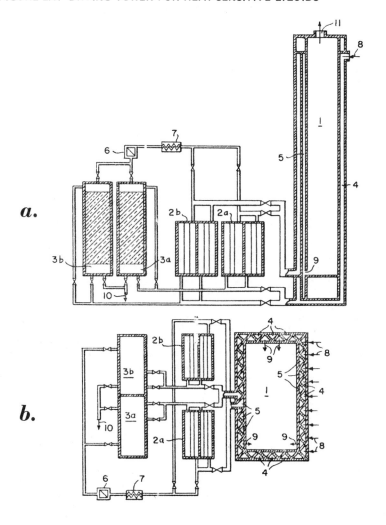

Source: P. Hussmann; U.S. Patent 3,190,343; June 22, 1965

therebetween. This space is divided into outer cells **4** and inner cells **5** by an intermediate zigzag wall which has a heat-insulating lining facing the interior of the tower. It is preferred to make all the walls of cement which has a considerable heat-storing capacity. The fresh air for drying as well as for regeneration of the adsorption medium is passed into the plant from the atmosphere through the outer cells **4** of the tower jacket. The header is subdivided into two branch conduits. In one of the operating cycles, the fresh air, which has been preheated by passage through jacket cells **4**, is led through the upper branch conduit into a regenerated adsorption medium packing **2a** consisting of two vertical packings arranged in parallel. The other portion of the preheated fresh air is conducted into a heat storage device **3a**. The entry valve of the lower branch conduit then remains closed. In the following operating cycle, a portion of the fresh air coming from cells **4** is conducted by the lower branch conduit to adsorption medium packing **2b** which is similar in structure to packing **2a**. The other portion of the fresh air is then led from the lower branch conduit to heat storage device **3b**. At this cycle, the entry valve of the upper branch circuit remains closed.

Depending on the cycle, the dehumidified air coming from packing **2a** or **2b** is conducted to the inner cells **5** of the tower jacket and is led up and down these cells by variously placed fans. Finally, the dehumidified and conditioned drying air is introduced into the tower through opening **9**. The fresh air further heated in heat storage devices **3a** or **3b** is led through the flue of the heating element of another heat exchanger **6**, where it is additionally heated, and is finally directed into a heat exchanger **7** where it is brought to the required regeneration temperature. This hot air is then directed to the adsorption medium packing requiring regeneration, whence it is passed into one of heat storage devices **3a** or **3b** to heat the same. The regenerating air then passes from the heat storage devices into the atmosphere through flue **10**. The inlet openings for cells **4**, which admit the fresh air thereto, and for cells **5**, admitting the dehumidified and conditioned drying air, are provided with suitable filters.

If the temperature of the adsorption packing used for dehumidifying the air exceeds a predetermined value, for instance 50°C., a thermosensitive relay may automatically operate suitable valves in the air conduits to direct the fresh air to a previously regenerated adsorption packing and to direct the regenerating air to the exhausted packing. The control of the next cycle may then be taken over by a thermostat in the regenerated packing. However, the cycle control may also be effected in an obvious manner by a thermostat in the conduit directing the conditioned air to cells **5**.

By using a tower wall jacket as heat storage and balancing device, it is possible to equalize the air temperatures so as to obtain a drying air of nearly constant temperature throughout the operation of the tower. The entire air circulation through the drying plant may be effected by fans mounted in the fresh air inlet openings of the outer wall of the drying tower.

Gasification Prior to Dehydration

Another process for the dehydration of fruit juices is described by *S.I. Strashun and W.F. Talburt; U.S. Patent 3,241,981; March 22, 1966; assigned to the U.S. Secretary of Agriculture.* The principles of this process are primarily concerned with control of the factors of dehydration whereby to ensure extensive expansion (that is, from 10 to 25 times in volume) of the liquid during dehydration and to maintain this expansion throughout the dehydration. It has been found that a primary factor in ensuring extensive expansion is the step of gasifying the liquid prior to application of dehydrating conditions. This gasification greatly enhances expansion and is so effective in this regard that it will cause extensive expansion hence successful dehydration of liquids which could not otherwise be dehydrated under the same conditions, or which could only be dehydrated by using high levels of vacuum or by adding drying aids.

By utilizing this step of gasification and control of other factors in accordance with this process, successful dehydration of fruit juices and other alimentary liquids is attained in that the film of liquid will consistently expand to a large degree and remain in an expanded condition throughout the dehydration procedure. As a result, the dehydration proceeds rapidly and efficiently and yields a free-flowing, porous product which exhibits a very high rate of rehydration when contacted with water to make a reconstituted juice. In addition, no damage to the color, flavor or vitamin content of the material is involved.

The liquid foodstuff to be dehydrated is first prepared by reaming, pressing, macerating, crushing, comminuting or extracting with water the edible portions of fruit or vegetables as, for example, orange, grapefruit, lemon, lime, apple, pear, apricot, strawberry, raspberry, pineapple, grape, prune, plum, peach, cherry, tomato, celery, carrot, spinach, lettuce, water cress, and so forth. The liquid preparation may be clear, contain suspended pulp, or may even be thick like a puree. Besides liquid foods of vegetative origin the process may be applied to animal products as for example meat juices, meat extracts, soups or lacteal products such as whole milk, skim milk, buttermilk, whey, cream or milk containing added flavorings or nutrients such as sugar, chocolate, fruit juices, fruit pulps, and so forth.

The liquid preparation is introduced into an evaporator where it is concentrated so that it will be in proper condition for the subsequent dehydration step. The concentrate is then introduced into an agitator. A gas is also introduced into the agitator, this gas being thoroughly whipped into the concentrate to form an intimate dispersion of the gas in the concentrate. Although air is the most convenient gas to use, it is often preferred to use nitrogen, carbon dioxide or other inert, nontoxic gas whereby to minimize oxidative or other deleterious effects. To reduce the size of the gas particles in the concentrate, the concentrate (after having the gas dispersed by the use of agitators or the like) may be passed through a colloid mill.

The gasified concentrate is introduced into the feeding vessel of a dehydrator connected to a source of vacuum to maintain the interior of the dehydrator at a pressure of 1 to 3 millimeters Hg. There is also provided a flexible metallic belt which traverses over a heated drum and a cooled drum. As the gasified concentrate enters the feeding vessel, it is exposed to the vacuum whereby frothing occurs as some of the gas in the concentrate is liberated. The concentrate now in the form of a liquid froth or foam is applied in a thin film, having a thickness on the order of 0.005 to 0.1 inch, to the underside of the belt. The optimum thickness of film to be employed in any particular instance will depend upon many factors such as the nature of the material being dried, the moisture content of the film, the speed of traversal of the belt, the temperature applied by the drum, and so forth. With many fruit juice concentrates, a film thickness of 0.006 to 0.20 inch gives efficient results. The means for applying this film comprises a roller which is positively rotated in a counterclockwise direction and which is spaced from the belt a distance equal to the film thickness desired. Wipers or scrapers may be provided to accurately define the thickness of the film.

The thin film of frothy concentrate applied to the underside of the belt is moved toward the drum which is hollow and through which steam or other heating medium is circulated to maintain the drum at a dehydrating temperature. In the case of orange juice and other fruit juice concentrate, excellent results are obtained with a temperature of $175°$ to $300°F$.

Before arriving at the dehydrating drum, the applied film is preferably subjected to what may be termed a predrying. This takes place where the film is subjected to irradiation from radiant heaters which are metallic rods heated to glowing temperature by electrical resistance coils embedded therein or which may be infrared lamps or the like. The significance of this predrying can be explained as follows. To obtain complete dehydration in the short time that the belt is in contact with the drum, it is necessary to maintain this drum at a high temperature, on the order of $175°$ to $300°F$. If the film without predrying is applied to the hot drum, unfavorable results are often obtained. Thus, as the film is initially heated by the drum, it expands to a desirable degree but as the expanded film travels about the drum it may collapse, that is, shrink to about its volume before expansion. This phenomenon is caused by the expanded film assuming too high a temperature while its moisture content is still high. In effect, the expanded film melts and loses its vapor bubbles which theretofore gave it an expanded structure.

The predrying treatment has the effect of removing part of the moisture content of the film at a relatively low temperature, whereby when the film contacts the hot drum its moisture content is decreased and its viscosity is increased to such an extent that it will maintain its expanded structure even though subjected to the high temperature of the drum. In effect, the predrying stage has the effect of removing moisture from the film to increase what may be termed its pseudo-melting point, that is the temperature range in which the expanded film will collapse.

The belt carries the predried film about the drum whereby the principal dehydration takes place. The dehydrated film still in its expanded condition then passes about another drum where the film is cooled so that it will lose its plastic character and become relatively brittle and easy to remove. The cooled product is removed from the belt by a scraper which may be provided with means for oscillating it in a horizontal plane to give increased dislodging effect. The cooled product falls from the scraper into a hopper from whence it can be removed to a container, which is provided with a valved conduit for connection to the source of vacuum so that it can be evacuated prior to opening of the valve.

Radiant heaters are provided so that the surface of the film away from the belt is properly dehydrated. In some instances, where such heating is not provided, the upper surface of the film is dehydrated to a lesser extent than the bottom surface of the film, with the result that the final product tends to roll up on the scraper.

In the dehydration of some fruit juices, purees, etc., it may be necessary to make some provision for returning volatile flavoring materials which are vaporized during the concentration and/or dehydration. In the case of tomato and apricot products, such provisions are not necessary as the dehydrated product retains its natural flavor and odor. In the case or orange, apple, pineapple, strawberry, raspberry and many other fruit products, provision should be made to restore flavoring substances to obtain a high quality product.

The restoration of flavor may be carried out in several ways. In one technique, the volatile flavoring component is mixed with molten, supercooled sorbitol and the mixture allowed to crystallize. The sorbitol containing absorbed flavoring material is then incorporated with the dehydrated juice to furnish the approximately original amount of flavoring component. The use of sorbitol to absorb the flavoring component is preferred as thereby the flavor is stabilized and prevented from vaporizing.

Example: (A) Aeration of Concentrate — To a lot of orange juice concentrate was added 1.5% of its weight of sodium carboxymethylcellulose using vigorous agitation to disperse this drying aid into the concentrate and also to draw air into the mixture and disperse it thoroughly therein. The aerated concentrate (60°Brix) was then dehydrated in the apparatus previously described. Operating conditions were as follows:

Temperature of drum	233°F.
Belt speed	40 ft./min.
Contact time between belt and drum	10.2 sec.
Thickness of film (approximate)	0.010 inch
Pressure in dehydrator	2.8 mm. Hg
Production rate of dehydrated product	33 lbs./hr.

It was observed that the film or orange juice concentrate expanded 20 to 25 times in volume during dehydration and maintained such expanded volume throughout. Thus, dehydration proceeded rapidly and efficiently, the product was readily removed from the belt by the scraper, and the product was in a porous condition so that on agitation with water for a few seconds it formed a reconstituted juice.

(B) No Aeration of Concentrate — In a comparative experiment, the orange juice concentrate was mixed with a previously prepared solution of sodium carboxymethylcellulose using enough of this solution to add 1.5% of this drying aid. In this case, the mixing was gentle to avoid incorporating air into the concentrate. The resulting nonaerated concentrate (60°Brix) was applied to the dehydrator using the same conditions as in (A). It was observed that the dehydration was unsuccessful in that the concentrate film did not expand significantly, with the result that the product stuck to the belt and could not be removed with the scraper. Eventually the belt became fouled with a brown, hard layer of overheated material.

VACUUM DRYING

Pulp Removal Before Dehydration

It has been found that fruit juices and other liquids can be dehydrated by a process which involves concentrating the juice then dehydrating the liquid concentrate by maintaining it in contact with a heated surface while being exposed to vacuum, the conditions of temperature being controlled to get rapid dehydration without damage to the product. A primary advantage of the above process is that the drying under vacuum in contact with a hot surface results in a puffing or expansion of the material during the dehydration, this expansion being caused by the entrapment of a multitude of small steam bubbles throughout the mass. This expansion is very desirable, as the final product is then in a porous form due to the presence of the numerous small voids. The product thus is easy to remove from the trays, breaks up easily into small particles or flakes, and exhibits an extremely high rate of rehydration so that a reconstituted juice can be prepared by agitating with water for less than one minute.

The expansion of the product also has the advantage that it accelerates the rate of dehydration. Thus, when the material expands, moisture can diffuse out of the mass very readily so that dehydration is completed in a short time, an hour or less in many cases.

It appears that if too much pulp is present, the pulp in some way interferes with the expansion effect so that instead of the steam bubbles being trapped in the mass they escape with the result that the material undergoing dehydration remains constant or even decreases in volume. *S.I. Strashun and W.F. Talburt; U.S. Patent 2,959,486; November 8, 1960; assigned to the U.S. Secretary of Agriculture* have found that if all or part of the pulp is removed prior to dehydration, the problem is solved and the material will expand properly during dehydration.

One method of applying the principles of this process in practice involves removing all of the pulp prior to dehydration whereby complete expansion during dehydration will be achieved. However, in many cases it is not essential to remove all of the pulp as a satisfactory degree of expansion can be attained even though some of the pulp is left in the liquid. The amount of pulp which may be safely left in the liquid to obtain satisfactory expansion will vary depending on the nature of the food product in question, and will thus be different for tomato products, orange products, apricot products, etc.

In the case of tomato, for instance, it has been found that the juice should contain less than 6% of pulp by volume to obtain satisfactory expansion; the volume of pulp is determined by centrifuging. Ordinary tomato juice contains 20 to 30% pulp by volume and in this condition cannot be successfully dehydrated because it will not expand. Thus at least part of the pulp must first be removed to provide a juice of less than 6% pulp which is amenable to dehydration, that is, which will expand during dehydration. In the case of orange juice, successful dehydration can be accomplished with ordinary juice which usually contains about 12% pulp by volume. If, however, it is desired to dehydrate an orange puree or other liquid preparation containing more pulp then part of the pulp must first be removed so that the liquid being treated does not contain more than 12% pulp by volume. In such case the desired expansion will be obtained.

In applying this process in practice, suitable fruit or vegetable material is pressed, macerated, comminuted or otherwise treated by known techniques to produce a juice or other liquid preparation. It is obvious that the liquid preparation should be made from ripe, sound produce of high quality. The partly or completely depulped liquid is then subjected to concentration so that it will be in proper condition for the subsequent dehydration step. It is preferred to conduct the concentration under vacuum at a temperature not over 50° to 125°F. in order to avoid heat damage to the material. The concentrate as above prepared is then ready for dehydration to the solid state. This dehydration is preferably achieved by the application of vacuum to the concentrate while it is spread on a heated surface. To this end, the concentrate is poured on trays which are placed in a vacuum dryer equipped with hollow shelves through which heating or cooling media can be circulated. After inserting the trays containing concentrate into the dryer, the dryer is closed and vacuum applied, the vacuum being maintained until the dehydration is completed. Pressures of around 2 to 20 mm. of Hg are used. A heating medium is circulated through the hollow shelves so that the concentrate is heated by conduction through the tops of the shelves, the bottoms of the trays, and so to the product. The shelves are maintained at a temperature near or above the boiling point of water, i.e., 150° to 300°F.

When the product temperature rises to 110° to 175°F. (due to falling off of the rate of evaporation), the temperature of the circulating medium should be immediately decreased, as by circulating cold water, to abruptly decrease the shelf temperature; then a medium at 110° to 175°F. is circulated through the shelves until drying is complete. When the drying cycle is completed as indicated by the product reaching the same temperature as the shelves, thus signifying absence of evaporation, the temperature of the shelves is reduced by circulating cold water through the hollow shelves. After the product is cooled to 70° to 100°F., the vacuum is broken, the dryer opened and the trays removed. By applying a spatula to the trays the product is easily removed, the scraping action of the spatula breaking up the product into a mass of fine flakes.

Inasmuch as the dehydration of juices, purees, etc. in accordance with this process necessitates removal of all or part of the pulp prior to dehydration, the final dehydrated juice or puree may contain an inadequate amount of pulp for forming a reconstituted product of the desired consistency. To overcome this situation, the pulp which is removed from the original juice, puree, etc. may be dehydrated, then mixed with the product made by dehydrating the depulped juice or puree. The dehydration of the pulp may be easily done in many different types of apparatus. For example, it is preferred to dry it in a vacuum tray dryer using the same two-stage temperature heating as explained above in connection with dehydration of the depulped liquid.

Example: A lot of orange juice having a pulp content of 12% by volume was concentrated to 60°Brix by high-vacuum, low-temperature evaporation. A second lot of orange juice having a pulp content of 18% by volume was concentrated to 60°Brix by high-vacuum, low-temperature evaporation.

The two samples of concentrate were then subjected to dehydration in a vacuum tray dryer under the same conditions. Thus the trays of concentrate were inserted in the dryer and the vacuum applied to keep the dryer at a pressure of 2 mm. Hg. The initial shelf temperature was 200°F. and this temperature was maintained until the product temperature reached 120°F. (25 minutes). The temperature of the shelves was then dropped to 125°F. and maintained at this level until the 45th minute. After this time cool water was circulated through the shelves and the vacuum broken, and the products removed after 60 minutes of operation.

It was observed that in the case of sample A (made from juice containing 12% pulp), the concentrate expanded about 16 times in volume during the dehydration, thus to produce a final product which was porous, easy to remove from the trays and broke up readily into flakes which reconstituted by stirring with water for a few seconds. Further, the final product was properly dehydrated having a moisture content of 4.5%.

In the case of sample B (made from juice containing 18% pulp), the concentrate did not expand during the dehydration. Further, the product was not properly dehydrated and was moist and pasty. To obtain further dehydration, this product was dried overnight under vacuum with a shelf temperature of 100°F. The next day the dry product was observed and found to be hard and adhered tightly to the trays. The product would not readily dissolve in water but required intensive agitation with the water for more than 3 minutes to form a reconstituted juice.

Freeze Concentration Followed by Spray Drying in Vacuum

A process has been developed by *H.A. Toulmin, Jr.; U.S. Patent 2,957,773; October 25, 1960; assigned to Union Carbide Corporation* for making a powdered juice concentrate. In this process the raw juice is dehydrated through a suitable low temperature process.

This process may consist of conveying the juice through a plurality of freezer units, each one of which is maintained at a higher temperature differential than the preceding unit. The ice which is removed from the juice in each one of these units is separated after each freezing step. The dehydrated juice is then heated to room temperature, and the heated juice is sprayed into a chamber in the form of a fine mist. The chamber is maintained under subatmospheric pressure which is below the vapor pressure of water at room temperature. Consequently, the water will evaporate within the chamber and is drawn off by a suitable pump. The residue will comprise granular solids, some of which will be in crystalline form. The residue is then collected from the chamber and may be further comminuted to form a powdered juice concentrate having uniformly sized particles.

Figure 2.2a is a schematic arrangement of the system with the flow of juice between units being indicated. Figure 2.2b is an elevational view of one of the freezer tanks combined with a centrifuge with portions of each removed to show inner structural details. Figure 2.2c is a longitudinal sectional view of the heater unit, and Figure 2.2d is an elevational view of the vacuum chamber with a portion of the outer wall removed.

The processing of the juice will now be described, referring to the drawings. The fresh juice before being delivered to the freezer units is precooled to a temperature of about 20°F. The precooled juice is then unloaded into the freezer unit **10** which is maintained at a temperature of -20°F. The subjection of the precooled juice to such a low temperature at the inception of the dehydration process will result in maximum water removal at that time when there is maximum water content within the juice. The juice within the freezer unit must be vigorously agitated to prevent the occlusion of juice and other solid particles in the ice formed therein. By vigorous stirring of the juice during the process of freezing in this unit, large sized ice crystals are formed which essentially comprise pure water.

After the completion of the first freezing stage, the juice has a concentration of 18°Brix. The partially dehydrated juice is then delivered through the discharge line **13** into the centrifuge **19** where the crystals of ice are separated from the juice. The juice separated within the centrifuge **19** is then delivered to the freezer unit **11**.

The freezer unit **11** is similar in structure to the freezer unit **10**. However, this unit is maintained at a temperature of - 10°F. By increasing the temperature at which the juice is to be frozen, additional water within the partially dehydrated juice will be removed. Again, by vigorously stirring the juice within the freezer unit **11**, the water will be removed in the form of ice crystals which are somewhat smaller than the crystals produced in the freezer unit **10**. This is true because the water content of the juice within the freezer unit **11** is less than when the juice was processed at the first freezing stage.

When the juice within the freezer unit **11** attains a concentration of 25°Brix, it is delivered through the discharge line **14** into the centrifuge **20**. Similarly, the ice crystals formed within the freezer unit **11** are separated from the partially dehydrated juice and the juice is then delivered to the freezer unit **12**. The freezer unit **12** is maintained at 0°F. In this unit an additional quantity of water is removed and the resultant dehydrated juice has a concentration of 40°Brix. By vigorously agitating the juice within this unit, the crystals formed therein are, in turn, progressively smaller than the crystals formed in the preceding freezer units.

When this freezing step has been accomplished, the slushy mass produced within this freezing step is then delivered through the discharge line **15** into the centrifuge **21** where again the ice crystals are separated from the dehydrated juice.

FIGURE 2.2: FREEZE CONCENTRATION AND SPRAY DRYING

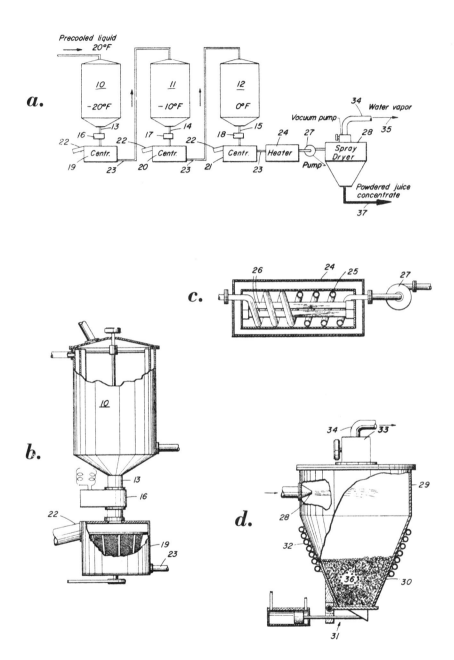

Source: H.A. Toulmin, Jr.; U.S. Patent 2,957,773; October 25, 1960

The dehydrated juice separated within the centrifuge is then delivered to the heating chamber **24** where it is gradually heated to room temperature or 70°F. The heated juice is then diffused into the chamber **29** through the action of the pump **27** and the spray nozzle **28**. The result is a fine mist of dehydrated juice which is at approximately 70°F. The vapor pressure of water at room temperature is approximately 1.8 cm. of Hg. Therefore, by maintaining the interior of the chamber **29** at subatmospheric pressure of 18 mm. of Hg or less, the water within the chamber will evaporate by applying a small amount of heat to the chamber by means of the heating coils **32**. The water vapors will then be drawn off via the conduit **34** under the action of the pump **33**. The residue remaining after the water vapors have been removed will consist of solid particles which will descend to the hopper **30** and accumulate as indicated at **36**. This residue will be in the form of solid or granular matter. Some of the residue may be crystalline in structure.

Upon removal of virtually all of the water vapor from the residue remaining in the chamber, the residue is then discharged, as shown at **37**, through the valve **31** for further processing. This processing may comprise the comminution of the residue to result in a powdered juice concentrate of uniformly sized grains.

Fruit juices comprise heat-sensitive materials which are adversely affected by the application of heat to the juice. By heating the juice to room temperature, the amount of heat applied thereto is insufficient to adversely affect the flavor and taste characteristics of the juice. In addition, by limiting the heating of the juice to room temperature, the volatilization of the aromatics from the juice is largely prevented.

Addition of Xylitol

A process is described by *L. Nobile; U.S. Patent 2,970,058; January 31, 1961; assigned to Ledoga SpA, Italy* for preparing fruit juice powders having an outstanding storage stability and which retain their organoleptic properties, such as flavor, taste and aroma, substantially unaltered with the course of time. It has been found that xylitol, a pentitol obtainable for example by reduction of xylose, is an edible, nontoxic, inexpensive substance which exhibits the outstanding property of strongly and longlastingly retaining the taste and flavor of the powdered fruit juices admixed therewith. Furthermore, xylitol permits the preparation of extremely storage stable powders.

Broadly stated, the process comprises the steps of preparing a fruit or vegetable in juice or pulp form and admixing therewith an amount of xylitol sufficient to obtain a free-flowing powder. The amount of xylitol is never critical of itself, especially in view of the fact that it is an edible, harmless and nontoxic substance. The process can be performed in several ways, that is, the xylitol may be admixed to the pressed juices and the mixture thus obtained concentrated in vacuo, or, for example, xylitol may be added to the vegetable or fruit concentrate (which can also be almost totally dried) subsequently. The only provision which is recommended for a long storage life of the powdered fruit juices obtained with the process is that of introducing the finished product in fluid-tight sealed cans; this provision is necessary to overcome the harmful effects of the hygroscopic salts possibly present. The process will be illustrated by a few operative examples.

Example 1: One kilogram of pineapple juice is admixed with 100 grams of xylitol. The mixture is concentrated under vacuum, at an absolute pressure adapted to preserve the taste and flavor of the fruit while removing the undesired water contents. The pressure is not critical in any way and can be selected according to the individual requirements and plant facilities. Once the desired degree of dryness is achieved, the mixture is powdered and packed in fluid-tight containers; if desired, certified flavoring and coloring substances may be added prior to sealing.

Example 2: One kilogram of apricot juice is admixed with 30 grams of xylitol. Xylitol is caused to be dissolved in the juice and the solution is concentrated in vacuo until dry. The dried mixture is added to 50 grams of xylitol, powdered and packed in sealed containers. The process is also applicable to nonacidic juices of fruits and vegetables.

Use of Drum and Belt Type Vacuum Dehydrator

In the dehydration of wet materials, moisture is evaporated at a high rate during the initial part of the drying process. However, the rate of drying falls off rapidly as the action progresses with only a small amount of moisture being given off during the latter part of the operation. As a consequence of this rapid initial drying, some materials have a tendency to boil and spatter in giving off the easily evaporated moisture at the beginning of the drying operation if heated too rapidly. Furthermore, during the final stages of drying, the material when nearly dry is liable to suffer deleterious overheating if the drying temperature is too high. Accordingly, it is desirable that the material being dried be heated gradually during the initial stage of drying, then more rapidly during the principal drying operation to effect dehydration as quickly as possible, and finally at a reduced rate to prevent damage to the nearly dry material during the final stages of drying.

A method for dehydrating liquid materials and an apparatus for carrying out the process were described by *G.K. Viall and W.E. Conley; U.S. Patent 3,085,018; April 9, 1963; assigned to Chain Belt Company*. The apparatus may be in the form of a vacuum dehydrator of the belt and drum type wherein the major drying action upon a thin film of liquid material carried by the belt is accomplished expeditiously and economically by heat from a heated drum as the belt runs over it. The heated drum is operated at the highest temperature that is feasible in effecting the major drying action whereby most of the moisture is removed in the shortest time with best efficiency and least detrimental effect upon the product.

To make possible such high temperature operation of the heated drying drum, the drying action of the drum is supplemented by controlled auxiliary heaters, some of which are arranged to preheat the material carried on the belt toward the drum to precondition it for drying on the drum, and other of which are arranged to provide further heating of the material on the run of the belt receding from the drum for completing the drying action without overheating the material. The drum is heated to the predetermined uniform high temperature preferably by steam in order to effect the greater part of the drying action most economically.

To provide for close control of the preheating and afterheating operations, the auxiliary heaters are constituted by independently controlled individual heating units that may be heated by steam or by electricity, although other sources of heat may be utilized both for the major heating effect and for operating the auxiliary heaters. For preheating the material, one group of auxiliary heaters is arranged to supply heat to the inner surface of the belt between the point of application of wet material to the belt and the point at which the belt runs onto the heating drum, the arrangement being such that the material is subjected to progressive controlled partial drying and preconditioning to facilitate the subsequent rapid drying by heat from the drum.

This auxiliary preheater is preferably in the form of a plurality of electrically operated radiant heaters controlled to provide regulated progressive heating of the wet material on the belt as it advances toward the zone of major drying on the drum. By this arrangement, the wet material may be warmed just sufficiently to maintain bubbles entrapped as foam within the film of material to expand and puff it, the heat being applied gradually and not rapidly enough to cause violent boiling that might otherwise result in dislodging the material from the belt. By this initial drying and expanding of the film of material, it is preconditioned to absorb heat more rapidly and to dry more expeditiously without boiling when subjected to the major drying action as the belt passes around the heating drum. With this preconditioning of the material, the drying temperature of the heating drum may be higher and the major drying action much more rapid and effective than could otherwise be the case without heat injury to the product.

By the time the material on the belt passes around the heating drum, it has been dried to such an extent that any further exposure to the relatively high temperature of the drum might result in overheating and damaging the nearly dry portions of the material that are not protected by the cooling effect of evaporating moisture. Whatever further drying of the material may be desirable is accomplished by the other group of auxiliary heaters that may also be electrically operated radiant heaters arranged to apply controlled heat to the outer surface of the nearly dry material on the run of the belt receding from the heating drum.

After the progressive drying action is completed, the belt passes around a cooling drum where the dried material is cooled and is then removed from the belt in the form of readily reconstituted discrete particles. The cooling drum also serves to cool the belt to prepare it for receiving the film of wet material that is continuously applied to it as it moves toward the heating drum. The dried material from the belt may be discharged from the apparatus directly into product-receiving containers or the particles may be retained for some time in an auxiliary drying compartment where they are agitated while subjected to further drying action of the vacuum in the system, together with moderate heating to reduce the residual moisture still further before being discharged into the containers.

Detailed drawings and descriptions of the process are contained in the patent.

DEACIDIFIED CITRUS JUICE

R.E. Berry and C.J. Wagner, Jr.; U.S. Patent 3,723,133; March 27, 1973; assigned to the U.S. Secretary of Agriculture have developed a process by which fruit juices are deacidified and converted to a dehydrated form. In this form the product can be mixed with other acid citrus products to produce a blend in either liquid or dehydrated powder form, thereby converting the unacceptable product to one with lower acidity and commercial acceptability.

Citrus fruit juices and products are manufactured by commercial methods from fruit which varies in maturity and flavor quality throughout the processing season. Thus, there are early fruit products which are rather high in acidity in October and lower in acidity in November and December, midseason varieties which are high in acidity in December and January and lower in acidity in subsequent months, and late season varieties which are high in acidity until about March or April and become progressively lower in acidity until June. As a consequence of these variations in natural characteristics of available fruit, the juices and products from these fruits vary in acid content during the season. High acidity products in general are not considered to be of the best quality for consumption. Quality is affected to a large extent by the Brix/acid ratio.

F.E. Nelson et al. devised a method whereby acidity in grapefruit products would be lowered by partial neutralization with calcium hydroxide, the precipitate removed, and the clarified partially deacidified juice would be blended with high acid grapefruit juice to adjust their acidity. Recent technology advances in dehydration methods, particularly foam-mat drying, have now made it possible and feasible to prepare dehydrated solids from high-sugar-containing materials such as citrus juices, in an economical manner commercially.

If deacidified juice could be converted to a dehydrated form, this would give it a higher degree of storage stability and lessen weight and storage requirements and cost. This dehydrated product then could be made available for blending with other juices and products to adjust their acidity without adding any unnatural ingredients. In all such cases the material blended would be of a natural consistency and only trace amounts of artificial or synthetic material would be included.

In general, this process can be described as a citrus product prepared from deacidified orange juice which is dehydrated and then blended with high-acid products. The method, briefly, consists of obtaining single strength orange juice or other citrus juice, centrifuging to remove suspended solids, neutralizing the clarified supernatant with calcium or other suitable metal hydroxide, recentrifuging to remove the precipitated calcium citrate or other metal citrate, recombining with originally removed suspended solids, concentrating this de-acidified whole juice, and dehydrating the resultant concentrate.   The resultant dried de-acidified powder is then blended with high acid orange juice, or other selected high acid juice, from early season fruit, immature fruit, orange concentrate of relatively low Brix to acid ratio, and high acid dehydrated powders.

For example, a single strength orange juice was prepared from frozen concentrate orange juice by the addition of water to form a 12.5°Brix solution (same density as aqueous 12.5% sucrose solution) with a Brix/acid ratio of 16.7:1. This juice was centrifuged at 4,000 rpm using a pilot model basket centrifuge with a three minute process cycle, ten seconds flush, operate 0.5 second and feed 25 seconds. After centrifugation, the serum was pasteurized by heating it to 191°F. at 0.5 gallon per minute. This pasteurized serum was stored in a cold wall tank. The sludge separated during the first centrifugation was stored in closed containers at 40°F.

Dry, finely divided calcium hydroxide was added to the centrifuged serum at the rate of 0.33 gram $Ca(OH)_2$ to 80 grams of single strength serum. This was allowed to stand with stirring for several hours, then the solution was screened through a 60 mesh (U.S. Sieve Size) screen to remove undissolved calcium hydroxide. Additional finely divided calcium hydroxide was added to adjust the pH to about 7.0. This was centrifuged under the same conditions as before. The precipitate of calcium citrate removed during this centrifugation was discarded. The neutralized clarified serum was mixed with the suspended solids sludge which had been separated during the first centrifugation. Before this mixing, the pH of the serum was 6.8, and after mixing the suspended solids back into the juice the pH was lowered to 6.5. This neutralized recombined citrus juice was then concentrated on a high vacuum falling film evaporator to 51.5°Brix and mixed with the appropriate amount of foaming agent and dried by foam-mat drying.

The resultant dry product was evaluated in several different ways. A small portion of it was reconstituted with water to form a juice of about 12°Brix. This was tasted by a panel of five members and judged to be very low in acidity, mild and bland, and tasted somewhat like papaya juice or peach juice. In another test, a sample of orange juice powder prepared from a highly acid concentrate, when reconstituted with water to a 12.5°Brix juice, resulted in a solution having a pH of 3.44 and containing 1.427 grams citric acid/100 grams juice. The Brix/acid ratio of this material was 8.76:1.0.

When this dried powder was mixed with the deacidified orange juice powder in the ratio of two parts acid juice powder to one part deacidified juice powder, and the dry mixture was reconstituted to 12.5°Brix by the addition of water, the resultant juice had pH 3.9, contained 0.95 gram citric acid/100 grams juice and the Brix/acid ratio was 13.16:1. When this was compared to the reconstituted juice from the original powder by a taste panel, the sample containing the added deacidified juice powder was judged to be very greatly improved in organoleptic quality.

In another example, single strength fresh orange juice was prepared from immature Valencia oranges. The pH of this juice was 3.25, it contained 1.736 grams citric acid/100 grams juice, and had a Brix/acid ratio of 7.19:1. When this immature Valencia juice was mixed with deacidified instant orange juice in the ratio of two parts green Valencia juice to one part deacidified juice powder and the resultant solution adjusted by the addition of water to 12.5°Brix solution, the resultant juice had a pH of 3.73, contained 1,097 grams citric acid/100 grams juice, and had a Brix/acid ratio of 11.39:1. This sample was also judged by the taste panel to be much more palatable and an acceptable product whereas the original green Valencia juice from which it had been made had been judged entirely unpalatable.

Tests on the deacidified juice powder itself, when reconstituted to 12.5°Brix solution with the addition of water, indicated pH 6.57 and a citric acid content of 0.009 gram/100 grams juice.

SEPARATION OF JUICE INTO FRACTIONS BEFORE DEHYDRATION

R.J. Bouthilet; U.S. Patent 3,365,298; January 23, 1968 has developed a process for preparing a solid fruit beverage base which, when dissolved in water, provides the flavor characteristics of natural fruit. The base concentrate combines fruit oils, proteins and water-soluble fruit constituents held in pellet form by a binder.

While the previous methods of stabilization have merit, it would seem more sensible to separate the ingredients of the juice and to recombine the ingredients only just before consuming as one virtually does in a fresh squeezed drink. Recombination is a relatively simple matter. Of the vital constituents in a juice such as orange juice, many are cheaply available as pure compounds, for example, citric acid, potassium citrate, ascorbic acid, pectin and carotenes, these being identical in every respect to that in the fruit.

If one separates only a very few ingredients from the orange, a preparation can be made to duplicate in every way the ripe orange juice and improve it by providing a preparation which is more stable than the fresh juice. The flavors may be fixed by physical isolation in solids such as gums to give more lasting stability mixed in a dry powder. The most common way of accomplishing this is to simply mix the flavor with a concentrated solution of the gum and spray dry. The little coated particles are similar to the particles found in a whole fruit and are called oil sacs. When pectin is used as the gum for the fixing of flavor, the nature of the particles is similar almost to the fruit. It is obvious then, if the separate components of the fruit be separated before they are allowed to interact, the quality of the beverage is closer to fresh than otherwise possible.

In the dehydration of juices, it is not enough to isolate intact ingredients, but it is necessary as well to isolate the whole mixture from the air. Even in dried powdered products, ingredients react slowly with air to lose quality. In the preparation of pellets such as are used here, the technique of tablet making involves the compression of dry powdered ingredients into solid pellets. Two ingredients are essential to provide good tablets, one being a binder and the other a lubricant. The binder may be in the form of some of the well-known gums, while the lubricant may be in the form of a solid fat or oil such as are commonly known and used. In the event that the pellet is to be required to dissolve quickly, a disintegrant is used. Disintegrants are similarly commonly in use, the disintegrants usually being a hydrophilic substance which quickly swells when wet.

The steps in the method involve first, the inactivation of the enzymes of the fresh fruit. The second step comprises the separation of the ingredients into homogeneous compounds. The next step involves fixing the ingredients in gums, and the fourth step involves drying the stable ingredients (or suitable substitute identical ingredients from an outside source, such as the pure compounds previously described), and lastly compressing the stabilized ingredients into pellets preferably having a minimal practical surface. As an illustration, a sample of a dehydrated beverage concentrate may be prepared in the following manner.

Step 1 — 25 pounds of whole fresh washed oranges of mixed varieties were ground in a colloid mill to a fine suspension. This suspension was frozen to a slush in a Sweden Soft Ice Cream Machine, and then mixed intimately with two gallons of isopentane, chilled to dry ice temperature. The solvent was separated from the ice and when evaporated at low temperature yielded an orange oil with the odor and flavor of fresh orange juice. (For purposes of clarity in this illustration, this oil is designated as fraction 1.) This oil was mixed with a suspension of water-soluble gum and spray dried to yield an orange juice flavor powder, the type prepared by commercial flavor houses. (This orange juice flavor powder is designated as fraction 1-X.)

Step 2 — The residue left in the ice in the above fractionation was dried in vacuo and extracted with 20% alcohol. This was then redried by lyophilation to yield a mixture of the water-soluble constituents of orange juice. Upon analysis this mixture was found to contain organic acids and their salts, dextrose, sucrose, trace substances as a source of ash, and ascorbic acid. Sugar is the predominant ingredient. This fraction was called fraction 2.

Step 3 — The residue from the alcohol extraction in Step 2 was heated on a boiling water bath for 30 minutes to denature the proteins, then dried in vacuo to yield a mixture of proteins and gums (polysaccharides) essentially pectin. This when dry was ground to a fine powder. It is called fraction 3.

Step 4 — The preparations from the preceding steps were combined in the following ratio: 33 parts of fraction 2, 6 parts of fraction 3, and 3 parts of fraction 1. 20 parts of this mixture when mixed with 80 parts of water yielded a beverage similar in many respects to fresh orange juice.

Step 5 — A sample of the dry powder from Step 4 was compressed into cube shaped pellets. The resultant cubes when aged and stored retained their flavor and in all respects were an improvement on beverages prepared from frozen orange juice or any orange-type beverage drink.

CONCENTRATED FRUIT AND VEGETABLE JUICES

PREPARATION OF "FOUR-FOLD" CONCENTRATE

Addition of Fresh Juice

A standard commercial practice for preparing fruit juice concentrate was developed by *L.G. MacDowell, E.L. Moore and C.D. Atkins; U.S. Patent 2,453,109; November 9, 1948; assigned to the U.S. Secretary of Agriculture.* It was found that a concentrated fruit juice containing a substantial portion of the original aroma, flavor, and palatability could be made by adding a portion of fresh, single-strength juice to a relatively strong concentrate (however prepared) and thereby obtaining a concentrate of medium strength. The fresh juice returns much of the natural aroma, flavor, and palatability to the concentrate. When concentrates prepared in this manner were diluted to original concentration with water, the resulting product was hardly distinguishable from fresh juice. An example will illustrate the process.

Example: Valencia oranges were washed, allowed to dry, and halved. The juice was extracted on a revolving burr and screened of suspended pulp. Four gallons of this juice (12°Brix) were concentrated under vacuum at a temperature of 40° to 65°Brix (about 7-fold). Fresh deaerated single-strength juice was added to the concentrate until a Brix of 42° (about 4-fold) was obtained. The product was then sealed under vacuum and placed in cold storage and frozen storage.

Separation into Fractions for Concentration

A.F. Lund; U.S. Patent 3,053,668; September 11, 1962; assigned to Cherry-Burrell Corp. has developed a process for producing orange juice concentrate in which the juice is separated into fractions that can be easily handled. Each truck load of oranges received at the processing plant is sampled and the sample tested for sugar-acid ratio. On the basis of these tests, the select oranges are separated from those of lesser quality and stored in bins according to the determined grade. In the process the select fruit is withdrawn from the storage bins and processed separately through different steps than the fruit of lesser quality.

The juice from the select fruit is extracted and the juice and pulp is separated from the seeds and peel. The juice-pulp mixture is then passed through a finisher where it is divided into a fine finished juice and a pulpy juice fraction containing suspended particles of pulp that will not pass through a screen size selected from the range of 0.020 to 0.065" depending on the preference of the particular processor. The finisher is adjusted so that the pulpy juice fraction contains 60 to 90% pulp, and preferably 75% pulp. Thus, this pulpy juice is wet enough so that it can be handled easily in the processing equipment. If the pulpy juice fraction contains more than 90% pulp, it is too dry and very difficult to pump through the equipment. The pulpy juice fraction is then heated to a temperature in the range of 187° to 240°F., preferably 203°F., which deactivates the enzymes and prevents the gelation that causes separation.

The heated pulpy juice is thereafter cooled to a temperature in the range of 40° to 110°F. and recombined in blend tanks with the fine finished juice fraction that has not been heated but passed directly from the finisher to the tanks. Note that neither fraction of the juice from the select fruit has been concentrated and therefore the juice remains at its original concentration, usually around 13°Brix. Since only the pulpy fraction has been heated, the juice therefore will have excellent flavor comparable to that of the original fresh whole juice, and even though it contains some pulp that has not been heated, the enzyme activity is not sufficient to appreciably affect the stability of the juice.

The oranges of lesser quality are simultaneously processed by first being passed through extractors. The resulting pulp-juice mixture is then heated to a temperature in the range of 185° to 200°F. to reduce the enzyme activity and thereby increase the stability. The heated juice is thereafter concentrated, usually in double-affect evaporators, to a concentration in the range of 50° to 60°Brix, usually 58°Brix. After the concentration step, the whole juice has almost no flavor and therefore it is passed into the blend tanks where it is mixed with the juice processed from the select fruit.

The relative amounts of the over-concentrated juice from the oranges of lesser quality and the treated whole juice from the select fruit are regulated so that the final juice withdrawn from the blend tanks is at a concentration of approximately 42°Brix. This is a ratio of 3 to 1, the desired strength of the frozen juice concentrates presently on the market. The final concentrate is then subjected to the standard steps of chilling, canning, and freezing and then either stored or shipped.

This process can be easily adjusted to produce the desired amount of pulp in the final concentrate, the pulp obtained from the select fruit being handled gently so that the cells are not ruptured. The fine finished juice fraction from the select fruit is not heat treated and thereby retains all the flavor of the fresh fruit, and when it is blended with the over-concentrated juice from the poorer grade fruit results in a final juice of excellent quality and flavor and one that is also very stable.

Addition of Essence to Concentrated Juice After Freezing

It is common practice to concentrate orange juice under vacuum and freeze the concentrate for merchandising. However, in the course of concentrating the juice, much of the essence which gives the characteristic flavor to fresh orange juice is removed during evaporation. This essence consists generally of various volatile alcohols, esters, aldehydes and the like, most of which, because they are quite volatile, come off with the first part of the water removed in concentrating orange juice. Substantially all come off with the first 15 to 20% of such water.

Essence may be recovered from the juice in various ways. It can be taken off during juice concentration and the accompanying water vapors may be eliminated by appropriate condensation procedures, after which the essence may be condensed at low temperatures. Essence may also be recovered in reduced volume by appropriate stripping columns or the like from the vapors given off during juice concentration. The essence thus recovered may be returned to the concentrated juice. However, if the essence is added back to the concentrate and the concentrate is stored under conditions familiar to the art, the flavor imparted by the essence is gradually lost. This may be due to hydrolysis aided by the natural acidity of the juice or may be the result of enzyme activity. If the essence, however, is separated and stored separately at 40°F. or below, it is very stable toward flavor deterioration for months and even for more than a year.

J.A. Brent; U.S. Patent 3,140,187; July 7, 1964; assigned to The Coca-Cola Company
describes a process for recovering the essence of orange juice and adding it back to the
orange concentrate in such a way that the concentrate and essence do not mix. This preserves against flavor deterioration. Since the eseence freezes at approximately 30° to 32°F.
and the orange concentrate of normal Brix, such as 42° to 50°Brix, will not freeze until
the temperature is reduced to approximately 20°F., in carrying out this process, the essence
is frozen and is then added to the orange concentrate when it is at a temperature below
the freezing point of the essence. The essence is thus maintained in discrete frozen form
and does not become intimately mixed with the orange concentrate as is the case when
the essence is returned to the concentrate without maintaining its discrete nature by the
freezing manipulation here employed.

In carrying out the process fresh juice is stripped of the essence. This may be done in a
Mojonnier evaporator or in any suitable flash equipment. The fresh juice is heated to a
temperature of approximately 195°F. It is then sprayed into an enclosed space where it
is subjected to vacuum evaporation. The fresh juice which may be at 12°Brix is stripped
of approximately 15 to 20% of the liquid. In the flash evaporator substantially all of the
essence and approximately 15 to 20% of the water of the original juice are vaporized. The
essence is recovered from these vapors by separating it from the majority of the water
vapor. This may be done by selectively condensing the water vapors and the essence in
suitable condensers. It is usually accomplished by passing the condensate through appropriate strippers or scrubbers to reduce the volume to approximately one-twentieth of the
original volume of the condensate.

The juice which has been stripped in the Majonnier evaporator or in the flash equipment
is taken to the regular evaporators where it is concentrated to any desired concentration.
It is quite common to concentrate it to 42° to 50°Brix, after which it is cooled to the
desired temperature. In general, the orange concentrate is taken down to a temperature
of from 20° to 25°F. where it is still fluid and unfrozen. In this respect, the procedure
is similar to that involved in normal concentration of orange juice.

The frozen essence, which is at a temperature of 30°F., or just under the freezing point,
is added to the orange concentrate which is at a temperature below the freezing point of
the essence. The frozen essence which may be in snow form or in frozen pellet form, may
be added to the colder orange concentrate. Since the concentrate is at a temperature
below the freezing point of the essence, the essence will remain in discrete form. Thereafter, the mixture of concentrate and frozen essence are placed in containers for sealing
and freezing as is cutomary in handling frozen orange juice concentrate.

The product thus produced consists of concentrated orange juice having therein discrete
pieces or a portion of frozen essence of the fresh juice, the product being at a temperature
below the melting point of the frozen essence so that the frozen essence will remain
discrete from the concentrate. This product has many advantages over other concentrated
citrus or other fruit juices in that it does not tend to lose the flavor characteristics imparted
by the essence on storage as would be the case if the concentrate and the essence were
not maintained in discrete form. The product may be stored for long periods of time
during merchandising and since the deterioration of the flavor does not take place when
the juice is reconstituted by adding water, the reconstituted juice will have a flavor substantially the same as fresh juice.

Concentration by Freezing

Frozen orange juice concentrates adapted to be reconstituted for beverage purposes by the
addition of water have enjoyed rapidly increasing popularity and have come into widespread
use within the past few years. For most purposes a four-fold concentration (with respect
to Brix values) is employed, the concentrate being reconstituted by adding three parts of
water. Generally speaking, two methods of preparing such concentrates have been used.
The first method, in which juice is evaporated under vacuum to a 6- or 7-fold concentration, and fresh juice is added to dilute the concentration to a 4-fold concentration has

been described in a previous patent. A second method of preparing such concentrates comprises concentrating the fresh juice by freezing and then separating the resulting concentrate from the ice, usually by centrifuging although other methods such as drainage under suction can be used. This method has the advantages that all adverse effects of heat on the taste of the concentrate are eliminated, and that loss of volatile flavor constituents such as occurs during vacuum evaporation is avoided, since practically all volatile constituents are recovered in the concentrate. However, when this process is operated to produce concentrations as high as 4-fold, the efficiency in terms of sugar recovery is low due to the fact that large amounts of soluble solids (mostly sugars) are occluded in the ice and cannot be recovered without reprocessing the ice.

A process developed by *G. Sperti; U.S. Patent 2,588,337; March 11, 1952; assigned to The Institutum Divi Thomae Foundation* preserves the advantages of freezing concentration with respect to retention of volatile flavor and elimination of cooked taste, and at the same time eliminates losses of soluble solids (i.e., sugars) in the ice and provides a 4-fold or even greater degree of concentration.

Broadly speaking, the process comprises first the concentration of the fresh juice by freezing and separation of the resulting concentrate therefrom, preferably by centrifuging the frozen juice mass. This concentrate contains a substantial part of the sugars of the fresh juice, and also practically all of its volatile flavor constituents because it has been found that these constituents are concentrated in the initial discharge from the centrifuge and hence are practically completely recovered although substantial amounts of sugars and pulp still remain in the residue ice.

The residue ice is then thawed at least enough to release substantially all of the occluded sugars and pulp in a liquid solution-suspension that is practically devoid of volatile flavor constituents. Since the sugars and pulp are not heat-labile, this liquor is concentrated by evaporation, preferably under vacuum at low temperatures, without objectionable deterioration of taste and flavor. After most of the water has been removed, the concentrated liquor is returned to and mixed with the centrifuged concentrate so that the mixture contains practically all of the valuable constituents of the fresh juice without material loss or deterioration of volatile flavor. If the ice is completely thawed and evaporated, the recovery of soluble solids will be practically 100%.

S.P. Cole, C.S. Walker and H.W. Reed; U.S. Patent 2,967,778; January 10, 1961; assigned to Institutum Divi Thomae Foundation describe a process for providing an improved, simplified and less expensive procedure for the formation of relatively high Brix concentrates.

The process comprises introducing and recirculating the orange juice in a closed recirculatory system while subjecting it to freezing temperatures. The liquid is maintained in a state of continuous agitation to convert it into a pumpable slurry having an ice crystal size of less than $1/32$". A portion of the so formed slurry is continuously removed from the recirculatory system and subjected to centrifugal separation to remove the ice crystals from a liquid constituting a first concentrate having a higher Brix value than the original juice and containing substantially all of the flavoring esters present in the original juice excepting for that portion which clings to the ice crystals.

The ice crystals are then subjected to a single washing treatment with water under centrifugal conditions, the washing treatment being conducted with the ice crystals at a temperature below the freezing point and with the washing water at a temperature above the freezing point, the quantity of water employed being proportioned to provide a second concentrate having a Brix value closely approaching the Brix value of the first concentrate, and an ice residue having substantially no Brix value.

An improved method for carrying out the Sperti process in a continuous manner is described by *C.S. Walker; U.S. Patent 3,156,571; November 10, 1964; assigned to Institutum Divi Thomae Foundation*. This is accomplished by circulating the beverage between a cooler and a holding tank so that the formation of ice crystals takes place in the beverage being processed. The cooling is not carried out to such an extent, however, that the mass of beverage is not in a flowable condition so that it cannot be pumped. While the beverage undergoing cooling is being circulated between the cooler and the holding tank, fresh beverage is introduced into the circulating stream and partially frozen beverage including unfrozen liquid and ice crystals are withdrawn from the circulating stream.

The material withdrawn can then be centrifuged conveniently to separate concentrated liquid from solids including ice crystals. The solids can then be thawed, the resulting liquid concentrated by evaporation, and the liquid thus formed admixed with the concentrated liquid from the centrifuge, in accordance with the procedure of the Sperti process. The circulating procedure of this process possesses the advantage that when it is employed it is possible to control accurately the formation of the ice crystals in the beverage under-going freezing, so that the separation of solids from concentrated beverage in a centrifuge can be conveniently and effectively accomplished with high productivity. The patent con-tains a diagrammatic drawing of the apparatus which can be used in the process and de-tailed description of the controls necessary for proper functioning of the process.

Increasing Size of Ice Crystals

Freeze concentration processes for fruit juices have been developed to overcome the problems of flavor and aroma loss which result from the use of heat in evaporation processes. Unfor-tunately, freeze concentration processes also suffer from difficulties including (a) loss of concentrate when the liquid adheres to the ice crystals by capillary action; (b) extremely small ice crystals which make separation from the liquid fraction difficult when continuous concentration is practiced; and (c) expense caused by poor heat balance in the process.

Attempts to solve these difficulties have been made in a process developed by *A.F. Lund; U.S. Patent 3,205,078; September 7, 1965; assigned to Cherry-Burrell Corporation* which is explained with reference to Figure 3.1. After being extracted from the fruit and processed through a finisher **8** to remove the seeds and rag, the whole juice is passed through line **9** into a separator **10** of suitable design and type. The separator separates the whole juice into a pulp fraction and a liquid fraction, reducing the pulp fraction remaining in the liquid to less than 6% insoluble solids. The liquid fraction, which has a concentration of 9° to 15°Brix, is discharged through line **12** into a heat exchanger **14** and the pulp fraction is drawn off from the separator for heat treatment or other processing.

The heat exchanger serves to precool the liquid fraction before it is passed through line **16** and mixed with a concentrated, cold liquid that is drawn off a separating vessel **18** through line **20**. This liquid from vessel **18** is at a temperature of 24° to 29°F. and has been concentrated to 18° to 30°Brix. Since it is at a temperature below the freezing point of the water phase of the incoming liquid in line **16**, seed crystals of ice will be formed instantaneously in the mixture. The mixture is then pumped by pump **22** into one or more chillers **24** preferably of the swept surface type. The minute seed crystals of ice are rapidly increased in size by the fast cooling in chillers **24** resulting in a larger, more durable ice crystal than can be produced by other known continuous methods. Also, the addition of the higher concentrated liquid from vessel **18** minimizes the formation of ice on the heat transfer surfaces of the chillers because of the increased soluble solids content of the mixture.

After discharge from the chillers the ice crystals and liquid are separated by passing the mixture into separating vessel **18** where the ice crystals are allowed to rise naturally to the top level of the liquid. As already mentioned, liquid of a concentration of 18° to 30°Brix is drawn off the vessel at a level below line **26**, through line **20**, and mixed with the incoming liquid in line **16**. A strainer **28** preferably is provided at the connection of line **20** with vessel **18** to prevent ice crystals from being carried out at this point. The ice mass formed at the top of the vessel is then removed either mechanically or by gravity. With some products, such as citrus juice, the ice crystals withdrawn from the top of the vessel carry with them some of the concentrated liquid. This is because capillary action

FIGURE 3.1: METHOD OF FREEZE CONCENTRATING CITRUS JUICE

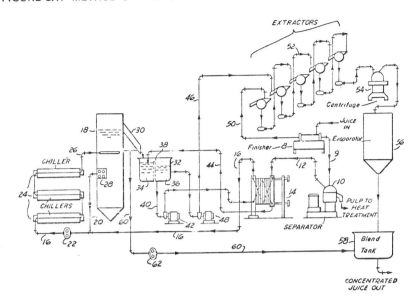

Source: A.F. Lund; U.S. Patent 3,205,078; September 7, 1965

holds the liquid on the surfaces of the ice crystals, and as these ice crystals become joined together in a soft mass, they will entrain the liquid. Therefore, the mixture is passed through a vacuum strainer 30 or into a basket-type centrifuge (not shown) for separation into ice and concentrated liquid, the ice fraction being discharged into a tank 32.

The tank is divided into two compartments 34 and 36 by a screen 38. The ice crystals from vessel 18 are discharged into compartment 34 where those that do not melt will remain. The ice water formed from the melting ice will pass through the screen into compartment 36 from where it is withdrawn through a line 40 and circulated through the heat exchanger 14. A pump 42 circulates the ice water through the heat exchanger 14 to precool the juice fraction from separator 10 and returns the water to compartment 34 of tank 32 through line 44. Here the water is used to melt the ice in compartment 34, the cold ice water then passing through the screen into the compartment 36. This arrangement utilizes available refrigeration to lessen the load on the chillers. Generally, there is refrigeration in excess of that required to precool the incoming juice in the heat exchanger and this excess can be utilized for other purposes in the processing plant.

The ice separated from the liquid juice fraction is also utilized for a second important purpose. The ice water in compartment 36, in addition to being used to precool the incoming juice, is withdrawn through line 46 by pump 48 and used for water extraction of the pulp originally separated from the juice by finisher 8. This pulp is withdrawn from finisher 8 through line 50 and passed into suitable extraction apparatus 52, the ice water being pumped into apparatus 52 through line 46. This is a highly desirable arrangement for at least two reasons. First, since the ice water from compartment 36 contains some juice, even though a very small amount, loss of soluble solids is thereby reduced to almost zero. Also, some of the disadvantages of water extraction of the pulp with tap water are avoided since this arrangement makes the system a completely closed one, nothing being introduced into the system except the product to be processed.

The liquid extracted from the pulp is withdrawn from the apparatus 52 and insoluble solids are removed by centrifuge 54. The liquid fraction discharged from the centrifuge is at a concentration of 3° to 7°Brix and therefore is passed into a suitable evaporator 56 where it is concentrated. The heat utilized in the evaporation process unfortunately destroys much of the flavor and, therefore, the concentrate is passed into a blend tank 58 where it is mixed with the liquid juice fraction that has been concentrated by the freeze method and pumped through line 60 from the vacuum strainer 30 by pump 62. These two fractions are blended in suitable amounts to obtain the desired concentration, which for commercial frozen orange juice concentrate is above 42°Brix.

Concentration by Freezing of Juices Other than Citrus

E.B. Huber; U.S. Patent 3,023,111; February 27, 1962 describes a process for the concentration of a wide variety of fruit and vegetable juices. In general the method involves the extraction of a juice from raw fresh fruit or vegetable by procedures which avoid the use of elevated temperatures, oxidation, or additions of objectionable chemicals. This juice is then subjected to a special kind of freezing capable of producing a semifrozen material containing separable aged ice crystals. This slurry is then subjected to centrifuging whereby the ice crystals are removed from the remaining concentrate. One or more additional steps of freezing and the centrifugal separation can be applied, until a concentrate is obtained having the desired solids content.

The preferred procedure is as follows. Fresh raw fruit is supplied to a disintegrating operation where it is macerated by means of a high speed disintegrator having impacting hammers. At this point it is desirable to introduce a small amount of enzyme, such as pectinol for the purpose of breaking down the pectin content (i.e., depectinizing). The disintegrated material is held for a period of time, such as from 4 to 8 hours, during which time enzymatic action is completed. At the end of this holding period, separable sludge solids are present.

The material is then subjected to desludging, as by passing the material through a suitable desludging centrifuge. Removed sludge solids can be subjected to filtering or to pressing, for the recovery of juice contained therein. The juice is subjected to a special freezing operation which serves to develop aged ice crystals. A suitable procedure to follow is to drip or spray the juice upon a refrigerated surface, whereby it is chilled to a temperature of the order of 0°F., over a period of the order of three minutes. The material is then removed from the refrigerated surface in the form of a slurry. This slurry consists of a liquid faction together with what are termed aged ice crystals. By aged ice crystals is meant crystals which are formed over a substantial time and which are of such size and separating characteristics as to permit ready separation by centrifuging. In particular the formation of large amounts of microscopic crystals which cannot be removed by centrifuging is avoided.

The relatively heavy slurry obtained is then subjected to centrifuging, whereby the crystals are separated from the liquid faction. The liquid faction can be referred to as the No. 1 concentrate. Assuming that a higher degree of concentration is required, this material is subjected to a second controlled freezing operation, carried out in the same way as before, to produce a viscous slurry containing ice crystals, and this material is subjected to centrifuging to produce a No. 2 concentrate. In most instances two such operations will produce a No. 2 concentrate having a sufficiently high solids content for all practical requirements.

The freezing can be preceded by cooling the juice to a temperature of the order of 32°F., after which it is applied as a thin layer to a refrigerated metal surface. Conventional ice machines of the type in which liquid is applied as a thin layer to a refrigerated surface, and then removed by scrapers, can be employed. Preferably an endless metal belt is employed, which moves at a predetermined speed, and is subjected to refrigeration as by enclosing it in a refrigerated tunnel. The liquid is applied (as by dripping or spraying) as a thin film on the belt, and the ice crystals are removed by suitable scraping means or blades. As previously stated, it is important for this operation to be carried out in such a manner that the resulting crystals are aged, thus avoiding production of microscopic or impalpable crystals which cannot be removed by centrifuging. Thus freezing should not be too rapid. In practice the material can remain on a refrigerated surface (at 0°F.) for a period of 3 to 5 minutes. Freezing in the manner just described not only produces crystals which can be separated by centrifuging, but also crystals having a maximum amount of entrained solids, skin fragments etc. from the liquid faction. In aging, there is a growth of the crystals from various centers, forming a filter bed of interlocking needle-like crystals similar to snow flakes.

The centrifuging operation can be carried out by the use of centrifuges of the basket type, or so-called continuous centrifuges. This equipment employs a rotor carrying a conical shaped screen, nested within a cone-shaped body, and surrounding an inner conical shaped body provided with a helicoidal rib. Material fed into the upper smaller end of the screen, travels downwardly, with progressive discharge of liquid through the screen, and with progressive discharge of separated solids from the bottom of the screen. Centrifugal apparatus of this type is desirable in that it permits continuous centrifuging operations, as distinguished from the batch operation of a basket type centrifuge. This method can be applied to juices derived from a wide variety of fruits such as pears, apples, prune plums, grapes, watermelon, cherries, citrus fruit and the like. Also it can be applied to juices derived from various vegetables, such as carrots. A particular example of the method is as follows.

Example: Raw whole apples were supplied to a Rietz disintegrator of the type previously described, the disintegrator being provided with a screen having a $3/16$" opening. Small amounts of pectinol were added as an enzyme. The resulting pulp was stored for a period of 8 hours, after which it was passed through a desludging centrifuge, to remove from 50 to 60% of the solids. These solids were pressed and the recovered juice merged with the juice from the desludging centrifuge.

The juice was then chilled to 32°F., and applied as a thin film to a moving metal belt. The belt was refrigerated to 0°F., and the material was applied to the belt to produce a layer of approximately $1/8$" thickness. The semifrozen material on the belt was permitted to remain thereon for a period of 3 minutes, after which it was removed by a scraper. The resulting semifrozen viscous slurry was supplied continuously to a centrifuge of the Mercone type, operated to discharge the ice crystals in a snow-like overflow.

As a result of the first centrifuging, substantially 50% of the water content of the juice was removed. The freezing and centrifuging operations were repeated, whereby the water content of the No. 1 concentrate was reduced by 50%. Thus the No. 2 concentrate contained only 25% of its original water content. This concentrate had a flavor and freshness almost identical with that of the original raw apples. It was preserved for a substantial period of time at ordinary refrigerating temperatures, and was preserved indefinitely by freezing.

USE OF INERT GAS

Stripping of Cut Back Juice by Nitrogen

Fruit juices generally deteriorate in color, flavor, and nutritive value on exposure to air. The rate of deterioration is quite rapid during stages of processing where high temperatures are employed. For this reason it is desirable to minimize exposure of the juice to air during extraction, straining and other treatment. Because of the presence of air, possibly in the intercellular spaces of the fruit, and because of the difficulty of extracting the juice without some aeration, the juice after extraction by conventional commercial techniques usually contains appreciable quantities of oxygen. This oxygen is conventionally removed from the juices, preferably as soon as possible, by vacuum deaeration methods. Such practice has been widely adopted in the commercial field of fruit juice processing. Nitrogen stripping of the material to be treated and nitrogen purging of the processing equipment offer many advantages over vacuum deaeration for the removal of oxygen and protection of the material from subsequent contamination. Among these advantages are (1) minimum removal of desirable dissolved volatile constituents such as flavor essence and aroma, (2) protection of the material being treated after leaving the deoxygenation operation, and (3) lessening the time during which the material being treated is in contact with oxygen.

Conventionally fresh juice is concentrated by high vacuum evaporation after or during which the juice is vacuum deaerated. A portion of single strength or specially processed concentrated juice, known as "cut back" juice is then added to the concentrate to restore flavor components. This specially processed concentrated juice has been heated so as to insure the retention by it of a substantial portion of its flavoring constituents. It has been found that a small proportion of such flavored or cut back juice is usually adequate to supply a marketable flavor to the blend of concentrate and cut back. In the conventional process of extraction the natural juice becomes aerated. The cut back juice thus contains oxygen which contaminates the concentrate and detracts from its stability during storage.

P.L. Smith and A.L. Bayes; U.S. Patent 3,044,887; July 17, 1962; assigned to Union Carbide Corporation have developed a process in which the oxygen is removed from the cut back juice by subjecting it to a stripping operation by an inert gas such as nitrogen, prior to the blending with the concentrate. Preferably the blend is protected from pick up of oxygen in subsequent operation by a purging operation of the head space with inert gas. Other inert gases may be employed such as argon and helium. Carbon dioxide may also be used in cases where a carbonated flavor is not objectionable. The cut back juice which has been subjected to inert gas stripping can also be concentrated under conditions which preserve its flavor constituents before being blended with the concentrated juice. The cut back juice is not ordinarily concentrated to the extent that the concentrated juice is. A flow sheet and detailed description of the process is given in the patent.

Totally Oxygen-Free System

In another process by P.L. Smith; U.S. Patent 3,102,036; August 27, 1963; assigned to Union Carbide Corporation a freeze concentration method is described in which oxygen is removed from the juice initially and the oxygen-free solution is maintained in a nonoxidizing atmosphere throughout the concentration process.

The process comprises subjecting the solution to a stripping operation wherein a substantially inert gas is passed through the solution so as to preferentially displace substantially all the oxygen present in the solution; decreasing the temperature of the solution to at least the freezing point of the solvent of the solution; and removing frozen solvent from the resulting concentrated solution, while continuously maintaining the solution in a nonoxidizing atmosphere after the stripping operation.

Example: Orange juice was passed through a stripping column prior to being concentrated by a freeze concentration process. The feed juice was at a temperature of 38° to 45°F. and was processed at a flow rate of 3,000 to 4,000 gal./hr. Nitrogen gas was passed through the juice in the column at a rate of 200 to 300 ft.^{3}/hr. Juice leaving the stripping column was continuously maintained in an atmosphere of nitrogen. The concentration of oxygen in the juice was checked at various points and found to be: 4 to 8 parts per million before stripping; 0 to 3 parts per million after stripping but before freeze concentration; and 0 to 3 parts per million after freeze concentration in an atmosphere of nitrogen.

Samples of the nitrogen-processed concentrate and samples of air-processed concentrate were stored for varying periods of time and then subjected to taste tests of a subjective nature. The results of the tests were consistently in favor of the nitrogen-processed juice, and the degree of preference increased as the storage times of the samples increased.

Recovery and Preservation of Essence

In the process of concentrating heat-sensitive materials such as fruit juices by evaporation, noncondensable gases and volatile flavoring constituents are often removed, thus making the concentrate less palatable.

P.L. Smith, R.E. Cornwell and T.W. Harwell; U.S. Patent 3,113,876; December 10, 1963; assigned to Union Carbide Corporation describe a process in which these constituents are recovered and preserved. The process consists of these steps: (1) removal of the essence from the solution; (2) concentration of the solution from which the essence was removed; and (3) recombination of the essence and the solution. The oxygen is removed from the essence and the solution and they are maintained in a nonoxidizing atmosphere.

In order to preserve the essence removed from the solution to be concentrated, the process removes oxygen from the essence by subjecting either the fresh single-strength solution or the removed essence to a stripping operation wherein a substantially inert gas, such as nitrogen, is passed through the essence or the solution so as to preferentially displace at least a portion of oxygen present in the essence or solution. This stripping operation removes only the oxygen from the solution, whereas the conventional vacuum deaeration process for removing oxygen from a solution by subjecting the solution to a vacuum removes a substantial portion of the essence of the solution along with the oxygen. Thus, the oxygen stripping column employed in the process removes only oxygen from the solution being treated, and the essence remains in the solution.

Inert gases other than nitrogen, such as argon, hydrogen, and helium, may be employed in the stripping operation, and carbon dioxide may be used in cases where a carbonated flavor is not objectionable. To provide adequate flavor protection for extended storage periods, it is preferable to reduce the oxygen content of the solution as low as possible, and a concentration below 0.4 part per million has been found to be acceptable in most solutions. After the oxygen has been removed, both the solution and the essence are continuously maintained in a nonoxidizing atmosphere so as to minimize subsequent contamination.

Figure 3.2 is a flow sheet of the preferred continuous concentration process. Referring to Figure 3.2, single-strength solution is supplied through a line 10 to a stripping column 12. An inert stripping gas, preferably nitrogen, is fed into the bottom of the stripping column through a line 14 and removes oxygen and carbon dioxide from the solution. The nitrogen gas and entrained oxygen and carbon dioxide are discharged through an exhaust line 16. Since the oxygen is removed from the solution by the stripping operation rather than by a vacuum deaeration process, most of the essence remains in the solution.

From the stripping column, the nitrogen-stripped solution is passed through a line 18 into a vacuum deaerator 20 which preferably operates at a temperature between 50° and 55°F. The vacuum deaerator removes the essence from the solution along with 1 to 5% of the solvent. This essence is discharged from the deaerator through line 22 into an essence recovery system 24 which recovers the vaporized essence. Any convenient heat exchanger may be used in the essence recovery system, but the recovery is preferably achieved by passing the essence into a body of the solution being treated, either single strength or concentrated, at a temperature sufficiently low to condense, absorb, and/or dissolve substantially all of the essence. From the essence recovery system, the recovered essence passes through line 26 into an essence storage tank 28, and then on through line 29 into the blend tank 44.

The solution remaining in the deaerator is discharged through line 30 into a high vacuum evaporator 32 wherein it is concentrated. Solvent, noncondensable gases, and some of the remaining essence are removed from the solution by the evaporator and are discharged through a line 34. The residue passes through a line 36 to heat stabilizer 38 and on to a second stage high vacuum evaporator 39. An exhaust line 40 from the second evaporator 39 joins the line 34 to take off the same kind of vapors. The residue solution in evaporator

FIGURE 3.2: PROCESS FOR PREPARING CONCENTRATED SOLUTIONS OF HEAT-
SENSITIVE MATERIAL

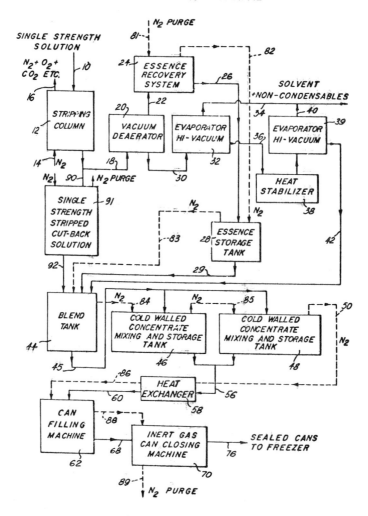

Source: P.L. Smith, R.E. Cornwell and T.W. Harwell; U.S. Patent 3,113,876; Dec. 10, 1963

39 passes through a line **42** to a blend tank **44**. From the blend tank, a line **45** conducts
the blended product to cold-walled concentrate mixing and storage tanks **46** and **48**. The
concentrate from the tanks passes through a line **56** to a heat exchanger **58**, wherein it is
further cooled, and on through a line **60** to a can-filling machine **62**. From the can-filling
machine, the filled cans pass by conveyor **68** to a can-closing machine **70** and on into a
freezer as at **76**.

After the oxygen has been removed from the feed solution in the stripping column, both
the essence and the solution must be continuously maintained in a nonoxidizing atmosphere
in order to minimize subsequent contamination. Thus, a nonoxidizing purge gas, preferably
nitrogen, is supplied to the essence recovery system through a line **81**; then on through

line **82** into the essence storage tank; then through line **83** into the blend tank; through line **84** into the cold-walled mixing and storage tank **46**; through line **85** into the tank **48**; through line **50** into the heat exchanger; through line **86** into the can-filling machine; and then through line **88** into the can closing machine, from which the purge gas and its entrainments are exhausted through a line **89**. Although the flow rate of the purge gas in the process system varies with the size and type of equipment employed, the flow rate should always be such as to provide sufficient pressure to insure that any leakage is outward from the system.

Since some small amount of essence is removed by the high vacuum evaporators, the exhaust vapors in line **34** may be passed into fractionating and distillation columns to separate the essence from the solvent vapors. Similarly, the exhaust vapors in line **34** may be completely condensed so that the essence can be removed therefrom by any of the essence-removal processes, i.e., by vacuum deaeration or an essence stripping operation. An alternate procedure would be to dispense with the vacuum deaerator altogether and to use the evaporators to perform the double function of concentration and essence removal. As described above, the exhaust vapors from the evaporators could then be passed into fractionating and distillation columns to separate the essence from the solvent vapors, or the exhaust vapors could be completely condensed and passed into a vacuum deaerator or an essence stripping column to remove the essence.

As indicated in Figure 3.2, the single-strength, stripped, cut-back solution is taken from line **18**, and passed through a line **90** into a storage tank **91**, and then through line **92** into the blend tank **44**. The main purpose of this cut-back solution is to control the concentration level of the solution in blend tank **44**. Since essence is added separately to the blend tank through line **29**, the cut-back solution is more important for controlling the concentration level of the concentrate than for adding flavor to the concentrate.

HIGHLY CONCENTRATED ORANGE JUICE

Concentration by Dielectric Heating

Citrus juices are commonly packaged at a density of 42°Brix, thereby yielding a four-fold concentrate. A density of 72°Brix would yield an eight-fold concentrate and would be desirable for such things as export to foreign countries, since it would result in a big saving in freight costs. Moreover, while the 42°Brix concentrate has to be stored at a temperature close to zero, the 65°Brix concentrate will stand storage temperatures as high as 20° to 30°F. without deterioration.

A process and apparatus is described by *R.G. Sargeant; U.S. Patent 3,072,490; January 8, 1963* in which there is shown apparatus for and a method of producing high density, low viscosity citrus juice concentrate, involving subjecting the juice to the action of radio frequency electrical energy. The juice is circulated by means of a pump from the bottom of a vertical evaporating chamber through a conduit and a high frequency electrical device, and thence through a spray head into the top of the evaporating chamber again.

One of the problems encountered in the concentrated fruit juice industry is caused by the well-known fact that, when a pectin containing juice is heated and then cooled, jellification occurs. This is particularly noticeable where the juice is concentrated by conventional steam evaporators, in which the juice is subjected to relatively high temperatures. In many cases, where an attempt has been made to run the density of orange juice concentrate, for example, up to 65° to 70°Brix with steam evaporators, the concentrate becomes very viscous, and tends to gel, and when placed in the cans and cooled, was found to be practically solid jelly. As a result of such jellification, the product, when reconstituted by the addition of water to produce a juice suitable for drinking, tended to separate, upon standing, into different strata or layers, instead of remaining a uniform mixture, and this separating tendency seriously detracts from the commercial acceptability of the product.

The process described by *R.G. Sargeant; U.S. Patent 3,366,497; January 30, 1968; assigned to Pet Incorporated* is directed particularly to the problem of producing acceptable high density concentrates from fruit juices containing substantial amounts of pectin or pectin compounds, such, for example, as apple juice, grape juice, and citrus fruit juices. Some of the citrus fruits, as for example, the popular variety of oranges, known as pineapple oranges, contain particularly large amounts of pectin.

The problem of jellification and high viscosity can be partially solved by the following method. The whole juice, before being introduced into the evaporator, is first run through a suitable centrifuge to separate it into two parts. One of these parts is a heavier portion containing most of the water, acids, sugars, etc., and containing about 80% of the entire juice. The other part is a lighter portion containing the pectin compounds and complexes, as well as other ingredients such as cellulose fibers, lipids, etc., and constituting about 20% of the entire juice. This pectin containing portion is then stored in a tank or the like, while the watery portion only is evaporated to the desired concentration. By thus first removing the pectin complexes, they are not subjected to the heat of the evaporator.

Separation of the pectin containing portion from the other portion is not absolutely complete. Thus, the pectin containing portion will contain small amounts of water, sugars, etc., and the watery portion will necessarily contain a certain amount of pectin complexes, cellulose, etc. But the centrifuge effects the separation of the major amounts of the several ingredients. Because of the fact that the watery portion of the juice necessarily contains a certain amount of pectin, it is highly desirable, when evaporating this watery portion, to maintain the temperature very low, preferably not over 85°F., the same as when evaporating whole juice, in order to hold down the viscosity. Moreover, much of the pectin complexes are contained in the cellulose or fibrous material. After having substantially separated the juice into the two portions described, and evaporated the watery portion to the desired concentration, several alternate procedures may be followed:

(a) The pectin containing portion, in its natural state, may be at once recombined with the concentrated portion to produce the final product.

(b) The pectin containing portion may be heated or flash-pasteurized to inactivate the pectin complexes and any enzymes present before being recombined with the concentrated portion.

(c) A small amount of water may be added to the fibrous mass containing the pectin complexes to thin it, and then the fibrous material may be strained out and discarded, thus getting rid of the pectin complexes which the fibrous material contained. The liquid passing through the strainer may be added directly back to the concentrated portion, or it may be first heat-treated or flash-pasteurized.

In any event, the separating out of the major part of the pectin complexes, concentrating only the remaining, watery portion of the juice, and thereafter recombining the two portions, results in a product of lower viscosity and less tendency to separate, whatever the method of evaporating employed. Fruit juices may be evaporated by means of high frequency electrical energy.

R.G. Sargeant; U.S. Patent 3,428,463; February 18, 1969 gives improvements which the author has made for the process of preparing orange or other fruit juice concentrates by means of radio or other high frequency electrical energy. Referring to Figure 3.3, a centrifuge is shown at **1**. The juice to be treated is delivered from the finisher through pipe **2**, and is separated by the centrifuge into two portions, one of which is designated as the water-containing portion, and the other the ester-carrying portion. This latter contains the major portion of the pulp and pectin complexes, while the watery portion contains most of the sugars and acids.

The water-containing portion is delivered through a pipe **3** to a vertically disposed, elongated deaerating chamber **4**. From the bottom of this chamber extends a pipe **5** to a centrifugal pump **6**, from the discharge side of which extends a pipe **7** which enters the chamber at

FIGURE 3.3: METHOD OF PRODUCING HIGH DENSITY, LOW VISCOSITY FRUIT
JUICE CONCENTRATE

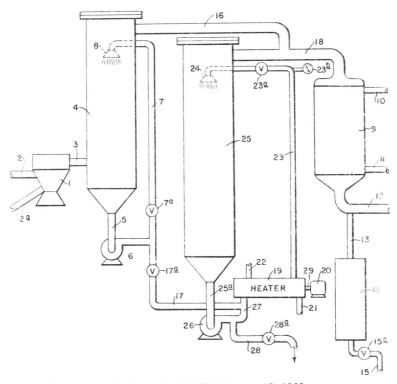

Source: R.G. Sargeant; U.S. Patent 3,428,463; February 18, 1969

a point near the top, where it terminates in a spray head **8**. Thus, the watery juice is
recycled by the pump through the pipe and spray head from which it falls to the bottom
of the chamber. Meanwhile, the deaerating chamber is maintained under a partial vacuum
by means of a conduit **16**, which, through a conduit **18** and condenser **9**, communicates
with a pipe **12** connected with any suitable type of vacuum pump (not shown). Thus,
any air contained in the juice is effectively removed.

From the discharge side of the pump **6** extends another pipe **17**, which, through a third
pipe **27**, connects with a heating device **19**, the construction of which will be hereinafter
described in detail. A heating medium, such as steam or hot water, is supplied through
pipes **21** and **22**, and the heater includes a shaft **29**, driven by a motor **20**. From the
heater **19** extends a pipe **23** into the upper end of a vertical, elongated evaporating chamber
25, where it terminates in a spray head **24**. This chamber is preferably jacketed. From
the bottom of this evaporating chamber extends a pipe **25a** to a centrifugal pump **26**,
which delivers into a pipe **28**, connected with the pipe **27**. Valves **7a, 17a, 23a** and **28a**
are interposed in the pipes **7, 17, 23** and **28**, respectively, to control the flow of liquid
therethrough, and a pressure gauge **23b** is preferably mounted on the pipe **23**.

From the top of the evaporating chamber extends a conduit **18** to a condenser **9**, a cooling
medium being supplied through pipes **10** and **11**. From the vacuum pump pipe **12** at the
bottom of the condenser extends a pipe **13** to a condensate receptacle **14**, from which the
condensate may be drawn off through a discharge pipe **15**, controlled by a valve **15a**.

With valve **7a** open and valve **17a** closed, the pump **6** continues to recycle the juice through the deaerating chamber until all of the contained air is completely removed. Then, when valve **17a** is opened, the pump **6** forces the juice from the deaerating chamber through pipe **17** to pipe **27**, where it joins the juice being concentrated and recycled by the pump **26** through the evaporating chamber **25**. This latter pump forces the juice through the heater **19** and pipe **23** to the spray head **24**, when the valve **23a** is open. The concentrated juice may be withdrawn from the pipe **28**. It will be noted that the juice is forced through the heater under substantial hydraulic pressure. By way of example, pressure of from 10 to 40 or more psi has been successfully used.

The heater **19** is of a very special kind known as the swept surface type. It comprises a cylindrical casing or housing through the center of which extends a cylindrical drum supported on a shaft. Surrounding this drum and separated therefrom by an annular space is a cylinder. Mounted on the surface of the drum are a pair of diametrically disposed scraper blades extending the length of the cylinder and having free edges adapted to engage and sweep over the inner surface of this cylinder as the shaft is driven by the motor.

Surrounding the cylinder and separated from it by a space is another cylindrical casing, this second cylinder being covered by suitable insulating material, the latter being enclosed by a metal shell. The juice being treated circulates through the annular space and flows from end to end in contact with the outside of the metal cylinder and is thus heated. Either steam or hot water may be used. The temperature of the water may be as high as 140° to 150°F., or more. By the use of a swept surface heater, such as described, the density can be carried to more than 72°Brix, without any difficulty.

In operation, the juice from the finisher comes in through pipe **2** to the centrifuge. From this, the esters and the major portion of the pulp and pectin complexes are discharged from the pipe **2a**, while the part designated as the watery portion is delivered through pipe **3**. This watery portion, which includes most of the sugars and acids, contains sufficient pulp so that the completed 72°Brix concentrate, drawn off from the pipe **28**, will have, when reconstituted, a pulp content of 5 to 13%. It is best to use an 0.020 screen on the finisher, set loosely. This usually would allow too much pulp to pass through, so, to meet some specifications, a centrifuge is used to reduce the amount of pulp in the juice portion going to the evaporator.

Another figure included in the patent shows apparatus for subjecting the juice to radio or other high frequency electrical energy before it enters the evaporating chamber. With the above described equipment and method, juice concentrates of a density of 72°Brix, and higher, and of a viscosity of 4,000 to 10,000 cp. at 75°F. have been produced. These concentrates, having a pulp content of around 7%, can be stored indefinitely without deterioration at up to 30°F., and when stored at 0°F. do not gel. Moreover, when reconstituted by mixing with water, they show no tendency to separate. It will be understood that satisfactory products of this character can be made either with or without the high frequency electrical treatment. The use of the electrical energy, however, produces a concentrate of better flavor and stability, and of lower viscosity, as well as a lower bacteria count.

Stabilization by Adding Orange Oil

Orange juice concentrations over 42°Brix would be desirable. However, orange juice concentrates of 60°Brix made in accordance with conventional processes generally become gelatinous or set upon storage and are not easily reconstituted with water. Further, orange juice concentrates of above 60°Brix made in accordance with conventional processes are highly viscous and cannot be easily processed. Various attempts have been made to provide highly concentrated fruit juice concentrates which are nongelatinous and which have a low viscosity. Known processes for producing highly concentrated fruit juices which can be reconstituted satisfactorily, for example, an orange juice concentrate which is eightfold concentrated (7 plus 1), do not generally provide high yields.

E.R. Fehlberg, G.H. Kraft and W.A. Gorman; U.S. Patent 3,391,009; July 2, 1968; assigned to National Dairy Products Corporation have developed a method for the manufacture of concentrated fruit juice which provides an increased yield of a highly concentrated fruit juice of low viscosity and does not substantially increase in viscosity upon extended storage. The 60°Brix fruit juice is nongelatinous and is easily reconstituted with water. The concentrate has an acceptable flavor and aroma without employing cut back juice.

The method comprises extracting orange juice from oranges, deaerating and heating the juice, adjusting the pulp content of the juice to less than 8% pulp and usually to between 4 and 8%, concentrating the juice to provide an orange juice concentrate of at least 50°Brix and stabilizing the juice concentrate by adding a minor amount of orange oil, or emulsion thereof, to the concentrate.

Orange oil is a commercially available cold pressed product, which has, in the past, been conventionally added to orange juice concentrates of low Brix to increase their flavor. Orange oil comprises 98% d-limonene and 1.65% aldehydes. The orange oil is added in an amount between 0.15 and 0.25% by volume of the orange juice concentrate. It has been observed that the presence of the orange oil in the orange juice concentrate acts to maintain low viscosity in the concentrate and provides a surfactant effect when added to orange juice concentrates of high Brix value.

Example: A 72°Brix orange juice concentrate is manufactured in a continuous process by extracting the juice from oranges in a conventional fruit juice extractor. The extractor is operated at 90 psig to provide 33 gallons of single-strength orange juice per minute. The orange juice obtained from the extractor is finished in a finisher having 0.020 inch finisher screens at a pressure of 95 psig. The orange juice obtained from the finisher contains 12% pulp.

The orange juice is passed into a deaerator which is maintained at 3 inches of mercury absolute and air is removed from the orange juice. The deaerated orange juice is passed into a plate and frame pasteurizer and is pasteurized by heating the juice to a temperature of 195° to 200°F. for 3 seconds. The juice is cooled to 110°F. in the cooling section of the pasteurizer and the pasteurized juice is introduced into a continuous centrifuge which is operated at 40 psig back pressure. About 32.6 gallons per minute of orange juice is obtained from the centrifuge which contains 5 to 6% pulp.

The pulp sludge from the separated orange juice in the centrifuge is removed from the centrifuge and mixed with 3 volumes of water per volume of pulp sludge in a mixing tank. The water and pulp slurry is centrifuged in a centrifuge operated at a back pressure of 40 psig to provide about 1.2 gallons per minute of pulp-wash liquor containing 5 to 6% pulp which is mixed with the orange juice in a storage tank. The mixed orange juice and pulp-wash liquor are passed into a deaerator which is operated at a pressure of 1 inch of mercury absolute. The deaerated orange juice mixture is then introduced into a double effect evaporator having a single stage in the first effect and two stages in the second effect. The flow rate of the orange juice mixture to the evaporator is 33 gallons per minute.

The orange juice mixture is concentrated to 27°Brix in the first effect of the evaporator at a temperature of 76°F. The orange juice mixture is further concentrated to 50°Brix at a temperature of 75°F. in the first stage of the second effect after which it is introduced into the second stage of the second effect of the evaporator where it is concentrated to 72°Brix at a temperature of 83°F. The evaporator withdraws 11,000 lbs. of water vapor per hour from the orange juice mixture.

The orange juice concentrate obtained from the evaporator is introduced into a cold wall mixing tank where it is cooled to 50°F. To the concentrate is added 0.20% of orange oil with agitation in the mixing tank to stabilize orange juice against gelling and setting. The stabilized orange juice concentrate is withdrawn from the cold wall mixing tank and is transferred to conventional canning apparatus where it is canned. The orange juice concentrate may be stored at 10°F. for a period of as much as 12 months and reconstituted with 7 parts of water per part of concentrate. The orange juice concentrate is readily reconstituted with water and provides a reconstituted orange juice which has an acceptable flavor and aroma.

VOLATILE CONSTITUENTS RECOVERED BY CONTINUOUS CONDENSATION

E.M. Byer and A.A. Lang; U.S. Patent 3,118,775; and U.S. Patent 3,118,776; January 21, 1964 describe procedures for recovering volatile fruit flavor constituents from juices which are being concentrated. U.S. Patent 3,118,775 describes the process as it applies to fruit juices typified by those from tomatoes, strawberries, boysenberries, grapes, prunes, apples and the like. U.S. Patent 3,118,776 describes the process as it applies to citrus fruit juices, and will be used for exemplification. The process recovers the specific fraction of volatile constituents from citrus fruit juices by a series of steps, which comprise the following.

(a) Causing the citrus fruit juice to flow rapidly, preferably in a thin, continuous film, over a heat exchange surface which is under a substantially reduced subatmospheric pressure, typically less than 1½" of mercury, in a closed system to partially concentrate the juice by separating it into a major juice concentrate portion and a minor volatile portion condensed to include flavor-producing volatile constituents boiling at temperatures higher and lower than that of water and to exclude noncondensible gaseous constituents. The temperatures employed to effect such separation will be dependent in great measure upon the identity of the citrus juice being processed. The juice temperature generally does not exceed 120°F. at any point in its flow over the heat exchange surface and, more ideally, the citrus juice is caused to travel at temperatures above 40° and below 80°F.

(b) The aforesaid volatile fraction produced by the first step is thereafter subjected to condensation continuously, at such temperatures that the volatile flavor-producing constituents are collected as a two-phase oily-watery mixture. Advantageously, the temperatures at which this oily-watery mixture may be collected by condensation may be only moderately reduced, typically 30° to 70°F. Usually, it is found that the desired condensate collected is in the neighborhood of 5 to 15% by weight of the juice initially introduced to the evaporation system or third effect vaporization chamber.

(c) The aforesaid two-phase mixture is thereafter separated into a minor subfraction containing desirable high as well as low boiling flavor-producing constituents and a major subfraction containing mainly previously condensed water and an undesirable quantity of oily flavor-producing constituents. It is this minor subfraction which has been found offers the desirable flavor enhancement to the juice concentrate. Such separation can be carried out by azeotropic distillation of the two-phase mixture under absolute pressures of less than 1½" of mercury at temperatures of 50° to 100°F. or by any other process yielding substantially the same minor subfraction, as will be described hereinafter.

The desirable subfraction which is collected comprises a major proportion of a watery phase and a minor proportion of an oily phase. This minor subfraction may be collected by condensation at temperatures of 30° to 70°F. Preferably the minor subfraction is further subdivided by allowing the watery and oily phases to separate one from another upon standing, the watery phase being continuously removed from the bottom of this separation while accumulating an oil level on the surface; upon sufficient oil accumulation, this oil is drained off separately and subsequently mixed with a quantity of suitable high boiling organic material, typically cold press citrus oil. It is preferred that this minor subfraction be collected by condensation in a closed system, that is one where no external vapors are introduced to the minor subfraction.

The minor subfraction of use generally represents about 0.5 to 1.5% by weight of the whole juice. The specific subfraction of citrus juices like orange and grapefruit is relatively colorless and has a light, cloudy appearance stemming from the emulsification therein of the minute quantity of oily material associated therewith, which minute quantity makes a desirable flavor contribution. The specific subfraction of use is ideally recoverable by evaporation at only moderately elevated temperatures and by condensation at temperatures ranging above 30°F. and upwardly to 70°F. when employing absolute pressures in the neighborhood of less than ½" of mercury; however, this fraction may also be condensed at temperatures below 30°F. by means of a brine solution or other refrigerating means which, depending on the temperature of condensation, may cause icing in the collection vessel and which permit recovery of the fraction as a snow rather than liquid.

The foregoing process for recovering the volatile fruit flavor fraction desired may be carried out under subatmospheric pressures at only moderately elevated temperatures in the initial whole juice concentration step; the juice concentrate portion may be subsequently introduced to further heat exchange equipment wherein it may be caused to again travel in the form of a thin film over one or more heat exchange surface also maintained under subatmospheric pressures but at higher temperatures whereby the more concentrated juices will be reduced in viscosity and thereby more effectively concentrated. Thus, most citrus juices call for moderate evaporating temperatures at least in the initial stages of concentration and, hence, it has been found useful in the process to employ highly volatile so-called refrigerant gases, typically ammonia, which when compressed contain sufficient latent as well as sensible heat to boil the desired volatile constituents.

Since the desired flavor fraction is collectable by condensation at temperatures above 30°F., a continuous heat exchange cycle employing such refrigerant gases is ideally suited to the process; thus, after the compressed refrigerant gas has surrendered its heat to the juice or juice concentrate, it can be employed in its liquid state to remove sensible heat as well as the heat of condensation from the distilled fraction, the heat of which can be reused to heat the liquid refrigerant for subsequent cycles. The heat of a compressed refrigerant gas may be utilized directly to boil further quantities of juice or it may be employed indirectly through transfer to another medium such as water which could serve in boiling the juice or the concentrate.

For citrus juice concentration it is preferred that the noncondensibles not be collected so that the redistilled condensate employed for flavor enhancement is relatively free of those materials like carbon dioxide and oxygen which impair the flavor values of the concentrate even at temperatures below 0°C. Thus, in the case of orange and grapefruit juice it is preferred that the noncondensibles and the difficultly condensible volatile vapors be evacuated from the initial concentration of the juice in a nonoxidizing atmosphere and discarded, thereby freeing the first two-phase oily-watery mixture collected of interference from such noncondensibles and highly volatile constituents stemming from their high gas velocities and resulting in more stable juice concentration fortified with the specific flavor fraction of the process. In this connection it is noteworthy that the citrus juice concentrates fortified with this flavor fraction have been characterized by their improved freedom from oxidative changes and the accompanying ability to avoid use of costly nitrogen packaging.

Ideally, the specific flavor fraction of use may be mixed in liquid form with the juice concentrate thereby offering the advantages of a simple plant recirculation. However, it has been found that a rather prolonged shelf life for flavor-enhanced concentrate is achieved when the specific flavor fraction of use is frozen into individual portions or pieces, typically cubes or discs, which are introduced to the juice concentrate in the can or other package just prior to freezing. It appears that, by maintaining the specific flavor fraction in a frozen condition separate from the frozen concentrate, the flavor values of the product are maintained over an unusually long period of time.

CONCENTRATION BY ULTRASONIC HIGH FREQUENCY VIBRATION

A process for concentrating citrus juice has been developed by *Z. Berk; U.S. Patent 3,352,693; November 14, 1967; assigned to Technion Research and Development Foundation Ltd., Israel* which uses ultrasonic high frequency vibration, thus minimizing the likelihood of impairment of flavor. A thixotropic composition is concentrated here by a process which comprises the steps of submitting the composition to an ultrasonic treatment and to solvent evaporation. Where the concentration is effected in stages it is possible to precede each individual evaporation stage by an ultrasonic treatment. The process may also be carried out by continuously recycling the product withdrawn from the evaporator back through the ultrasonic treatment unit into the evaporator. In this manner the concentration is increased gradually up to the desired level.

Fruit juices of a concentration of 75° to 80°Brix can be prepared. The preparation of juices or rather purees of such a high concentration is very desirable since at this level of concentration no special measures for preservation such as pasteurization or freezing have to be taken. All that is required with such a product is to store it under refrigeration so as to prevent nonenzymatic browning. Such a degree of concentration could hitherto not be achieved even with modern evaporators containing heat exchangers designed to cause a high turbulence of the concentrate.

The highly concentrated juices obtained with this process are considerably less viscous than would be the case if concentrates of the same degree of concentration were prepared in a conventional manner, and this state of comparatively reduced concentration appears to be permanent. Thus, for example, the viscosity of a 70°Brix orange juice concentrate prepared in accordance with this process was only 1.5 times that of the original 60°Brix concentrate. Against this, the viscosity of a 70°Brix concentrate obtained by evaporation without ultrasonic treatment was almost 8 times that of this original 60°Brix concentrate.

The process also consists in an apparatus for the concentration of a thixotropic composition comprising means for the ultrasonic treatment of the composition, an evaporator, and means for conducting the composition from the ultrasonic treatment means to the evaporator. The process is illustrated by Figure 3.4 which is a diagrammatic representation of

FIGURE 3.4: CONCENTRATION OF THIXOTROPIC COMPOSITIONS

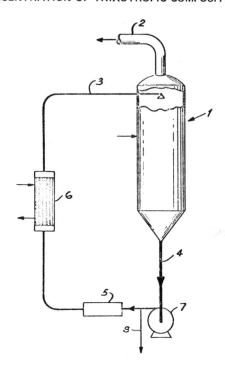

Source: Z. Berk; U.S. Patent 3,352,693; November 14, 1967

the apparatus which comprises an evaporator **1** having an adapter **2** for connection to a condenser, an inlet **3** for feeding the composition to be concentrated and outlet **4** for the discharge of the concentrated product. The apparatus further comprises an ultrasonic treatment unit **5**, a heater **6**, a pump **7** and an outlet **8** for the takeoff of the final product. During operation, the treated composition is circulated continuously from the evaporator through the ultrasonic treatment unit **5**, the heater **6** back into the evaporator and in this manner a gradual concentration takes place. As soon as the final degree of concentration is reached the product is discharged through outlet **8**.

The process is further illustrated by the following two working examples. In these examples the relative viscosity measurements were effected by comparing the time of flow in a No. 400 Oetwald Pipette. Where concentrates of different degrees of concentration were compared, correction was made for the differences in specific gravity.

Example 1: Fifty grams of commercial Shamouti orange juice concentrate with a soluble solids content of 60°Brix, was subjected to ultrasonic waves with a frequency of 20,000 cycles per second. The ultrasonic generator had a power output of 60 watts. The relative viscosity of the sample decreased as follows:

Period of Treatment, min.	Relative Viscosity
0	100
2	85
4	64
7	53
12	51
15	26

This example illustrates the very pronounced lowering of the viscosity by a suitable ultrasonic treatment. The applicability of this phenomenon in the concentration of orange juice is illustrated in the following example.

Example 2: A commercial sample of Shamouti orange juice concentrate, as described in Example 1, was subjected to ultrasonic treatment for ten minutes by means of the apparatus described in Example 1, and then concentrated, under reduced pressure to a soluble solids content of 63°Brix. The material was given another ten minutes of ultrasonic treatment and further concentrated to 70°Brix. The viscosity of this 70°Brix concentrate was only 150% of that of the original 60°Brix concentrate. The viscosity of a 70°Brix concentrate obtained without ultrasonic treatment was 700% that of the original 60°Brix concentrate. The sample concentrated in accordance with the process was capable of further concentration without scorching and browning.

CONCENTRATION BY REVERSE OSMOSIS

Controlling Degree of Liquid Concentration During Process

A process has been developed by *E. Lowe and E.L. Durkee; U.S. Patent 3,634,103; January 11, 1972; assigned to the Secretary of Agriculture* relating to concentrating liquids by reverse osmosis, and particularly to an apparatus and process for controlling the degree of concentration attained in reverse osmosis operations, and for discharging concentrated product from such operations without damage to the product.

In recent years it has been shown that reverse osmosis is useful for concentrating various liquid foods such as fruit juices, vegetable juices, milk, egg white, and the like. In conducting reverse osmosis the liquid to be concentrated is applied under superatmospheric pressure against a suitable membrane. Water (from the liquid) passes through the membrane, leaving a residue of concentrated liquid on the upstream side of the membrane. The permeate (water) and concentrate are separately discharged from the system. In conducting reverse osmosis operations, various problems are encountered. One involves the matter of controlling the degree of concentration achieved. For example, the operator may require a product of two-fold concentration or of four-fold concentration, etc., depending on particular circumstances. Because of the fact that fruit and vegetable juices are non-Newtonian liquids whose viscosity is a function of the shear rate imposed upon them, the usual techniques for controlling the degree of concentration are inoperative.

The apparatus and process are explained by having reference to Figure 3.5, where the liquid to be concentrated is held in reservoir 1. The liquid is forced under superatmospheric pressure by pump 2 into the reverse osmosis concentrator represented by block 3. This latter device may embody any of the known devices such as that of Lowe and Durkee, U.S. Patent 3,341,024.

Reverse osmosis unit 3 is provided with pipe 4 for discharge of water (permeate), and pipe 5 for discharge of the concentrated product. It will, of course, be understood by those skilled in the art that the concentrated product leaving unit 3 is under superatmospheric pressure, typically on the order of 100 to 2,000 psig. The illustrated apparatus includes an automatic syphon, generally designated as 6. This syphon meters the outflow of water (permeate) from the reverse osmosis unit, and also cooperates with a discharge metering unit (hereinafter described) to correlate the discharge of concentrate with the discharge of water (permeate), thereby providing means for controlling the degree of concentration attained in the reverse osmosis concentration. Details on the construction and operation of the syphon are provided below.

Syphon 6 includes an open-topped container 7 and a discharge pipe 8. The water leaving unit 3 via pipe 4 flows into container 7. When the water collecting in container 7 rises to the level indicated, syphon action takes over and the container is quickly emptied via pipe 8. During operation of the system, container 7 is repeatedly filled and emptied by this syphon action. A liquid-level switch 9 (which may take the form of a 0 current relay, for example) is provided near the top of container 7. Each time the water level rises to the point that syphon action takes over, the switch is actuated and it in turn actuates sequential motor 13.

Reference numeral 11 designates a rotary ball valve operated through shaft 12 by motor 13. This motor is of the sequential type, that is, when it is energized (by liquid-level switch 9 or by microswitch 14) it will rotate 180° and come to a dead stop. Thus, when motor 13 is energized by liquid-level switch 9 it rotates valve 11 to the position shown in Figure 3.5, that is, with communication between pipe 5 and pipe 15. When energized by microswitch 14, motor 13 rotates valve 11 180° so that there is communication between pipe 15 and pipe 16.

For controlling the discharge of concentrate in cooperation with syphon 6, there is provided a concentrate metering unit generally designated as 10. This unit includes baseplate 17 on which is mounted cylinder 18. A free piston 19 subdivides cylinder 18 into compartments 20 and 21. Attached to piston 19 is a piston rod 22 which passes (via a conventional gland or seal, not illustrated) to the exterior of cylinder 18. Microswitch 14 is mounted on threaded rod 23 so that the position of switch 14 can be adjusted for particular conditions. In operation, microswitch 14 cooperates with piston rod 22 as hereinafter explained.

At its right end cylinder 18 communicates with pipe 15 at its left end with pipe 24 equipped with needle valve (adjustable orifice) 25. A branch pipe 26 equipped with check valve 28 is connected to main 29 which provides a supply of water under usual house pressure, e.g., 35 to 75 psig.. During operation, compartment 20 is kept full of water at the pressure supplied by main 29. Needle valve 25 permits the discharge of excess water from compartment 20 as piston 19 traverses to the left, while maintaining the desired back pressure of water in compartment 20. Check valve 28 prevents water from backing up into the house line. A valve 30 is also included in the system so that one may control the rate at which water from main 29 will enter pipe 26 and other units connected therewith. During the operation, valve 30 remains in a set position. The operation of the controlled

FIGURE 3.5: CONTROLLING DEGREE OF LIQUID CONCENTRATION DURING
REVERSE OSMOSIS

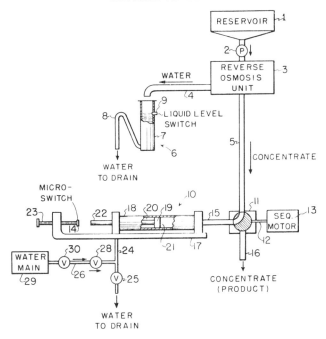

Source: E. Lowe and E.L. Durkee; U.S. Patent 3,634,103; January 11, 1972

discharge system is explained as follows. As the action in reverse osmosis unit **3** takes place, water (permeate) is discharged into cylinder **7**. When the water level rises to contact switch **9**, motor **13** is triggered to rotate valve **11** to the position shown in Figure 3.5. This allows concentrate to flow into compartment **21** where it meets the resistance offered by the water present in compartment **20**. The concentrate being at higher pressure, forces piston **19** to the left and this, in turn, forces excess water out of the system via needle valve **25**, this valve having been preset to provide a small orifice. By providing a back pressure of water in this manner the concentrate is gradually removed from the reverse osmosis unit **3** without subjecting it to excessive stress. At the same time a metering effect is attained.

Thus, when piston rod **22** moves far enough to the left to contact microswitch **14**, valve **11** is rotated to a second position whereby flow of concentrate into compartment **21** ceases. Instead, compartment **21** is now in communication with product discharge pipe **16**. Since the end of this pipe is open to the atmosphere, the concentrate in compartment **21** is instantly depressurized. The concentrated product is, however, not damaged because there is no flow, i.e., the product is not subjected to shear forces. It is thus a significant advantage of the process that the concentrated product is depressurized while in a static condition, whereby the pressure reduction is achieved without damage to its properties.

After the above described depressurization takes place, the pressure on the left side of piston **19** is the higher and consequently the influx of water through pipe **24** forces the concentrate from compartment **21** via valve **11** and pipe **16** out of the system. The rate at which the concentrate is discharged depends on the setting of valve **30**. In typical operations, this valve is set so that the concentrate is discharged gradually, rather than in a sudden burst, whereby to avoid subjecting it to undue stresses. Meanwhile, of course, additional water (permeate) has been flowing into container **7** and when the level of water reaches switch **9** the discharge cycle commences again. The following explanation illustrates how the system may be adjusted to secure a desired degree of concentration.

If, for example, it is desired to obtain a two-fold concentrate one proceeds as follows. The position of microswitch **14** is so set that piston rod **22** contacts this switch when compartment **21** takes on the same volume as the effective volume of container **7**. (By effective volume is meant the volume of container **7** considered from its base up to the level at which syphon action takes over.) In this way one obtains one volume of concentrate per volume of water removed from the starting liquid. If, on the other hand, a more highly concentrated product is desired, one would adjust the position of microswitch **14** so that it is contacted by piston rod **20** when the volume of compartment **21** is a predetermined fraction of the effective volume of container **7**. For example, if the strength of rod **23** is such that the switch **14** is triggered when compartment **21** has a volume one-half of the effective volume of container **7**, one will achieve a 2/1 concentration, or, expressed in other terms, a three-fold concentrate.

As explained above, in the system of the process the discharge of concentrate is controlled by metering the discharge of permeate. This system not only provides the desired result that the ratio of concentration can be set and maintained at any desired level, but also provides additional advantages of simplicity and economy. These latter advantages stem from the fact that the permeate is discharged from the reverse osmosis concentrator at essentially atmospheric pressure. Hence the metering of this liquid, the permeate, can be done with simple and inexpensive equipment, such as the syphon arrangement shown in Figure 3.5.

This process is of wide applicability and can be used for the concentration of liquid foods of all kinds. Typical liquids to which the process may be applied are juices, extracts, pulps, purees, and similar liquid products derived from fruits or vegetables such as orange, grapefruit, lemon, lime, apple, pear, apricot, strawberry, raspberry, cranberry, pineapple, grape, prune, plum, peach, cherry, tomato, celery, carrot, spinach, onion, lettuce, cabbage, potato, sweet potato, watercress, etc. The liquid products may be prepared in customary manner by subjecting edible portions of the produce to such operations as reaming, pressing, macerating, crushing, comminuting, extracting with water, cooking, steaming, etc. These operations may be applied to the fresh produce or to processed produce, that is, produce which has been subjected to such operations as cooking, blanching, freezing, canning, sun-drying, sulfiting, or preservation by application of chemical preservatives or ionizing radiations.

Blend of Vacuum and Reverse Osmosis Fruit Juice Concentrates

In the usual process of concentrating fruit juices by evaporation, most of their volatile flavoring compounds, generally referred to as volatile esters or essence are lost. The volatile flavoring compounds comprise various water-soluble alcohols, esters, aldehydes and the like, which are quite volatile and generally go off to a large extent with the first 15 to 20% of the liquid removed in concentrating the juice by vacuum concentration.

Another prior art technique involves first preparing an initial concentrate from fresh juice by evaporation under vacuum to produce a concentrate of 55° to 60°Brix. To impart flavoring to the juice, this concentrate is then passed along one side of a permeable membrane, while fresh juice is passed on the other side of the membrane so that the difference in osmotic pressure transfers the essence of the fresh juice to the concentrated juice, thus restoring some of its flavor qualities while maintaining its concentration. The fresh juice is then passed to the concentrator and cycled back along the other side of the membrane to produce a continuous process. Still other processes have relied solely on reverse osmosis to concentrate fresh juice to produce a fresh juice concentrate. Reverse osmosis processes in principle can be used to concentrate juice beyond 43% solids. However, it is a more expensive process than evaporation and as the concentration increases, the cost of further concentration by reverse osmosis increases very rapidly. The cost of reverse osmosis processes therefore has effectively limited it to the production of concentrates of no more than 45°Brix.

M.P. Tulin; U.S. Patent 3,743,513; July 3, 1973; assigned to Hydronautics, Incorporated has developed a process for providing a fruit juice concentrate capable of speedy reconstitution by addition of water to make a cold drink having a substantial portion of the original aroma, flavor and palatability of the fruit comprising concentrating whole fruit juice under vacuum and blending the vacuum concentrated juice with a concentrated fruit juice prepared by reverse osmosis to form a full flavored concentrate.

The fruit juice is preferably a citrus juice and more preferably is orange juice. The vacuum concentrated juice is preferably concentrated to between 60° to 72°Brix, and blended with a reverse osmosis concentrate of 20° to 45°Brix. The vacuum concentrated juice is blended with from 30 to 60% by volume of reverse osmosis concentrate based on the volume of the final concentrate to yield a final concentration of 55° to 60°Brix.

The first step of the process for preparing a Hibrix concentrated juice comprises concentrating whole juice of fruit under vacuum. The concentration by vacuum is effected in the usual vacuum concentration equipment in which the water is removed by applying vacuum to fresh juice in a suitable closed container. The concentrated juice thus obtained is generally quite free from the essence of flavoring compounds which, as pointed out above, generally comprise water-soluble alcohols, aldehydes and the like. At present most vacuum concentrations of juices are designed to produce a product having a concentration of from 65° to 75°Brix.

Juices that can be concentrated by the process include apple juice, grape juice, orange juice, tomato juice, tangerine juice, and other citrus juices. The concentrated fruit juice produced by vacuum concentration is blended with a concentrated juice prepared by reverse osmosis to produce a full flavored concentrate. The reverse osmosis concentrated juice can be prepared by any of the commonly used reverse osmosis techniques for juice concentrations.

Single strength fresh juice is concentrated by a reverse osmosis technique to a concentration of from 20° to 45°Brix. The fresh fruit juice for preparing the reverse osmosis concentrate can be from the same batch of juice used to prepare the vacuum concentrate, or can be from a different batch. Fruit juice concentrates prepared by reverse osmosis retain their volatile flavor compounds.

The concentrated fruit juice prepared by reverse osmosis is then blended with the concentrated juice prepared by vacuum evaporation and from which all of the flavor essences essentially have been removed. The blending of the vacuum concentrated juice with the reverse osmosis produced juice is accomplished in any suitable blending apparatus such as a Waring blendor. The reverse osmosis concentrate imparts its volatile flavoring compounds to the final blended concentrate.

About 30 to 60% by volume of reverse osmosis produced concentrated juice is blended with from 70 to 40% by volume of the vacuum concentrated juice, based on the volume of the final concentrate. Preferably, the concentrates are mixed to provide a final concentrate having a density of between 55° to 60°Brix. For example, when mixing a juice evaporated to a concentration of 72°Brix with an equal volume proportion of a juice concentrated to 45°Brix by reverse osmosis, a blend is produced having a concentration of 58°Brix. The concentrates of the process have a density higher than those obtained by the prior art processes of mixing single strength fresh juice with a vacuum concentrated juice, yet retain all of the flavor properties of such concentrates. Thus, the blends upon reconstituting with water, produce a product having a taste quality substantially similar to fresh juice. The following examples are given by way of illustration. All percentages referred to are by volume unless otherwise specifically indicated.

Example 1: Valencia oranges are washed, allowed to dry and halved. The juice is then extracted and screened of suspended pulp. A portion of the fresh juice having a concentration of 12°Brix is concentrated under vacuum and at a temperature of 40°F. to 72°Brix. The remaining portion of juice is concentrated by reverse osmosis to 45°Brix. The reverse osmosis produced concentrate is blended with the vacuum concentrated juice in equal proportions to obtain a final concentrate of 58°Brix. The product is then sealed under vacuum, and placed in cold or frozen storage.

Example 2: The process of Example 1 is repeated except that 60% of the reverse osmosis concentrate is blended with 40% of the vacuum concentrated juice to produce a final concentrate of 56°Brix.

Example 3: The process of Example 1 is repeated except that the juice concentrated by reverse osmosis is concentrated to 20°Brix. About 30% of the 20°Brix reverse osmosis concentrate is then blended with 70% of vacuum concentrated juice to produce a final concentrate of 56°Brix.

Concentrated Fruit Juice Prepared in Its Container by Dialysis

R.D. Scott; U.S. Patent 3,758,313; September 11, 1973 has developed a process to provide liquid foodstuff concentrates through the use of dialysis, in a rapid, commercial manner, which concentrates and packages the product in a single operation. The wall of the package comprises a dialysis membrane. These advantages accrue from the process:

   (a) there is reduced handling of the food product;
   (b) bacteria are destroyed by the osmotic pressure across the dialysis membrane,
       so the concentrate is sealed in a sanitary state;
   (c) production costs are minimized;
   (d) the consumer receives an easily handled package, ready for use;
   (e) heat exposure is minimal, so that delicate organoleptic factors are not lost
       or destoryed;
   (f) automated handling is facilitated; and
   (g) a great variety of different foodstuffs can be packaged on the same equip-
       ment with only minimal changes, if any.

The food product package comprises a sealed container having a dialytically responsive wall, such as a bag formed of semipermeable membranous film, and the food product concentrate therein consists of the nondialyzable portion of the food product. The bag typically has a sealable filler opening and means adjacent the opening for supporting the bag during filling operations.

Method is provided for preparing packaged food concentrates which includes sealing a liquid food portion to be concentrated in a container having a wall comprising a semipermeable membrane, and dialyzing the food product through the membrane against a dialyzing solution to remove water from the food product sealed within the container. A bath of dialyzing aqueous salt solution is maintained, e.g., at a temperature between 40° and 200°F. and preferably to 120°F., and the sealed container is immersed in it for food product dialysis. The container preferably takes the form of a bag comprising a flexible, semipermeable film having a sealable filler opening. The bag is then filled through its filler opening with the food portion to be concentrated, the bag opening sealed and the bag immersed in a dialyzing aqueous salt solution to concentrate the food portion contents of the bag.

In a specific embodiment of the process particularly adapted for high speed automated production, the method includes advancing a series of individual food containers having a wall portion comprising a semipermeable membrane to a fill station, filling the food product to be concentrated, e.g., orange juice into the containers in sequence at the fill station, sealing each food product portion in its individual container, immersing the containers in a bath comprising a dialyzing aqueous salt solution, e.g., a solution of sodium chloride in water at a temperature between 40° and 120°F. and a concentration between 1 and 10% by weight in a manner and for a time to dehydrate the food product through the dialysis membrane wall portion of the container, e.g., to remove from 50 to 90% of the product water and thereafter withdrawing the containers from the bath. The containers may be bags formed of flexible semipermeable film, e.g., cellulosic, proteinaceous or synthetic organic polymeric film which is differentially permeable to food product components. Apparatus for carrying out the process is illustrated and described in the patent.

NO APPLICATION OF HEAT

A process for concentrating citrus juices without the application of external heat is described by *H.W. Harrell; U.S. Patent 3,340,071; September 5, 1967.* This method minimizes the destruction of the nascent qualities of the juice. It has been found that by vigorous agitation of confined liquids by internal mechanical means, vaporization of certain constituents, emulsifications and like combinations of liquids will occur. Vigorous agitation or turbulence generated by internal mechanical means will produce liquid vapor which may include droplets, atomized liquid and free floating aresol particles, as well as volatilized portions of the lighter constitutents of liquid combinations, dispersals, or suspensions.

While the nature and amount of vapor produced by mechanical agitation and turbulence of the body of a liquid will, of course vary with the pressure applied to the liquid within a closed receptacle, as well as the type of agitation employed, it has been found that with the present apparatus considerable vaporization, both by volatilization and mechanical atomization will occur under atmospheric pressure. In the treatment of citrus juices the vigorous internal mechanical agitation and resultant turbulence produces a water vapor which may be drawn off from the body of the liquid to leave therein a more concentrated solution of the juices, the constituents of which are heavier and less volatile than the water content of the natural juices.

In this method and by use of the described apparatus internal heat is generated through the frictional action of contact with the liquid and the turbulence produced by rotating blades preferably operating between stationary blades. The relation of the blades being somewhat in the fashion of a maze to force the fluid in a tortuous turbulent path whereby the skin friction of the liquid in contact with such mechanical means produced the elevation of temperature of the liquid.

It is to be noted that in accordance with the well-known principle of fluid motivation mechanical impellers produce a localized reduction of pressure behind the blades. Thus, as the front or leading face of a rotating impeller pushes forwardly the fluid in front of the blade, substantially reduced pressure is effected behind the blade to provide for the inflow of fluid to compensate for that which has moved forward. In the present device this localized reduction of pressure in combination with the increased temperature produced by skin friction provides a localized continuously moving evaporation zone or zones where the heat and pressure are such as to induce distillation despite overall pressures and temperatures well below the temperatures and pressures required for conventional distillation. Thus the chamber pressure may be atmospheric and the mass temperature relatively low despite which volatilization will occur.

With respect to the concentration of citrus juices, deleterious effect upon the character of the juice such as taste, aroma and appearance, frequently ensues from evaporation by the application of external heat to elevate the mass temperature of the total liquid to the evaporation point, reduced pressure having been employed to minimize such required temperature. It has been suggested that such destruction of the nascent qualities of the juices by evaporation in a cooking method permits the enzymes and microorganisms to affect the pectin content whereby it may be precipitated, thus contributing to the disqualification of such concentrate from wide areas of the market. It has further been suggested that by a brief subjection of at least portions of the natural juices, while in motion to vaporizing temperatures by externally applied heat and the instant comingling of such portions with the body of the juice, will be effective to immunize the enzymes and microorganisms within acceptable limits. The method and the operation of this apparatus may produce a like inactivation, neutralization, or pacification of the enzyme and/or microorganism effect on the pectin to produce a citrus juice concentrate free of previously encountered objections. Drawings and a detailed description of the apparatus are included in the patent.

NONFROZEN CONCENTRATE

In the fruit juice concentrate industry, the relative sweetness-to-tartness relationship is known as the Brix-acid ratio. The Brix unit is a commonly used unit of measurement to express the concentration of dissolved solids in an aqueous solution. The acid unit is the citric acid concentration in the citrus juice. A Brix-acid ratio is obtained by dividing the Brix value by the acid value of a given product, and this gives a ratio compared with unity which forms a comparative scale for acceptability of such juice concentrates.

As examples of Brix-acid ratios of concentrated citrus fruit juices, high-grade fresh-frozen orange juice or grapefruit juice concentrate will usually have a Brix-acid ratio of 14:1. A range of Brix-acid ratios for many well-known commercial fruit juice drinks of citrus base ranges from 17:1 all the way to 54:1. In all such drinks, sugar is added to offset the sourness of the citrus juice and this may result in a syrupy final product. Also, if the solids content is too low, the final drink may be watery and insipid. Accordingly, careful formulation is needed in all such products in order to ensure acceptability of the final drink at consumer level as well as approval of the concentrated product at Government control level.

It is reiterated that the better quality products on the fruit juice market have at the lowest Brix-acid ratio of 14:1 and are more usually at 17:1 upwardly. Normally, Brix-acid ratios below 14:1 yield a drink which is excessively tart and is commercially unacceptable to the consumer. Even the addition of large quantities of sugar to offset the sour taste fails to make these low ratio drinks palatable since the result is a syrupy or cloying taste which is objectionable to the average consumer.

C.I. Houghtaling, F.S. Houghtaling, N.E. Houghtaling and R.W. Kilburn; U.S. Patent 3,227,562; January 4, 1966 describe a process for preparation of a citrus fruit concentrate which will provide a drink with a low Brix-acid ratio with an unusually tart taste and pleasant flavor.

A citrus fruit concentrate, preferably made up of several species of citrus fruit with or without the juices of other fruit, is formed in combination with an additive or additives which allow a low Brix-acid ratio to be attained, such as 11:1, and prevent the need of refrigeration of the finished concentrate or its original freezing, the juice or drink formed from the concentrate of the process being of a natural, tart, pleasing taste, flavor and appearance and, moreover, being capable of a variation of dilution, even dilutions of as much as 15:1, and more, without becoming thin, watery or otherwise unacceptable. The fruit juice concentrate can be stored on open shelves, i.e., without refrigeration, is not a frozen product and therefore does not have to be defrosted but is available for instant use, yet takes up little shelf room in store or home due to its concentrated condition.

The process generally comprises a mixture of sugar and concentrated fruit juices, some of which are from citrus fruits blended together in proportion to yield a concentration of soluble solids between 68° and 75°Brix and with a citric acid concentration between 6.2 and 6.8% by weight to which may be added essential oils from natural citrus fruits, together with an additive comprising an intimate mixture of the salts containing the cations sodium, magnesium, calcium, iron, potassium, manganese and aluminum and the anions chloride, oxide, silicate, iodide, bromide and sulfate. The patent contains formulas for such drinks and the additive mixture of minerals used in them.

CONCENTRATION OF GRAPE JUICE

Separation and Restoration of Essence

J.J. Mojonnier and M.A. Brna; U.S. Patent 3,061,448; October 30, 1962; assigned to Mojonnier Bros. Company have developed a process adaptable for the concentration of a variety of fruit juices whose aroma and flavor components are volatile and separable from the juice, but particularly advantageous in the concentration of grape juice.

The process for the concentration of grape juice comprises the steps of vaporizing a portion of the juice to provide vapor and an intermediate liquid concentrate, conducting the vapor to a distillation zone wherein the vapor is enriched in flavor and aroma components to provide essence, conducting the essence from the distillation zone and condensing the essence by transferring its heat of vaporization through an impervious wall to the intermediate liquid concentrate while the intermediate liquid concentrate is subjected to vacuum conditions and while the intermediate liquid concentrate and the essence are maintained separate from one another, thereby producing concentrated liquid product and essence condensate, returning a first portion of the essence condensate to the distillation zone to enrich the vapor in flavor and aroma components, and blending a second portion of the essence condensate with the concentrated liquid product. Complete drawings of the apparatus and description of the process are contained in the patent.

Reduction in Size and Amount of Equipment

A.N. Chirico; U.S. Patent 3,065,085; November 20, 1962; assigned to Chicago Bridge & Iron Company has developed a process for concentrating grape juice in which (1) a first fraction of the raw juice is vaporized and a relatively small amount of essence separated therefrom by rectification (2) the remaining juice is heated to evaporate water therefrom and provide a concentrated juice and (3) the separated essence is blended with the concentrated juice to produce a full flavor concentrate.

The process provides an unusually efficient method of producing a full flavor juice concentrate. By using the vapors from the rectifying column as a source of heat for heating the partially concentrated juice in the condenser-heater, the quantity of steam required to achieve a given increase in the concentration of a given quantity of juice is substantially reduced. Hence the production of concentrated full flavor juice by the process is more economical than with previous methods. An accompanying drawing schematically illustrates the system, and a full description of the process is given.

FRUITS

APPLES

Air Dried Segments for Pie

A number of methods of dehydrating apples are known to the art. A satisfactory method may be described as peeling, coring and segmenting apples which are then treated with a suitable antidiscoloration agent. The treated segmented apples are next subjected to a current of heated air to remove a major portion of the moisture from the apple segments leaving a residual moisture content by weight ranging from 16 to 30%. The apples are then further dried to an ultimate moisture content comprising less than 5% by weight of the apples. A major problem with regard to dehydrated apple segments resides in the rehydrating. The rate of rehydration varies according to the variety of apple used, but regardless of the variety, a substantial period of soaking has always been necessary.

W.R. Dorsey and S.I. Strashun; U.S. Patent 3,049,426; August 14, 1962; assigned to Vacu-Dry Company have developed a process in which segmented apples are initially dried to remove a major portion of the water from them and then are subjected to a perforating action to define a plurality of perforations extending through individual segments and where the cellular structure of ridges spacing the perforations is retained normally open and uncrushed. Further dehydration of the segments to an ultimate moisture content of less than 5% for purposes of preservation prior to end use is facilitated by the perforations, and upon rehydration the perforations enable the segments to readily receive water which materially shortens the time necessary for rehydration.

In making pies from dehydrated apple segments, it has previously been necessary to rehydrate the dried apples before baking the pie. Dehydrated apple segments perforated in accordance with this process may be placed in a pie shell, an aqueous solution which may include sugar and selected seasonings is poured over the segments in sufficient quantity to immerse them, and the pie is ready to be baked. The perforate apple segments thoroughly rehydrate during the baking to form a palatable pie. The necessity of preliminary rehydration of the apple segments is thereby eliminated with a consequent saving in time and money.

Figure 4.1a is an end view schematically illustrating apparatus for perforating the apple segments; Figure 4.1b is a fragmentary perspective of one of the rollers showing the perforations; and Figure 4.1c is a perspective of a dehydrated apple segment perforated in accordance with the process. A suitable perforate dehydrated apple segment has been obtained by initially sliding the apples into segments measuring about ¼ inch in thickness, sulfuring the apple segments with sodium bisulfite and then dehydrating the sulfured segments to a moisture content by weight of from 16 to 30%. The perforating apparatus comprised a pair of steel rollers 12 and 14 as shown in Figures 4.1a and 4.1b. Roller 12 was provided with a smooth rubber covering 20. Roller 14 was formed with a plurality of perforators or studs 22. Studs 22 were spaced at intervals of about 0.1 inch, projected radially from roller 14 $5/16$ inch, and defined a blunt end, oblong in plan measuring $1/8$ by $1/16$ inch. Rollers 12 and 14 were spaced slightly less than $5/16$ inch apart such that studs 22 impinged about 0.1 inch upon surface 20 of roller 12. The partially dehydrated apples were then passed through the oppositely rotating rollers forming partially dehydrated, perforate apple segments. The perforate segments were then further dehydrated to a moisture content of less than 5%.

FIGURE 4.1: METHOD FOR MAKING DEHYDRATED APPLE SEGMENTS

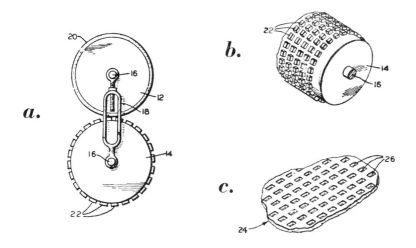

Source: W.R. Dorsey and S.I. Strashun; U.S. Patent 3,049,426; August 14, 1962

The perforating operation facilitated ultimate dehydration since the overall drying surface of the individual segments was increased, and provided an end product of a thin apple segment having a plurality of small perforations extending through the segment and wherein the structural integrity of individual segments was preserved with the cells of ridges spacing the perforations retained uncrushed and normally open. Figure 4.1c illustrates a dehydrated apple segment 24 having a plurality of perforations 26 in accordance with the process.

A satisfactory pie has been made from a recipe including the following ingredients:

Dehydrated apple slices perforated in accordance with this process	92 grams
Sugar	92 grams
National Starch Co. Instant clear gel #37	7 grams
Salt	1.3 grams
Cinnamon, ground	0.45 gram
Citric acid	1.00 gram
Lemon flavor, Fritzsche Bros. Aromalok (R) Lemojuice No. 30376	0.015 gram

The dry ingredients are placed in a 9 inch pie tin, known generally as a 309 pie tin, which is lined with a shell of pie dough. If desired, a top crust formed with a center hole is placed on the shell and two cups of tap water are poured slowly through the hole. Alternatively, the dry ingredients and water may be premixed or the dehydrated apples may be placed separately in the shell, the other ingredients being mixed with the water and then poured into the shell over the dehydrated apples. The pie is then baked 40 to 45 minutes at 425°F.

Since the perforate apple segments reconstitute during baking and without preliminary soaking, it will be apparent that the above enumerated steps may be performed at different points in space and time. This feature permits distribution to the consuming public of pies at different stages of production as the tastes of the consuming public may dictate.

Precooked Apples

W.R. Dorsey and S.I. Strashun; U.S. Patent 3,060,037; October 23, 1962; assigned to Vacu-Dry Company describe a process for precooking and dehydrating apples to produce a precooked product which is instantly reconstitutable for use.

As a first step of the process, mature apples are peeled, cored and then sliced to a convenient size, for example, ¼ to ⅜ inch in thickness. The sliced apples are then subjected to a sulfite treatment, well known in this art, for the purpose of preventing oxidation of the newly exposed cut surfaces. Subsequently, the sliced apples are precooked at a temperature of approximately 212°F. Preferably, the precooking is performed in a steam cooker or blancher. Throughout the precooking period apple samples are tested to assure proper control of the precook.

In performing these control tests on the product being precooked, the familiar fruit pressure tester is used. Such testers are normally utilized for determining the ripeness of fruit. One form of tester is described in Circular No. 627, United States Department of Agriculture. This pressure tester consists of a metal barrel within which a plunger is reciprocably attached by a coil spring. The barrel carries a longitudinal groove through which the restrained end of the plunger may be observed. The groove is calibrated in pounds of pressure required to force the plunger back into the barrel against the resistance of the spring. This type of tester is operated by placing the free end of the plunger against the fruit product to be tested, here an apple slice; forcing the plunger into the fruit flesh up to a predetermined depth; and observing the amount of force applied to obtain such penetration. The predetermined depth of penetration is established by a mark on the free tip of the plunger. The amount of force required is a measure of the relative firmness of the fruit.

In applying this pressure tester to the process, it has been found that a satisfactory final product (i.e., one that is not gummy or sticky and yet is sufficiently precooked) results only when precooking is controlled so that penetration tests on the precooked apple slices immediately prior to dehydration are within a critical range of between 1 and 5 pounds on a pressure tester of the foregoing type having a ⁷⁄₁₆ inch diameter plunger slightly concave at its free end. The slice being tested is placed against a firm backing and, in practice, the fruit resists penetration up to a certain force, whereupon the fruit suddenly yields allowing the plunger to penetrate as far as a disc of fruit compressed against the backing. The critical range refers to the force applied immediately prior to this yield point.

Immediately following the continuously monitored precooking step, the apple slices are dehydrated to about 1.5% moisture content on a dried basis either by air drying or in a vacuum dryer. Dehydration of the slices must be commenced immediately after the foregoing critical range is attained in order to prevent further internal cooking of the hot fruit. As a final step, the dehydrated and precooked apple slices are packaged in their condition as produced from the dryer and constitute a precooked sliced apple product which is instantly reconstitutable for use merely by the addition of water. The process may also be continued to produce applesauce or cake mix additive. The final product, whether in the form of slices, sauce or cake mix additive, instantly reconstitutes into an apple product merely upon the addition of water and by allowing the rehydrated product to stand for at least three minutes and preferably five minutes.

Applesauce from Precooked Apples

A process described by *W.R. Dorsey; U.S. Patent 3,235,391; February 15, 1966; assigned to Vacu-Dry Company* provides a process for dehydrating apples utilizing a flaking step to produce an improved applesauce product which retains a natural fruit flavor and still is smooth and without a grainy or gritty texture.

In producing the applesauce, mature fresh apples are first peeled and cored. The apple flesh is then sliced to a convenient size, for example, 1/4 to 3/8 inch in thickness. The sliced apples then are subjected to a sulfite treatment, as will be familiar to those skilled in this art, for the purpose of preventing oxidation of the newly exposed surfaces. Following the foregoing preparation, the sliced apples are precooked for about 3 to 7 minutes at a temperature of 212°F. This precooking step is preferably performed in a conventional steam cooker. For soft mature apples a time of 3¾ minutes produces a satisfactory final product, whereas for harder apples a precooking period for as long as 7 minutes has been found to be necessary to produce a satisfactory product.

After the apple slices have been treated as above, they are air dried to a moisture content in the range of 18 to 24% by weight. Seasoning is then added, if desired. For the apple product illustrated here, sugar is added in the approximate weight ratio of 25% sugar to the sliced apples on a dry weight basis. The sliced apples and sugar are thoroughly mixed.

The unseasoned fruit slices air dried to 18 to 24% moisture are diced to 1/8 to 3/8 inch cubes and then dehydrated by air-drying or vacuum-drying to a final moisture content of below 3.5%. After the diced fruit has been dried to a low moisture content, it is warmed until the cubes become pliable. This warming step has been performed satisfactorily on apples with radiant-type heating equipment and is important to render the fruit cubes pliable prior to the subsequent rolling step. Otherwise the cubes shatter and disintegrate when rolled.

After the diced fruit has been warmed, it is passed between rollers to produce an intermediate flaked product having a finished flake thickness within the range of 0.010 to 0.040 inch. This flaking step breaks down the fiber in the fruit without destroying the cellular structure, thereby eliminating any grainy or gritty texture. This step is very important to obtain a smooth product which retains a true fruit flavor. The flake thickness is critical. If the finished flake thickness for apples is under 0.010 inch, the product becomes cottony, loses some of its apple flavor, and tends to ball up on one's tongue when eaten. If the finished flake is over 0.040 inch in thickness, apple pieces are the result and not a sauce.

The flaked fruit flesh is then milled to granular size so that the granular particles pass through a U.S. 10 to 20 mesh screen. This form of instant sauce also may then be seasoned and packaged for commercial use. The described apple product forms an instant applesauce by the addition of approximately 8 ounces of cold water to from 1 to 1.5 ounces of dehydrated product. The mixture is stirred slightly until dispersed. It then is permitted to stand for 3 to 5 minutes, whereupon it is ready to be served and eaten without additional cooking.

Convoluted Apple Slices

R.F. Robinson, R. Fine, W.H. Lehmacher and D.R. Davis; U.S. Patent 3,384,496; May 21, 1968; assigned to Colgate Palmolive Company describes a process for preparing a crisp, dry apple product in which the apple slices are convoluted.

The process comprises heating apple slices having a biologically normal moisture content to reduce their moisture content and to render them pliant, convoluting the apple slices so that substantially convoluted, three-dimensional configurations are formed, and rapidly cooling the convoluted apple slices to impart a brittle, substantially nonpliant nature to them.

In the process an important feature resides in the step of convoluting the substantially pliant apple slices. By "convoluting" is meant the feature of distorting the surface of the apple slices to result in at least part of the perimeter and adjacent area of one side of the slices and the slices and at least part of the perimeter and adjacent area of the opposite side of the slices being bent at least proximate to one another. It has been found that this convolution can be effected most suitably in one of the following ways.

If apple slices are vertically suspended from a bar, hook, wire and the like and dried while in this vertically suspended condition, it was unexpectedly discovered that the apple slices assume a convoluted configuration resembling the integer 8. Apple slices may also be convoluted by drying the slices while tumbling them in a rotating, preferably perforated drum. Although a perforated rotating drum is preferred, any device may be used which would enable the apples to repeatedly free fall in a confined space before striking any of the confining surfaces of the device. The convolution of the apple slices may also be effected by allowing them to free fall in a wind tunnel which convolutes the apple slices while they are falling. Regardless of the method used to convolute the slices, it should be noted that the slices must be pliant and dry, that is, contain up to 3% moisture, during the convoluting step.

The drying of apple slices and the rendering of the slices pliant may be accomplished as follows. When drying the apple slices and rendering them pliant, the slices are placed in an oven or other suitable heating device and are dried by using hot, dry air at a temperature of 140° to 300°F. and a relative humidity of up to 5.5%. If desired, a temperature of up to 350°F. may be used for the first minute or so of drying to remove surface moisture from the slices.

During the drying, it is preferred that the hot, dry air flow past the slices at the rate of up to 700 linear feet per minute and preferably from 300 to 700 feet per minute. However, a static, hot, dry air system may also be used. The time required to dry the slices will generally vary from about twenty minutes to an hour if hot, dry air flows past the slices at a rate of 300 to 700 linear feet per minute. If, however, a static hot air system has been used, then the drying time will vary from four to eight hours.

Additionally, even flat apple slices which have been previously dried may be convoluted. This is accomplished by heating the dry, flat slices to at least 140°F. at which temperature they become pliant. Thereafter the dry, flat apple slices may be convoluted in any of the above described ways.

If the slices have been dried while in a horizontal position, either on a tray or conveyor belt, etc., they are then removed from their horizontal position and placed in a perforated, rotatable drum. Hot, dry air within the above temperature range and preferably at a velocity of from 300 to 700 linear feet per minute is forced through the perforations in the drum while rotating the drum and in order to maintain the pliant nature of the apple slices. After the drum containing the apple slices has been rotating under the above conditions for up to five minutes, the flow of hot, dry air is discontinued and cool, dry air is blown through the perforations in the rotating drum in order to fix the physical form of the apple slices. The apple slices become brittle when they have cooled to a temperature below about 130°F. The apples removed from the drum are crisp, convoluted, golden brown and dry.

A variety of edible apples may be used, such as Delicious, Winesap, McIntosh, Cortland, Rome, York, Staymen, Golden Delicious and the like. The apples are first washed with water to remove therefrom any residue of pesticide. Thereafter the apples are cored, peeled and sliced. Optionally, the peel may be left on the slices and the apples may be sliced in such a manner so that they do not have a hole therein. The thickness of the apple slices may vary from 0.07 to 0.25 inch. It has been found that the final product has optimum physical properties when the thickness of the apple slices is within this range.

Example 1: McIntosh apples are hand washed with water and then fed into a core and peeler machine. After the apples are peeled and cored they are sliced with a Hobart food slicer which is set so that the resultant slices are $\frac{1}{16}$ inch thick. The apples are sliced in such a manner that each slice has a hole in the center representing the portion of the core which was removed. The slices are then dipped in a mixture composed of 98% sugar and 2% cinnamon by weight.

A metal bar is then inserted through the hole in each cored slice. The slices are spaced from one another on the bar by a distance of $1\frac{1}{2}$ inches. The bars are then placed in a hot air dehydrator which has been preheated to a temperature of 300°F. The hot air is circulated through the oven at a rate of 300 linear feet per minute and allowed to contact the apple slices for 25 minutes. After 25 minutes the apple slices, which have formed a configuration resembling the integer 8, are removed from the oven and allowed to cool at room temperature. Upon cooling at room temperature the apple slices become crisp and permanently assume the integer 8 configuration.

Example 2: Greenings apples are washed and hand cored. Approximately one-half of the apples are hand peeled. All of the apples are sliced using a Hobart food slicer to form slices having a thickness of 0.09 inch each. Twenty-four slices are thus obtained, consisting of twelve peeled slices and twelve nonpeeled slices. Six peeled slices and six nonpeeled slices are dipped into a mixture of 98% sugar and 2% cinnamon by weight. The remaining slices are untreated. All of the slices are then placed in a forced air oven which has been preheated to a temperature of 250°F. by 350 linear feet per minute of hot, dry air. The slices remain in the oven at this temperature and rate of air flow for 20 minutes. During the 20 minutes, the slices are turned over periodically.

After 20 minutes the slices are removed from the oven and placed in a perforated basket having a handle. The basket containing the slices is then placed in the oven at the same temperature and air flow conditions as above. After remaining in the oven for 5 minutes, the door of the oven is opened and the basket, which remains in the oven, is hand rotated for 15 minutes causing the apples to tumble and convolute. The basket containing the still pliant convoluted slices is then removed from the oven and allowed to cool at room temperature. Upon cooling, the slices become brittle and thus permanently convoluted.

Applesauce by Pressing Before Dehydration

Dehydrated applesauce powder is a very desirable component of field rations for the Armed Forces and of rations for astronauts because of the generally widespread acceptability of applesauce. However, one serious drawback has militated against incorporating dehydrated applesauce powder in such field and space rations, namely its tendency to lump and cake, particularly as the result of storage of dehydrated applesauce in flexible packages under vacuum at relatively high temperatures, such as 100°F. or higher. In military operations such temperatures are frequently encountered. Hence, any food ration must be stable under such conditions in order to merit serious consideration as a component of field rations.

A.R. Rahman and T.R. Schmidt; U.S. Patent 3,535,127; October 20, 1970; assigned to the U.S. Secretary of the Army have developed a process for producing a dehydrated applesauce product which is readily rehydratable, is storage stable and does not cake during storage at elevated temperatures. The dehydrated applesauce product is made resistant to caking by removing a substantial proportion of the juice of the apple prior to dehydration thereof. Any loss in flavor and sweetness resulting from the removal of the juice is overcome by adding malic acid and a nonreducing sugar, such as sucrose, to the dehydrated applesauce powder at the time of packaging. If all, or substantially all, of the juice is removed, satisfactory flavor may be restored by packaging the dehydrated applesauce powder with a nonreducing sugar and apple essence, possibly augmented by malic acid.

Not less than 40% by weight of the apple pieces must be pressed out as juice prior to dehydration thereof in order to avoid caking of the final product when stored at temperatures of 100°F. or higher, especially when the final product is packaged in flexible containers under vacuum. Preferably, 45 to 55% of the juice based on the weight of the apple pieces is removed since the dehydrated applesauce product produced thereby retains a substantial amount of the normally present flavoring ingredients while it does not retain too much of the ingredients which tend to cause caking of the dehydrated powder at temperatures and times of storage experienced in the normal usage of this type of product by the Armed Forces.

As has been pointed out before, removal of substantial portions of the juice of the apple prior to dehydration results in a dehydrated applesauce product of reduced flavor and sweetness. To compensate for this, malic acid or other apple flavoring agents, such as apple essence or artificial flavoring, and granulated or powdered sucrose or other nonreducing sugar are added to the dehydrated applesauce powder.

The proportion of malic acid or other flavoring ingredient added and also the proportion of sucrose or other nonreducing type sugar added will depend to a considerable degree on the variety of apple being used, its growing conditions and the degree of maturity at the time of processing, as well as the degree of extraction of the juice during processing. In general, from 0.5 to 3.0% malic acid and from 100 to 200% sucrose or other nonreducing sugar, these proportions being based on the weight of the dehydrated applesauce powder, are added to the package at the time of hermetic packaging of the dehydrated applesauce powder, but they may be added during reconstitution of the applesauce or at other suitable times following dehydration of the apple pieces. These ingredients do not require thorough mixing with the applesauce powder at the time of packaging since they will normally be stirred into the rehydrated applesauce during reconstitution thereof immediately prior to eating the reconstituted applesauce.

It is desirable to dehydrate the apple pieces, after the removal of a portion of the juice therefrom, to a final moisture content of not more than about 3% by weight. For long-term storage purposes, it is preferred to reduce the moisture to not more than 2% by weight by dehydration of the pressed apple pieces.

BANANAS

Foam-Mat Process

A process developed by *R.C. Gunther; U.S. Patent 3,119,701; January 28, 1964; assigned to Gunther Products, Inc.* is concerned especially with the preparation of dried banana easily reconstituted by the addition of water by utilizing foam-mat drying. This process consists essentially of three steps, namely: (1) formation of a stable foam containing the product to be dried, (2) air drying of the foam to form a thin porous sheet or mat, and (3) compression of the dried mat followed by disintegration to yield a free-flowing powder.

The first step is accomplished by whipping the desired product with a whipping agent and a suitable stabilizer in a conventional food beater. The foam density and initial bulk density of the product can be carefully controlled by proper balance of the stabilizer, whipping agent and whipping time. The second step is normally carried out by metering the stabilized foam onto a continuous belt which runs through drying ovens maintained at predetermined temperatures. Drying times have run as long as one hour and temperatures have ranged from 130° to 190°F. The third step involves removal of the dried foam-mat intact from the continuous belt, compression to a high bulk density product and then disintegration to the desired powdered or other disintegrated form.

The foam-mat drying process is a very simple and inexpensive process. One difficulty that has previously been experienced with this process, however, is the lack of stability of the foam during the heating cycle. If the foam does not remain stable, cellular breakdown occurs causing serious impairment of the drying operation.

The substance used in this drying process is a composition of mashed bananas and a water-soluble cellulose lower alkyl ether or a water-soluble cellulose lower hydroxyalkyl ether, or a water-soluble cellulose in which the hydroxy groups of the cellulose are etherified with both lower alkyl and lower hydroxyalkyl groups. Examples of suitable alkyl groups which may be present in the preparation of the water-soluble cellulose ethers are methyl, ethyl, hydroxyethyl and/or hydroxypropyl groups. One specific example of a suitable water-soluble cellulose ether is the water-soluble cellulose methyl ether known as Methocel which is essentially the dimethyl ether of cellulose.

Especially good results have been obtained by employing either the dimethyl ether derivatives of cellulose (Methocel MC) or an alkyl ether of cellulose which contains both methoxyl substitution and hydroxypropoxyl substitution in the cellulose chain (Methocel HG). These cellulose derivatives are available in a wide variety of types and viscosities. All may be used under the right conditions and proper application because the predetermining factor in their use is the thermal gelation property.

The concentration of the cellulose ether based on the banana solids (dry basis) is sufficient to give a firm foam texture which is stable at about 160°F. The viscosity of the banana mix is also a factor to be considered, and cellulose ether concentrations are to be avoided which increase the viscosity of the mix to a point in which insufficient aeration is obtained in the foam-forming process because of poor beating properties. The minimum quantity of the cellulose ether, based on the banana solids (dry basis) should be 0.2% by weight in order to obtain a substantially firm foam texture. The maximum quantity is governed primarily by the increase in the viscosity of the banana mix to a point where insufficient aeration is obtained and ordinarily will not exceed 3% by weight on the same basis. The optimum quantity of the best cellulose ethers, such as the dimethyl ether of cellulose, is in the range of 0.4 to 1.5% by weight when considering both the quality of the foam texture and the foam density of the foamed mashed banana product.

It should be pointed out that the mashed bananas to be used in foam-mat drying should always be heat treated in absence of air to deactivate the many enzymes (amylase, sucrase, protease, lipase, raffinase and peroxidase) that are present in the banana. If the bananas are not so treated, darkening of the banana will occur during the foaming and drying stage. Commercially available mashed bananas are normally heat treated prior to canning, making further treatment unnecessary when this type is used in the process.

The foam density of the mashed banana product is a measure of the degree of aeration. A highly aerated product is desired because it produces a foam which dries more rapidly than lesser aerated foams and also yields a dried product which is more easily disintegrated into a free-flowing powder. The amount of the water-soluble ether of cellulose should produce a foam having a density less than 0.8 gram per cubic centimeter, and usually within the range of 0.20 to 0.80 gram per cubic centimeter. The following examples illustrate the production of a dried banana product by the foam-mat drying process.

Example 1: To 300 parts of mashed bananas was added 0.6 part of Methocel-65 HG (4,000 cp.) in the form of a 4% solution, a total of 15 grams of solution. The mixture was stirred for a short period to disperse the Methocel, after which it was whipped for 4 minutes with a wire wisk and a light, very firm foam weighing 0.29 gram/cc was produced. This foam was spread to approximately ⅜ inch thickness on a Teflon coated glass fiber mat and dried in an oven at 160° to 170°F. Drying time was 2 to 3 hours. The dry mat was very light in color and showed no sign of weep or breakdown. It could be easily crushed and the powder, when reconstituted with water, gave a mashed banana mix practically identical in color, odor and taste to the original material.

Example 2: To 300 grams of mashed bananas (prepared by blending ripe, whole bananas in a Waring Blendor and rapidly heating to 160°F. for 5 minutes in the absence of air, then chilling immediately to room temperature) was added 0.4 gram Methocel MC (4,000 cp.) (10 grams of 4% solution). After mixing well, the mix was whipped for 4 minutes in a Hobart Kitchen Aid beater. A very firm, light-colored fluff was obtained which weighed 0.27 gram/cc. The whipped foam was spread to ¼ inch thickness on a Teflon coated fiber glass mat and dried in the oven at 170°F. for 2½ hours. A fine celled, light colored mat was produced which could be easily crushed to a fine powder. When reconstituted with water, the material was nearly identical to the original banana mix.

Drum Drying

J. Aguirre, P.P. Noznick and R.H. Bundus; U.S. Patent 3,259,508; July 5, 1966 describe a process for drying bananas to a powder. The process is also useful for other high sugar-containing fruits having, for example, from 7 to 26% sugar on a liquid basis and either raw or processed in accordance with this process, such as grapes, strawberries, guava, caruba, pineapple, as well as starch-containing fruits having high sugar content, i.e., above 7% on a liquid basis, as stated.

The raw fruits are usually initially heat processed in a conventional manner to destroy pathogenic bacteria and undesirable enzymes. The bananas are peeled and the banana meat washed. Then the raw banana meat is macerated on a high shear machine such as an extrusion plate provided with perforations through which the banana substance is pressed and whereby it is made pumpable without, however, altering the physical composition but reducing the banana flesh to a shapeless, pulpy mass. The maceration step is carried out in a closed chamber and preferably an inert gas, such as nitrogen, is introduced to the chamber under sufficient pressure to drive out and replace the air, thereby to permit maceration to be conducted in an inert atmosphere, i.e., of nitrogen.

Optionally, the macerated banana pulp or puree is stabilized against oxidation by addition of an antioxidant, notably ascorbic acid. For example, about 0.15 to 0.50%, preferably 0.3% ascorbic acid on a dry weight basis of the banana puree is added, i.e., at a level of 50 to 60 mg./lb., preferably 200 mg./lb. (In another example, 400 mg./lb. was used of the banana puree.) This introduction of the antioxidant is made complete by intimate mixing, usually in the pulper.

The pulpy mass, with or without antioxidant, is now pumped in a closed conduit and subjected to a higher shear force in a homogenizer from under 1,000 to 3,000 lbs. pressure (psig), preferably 2,000 lbs. and sufficient or effective to be active to break up the starch and fibrous material, i.e., disrupt the same and simultaneously coat and entrap the relatively small amount of oil or oil-like material present in the mass, to thereby form a puree in which the ratio of solids content to water (considered as a total of 4 parts) is about 3:1, i.e., the banana having 25% solids and water being present as 75%.

The homogenization is a critical step which is essential to the success of the process. In some cases it precedes and in other cases it follows the heating and agitation step, and in some instances part of the heating and agitation is conducted before and part after homogenizing. By homogenizing a uniform product is continuously obtained, i.e., the film or sheet formed by drying on a drum dryer is free of variation in thickness, is devoid of perforated areas, and the web does not tear while being removed from the drying drum or drums.

The homogenized puree optionally containing the antioxidant is conveyed from the macerator by pumping in a closed conduit to the homogenizer and then still in a closed conduit to the heating and agitating operation and holding means such as a coil. The material in some cases is heated by flowing in a conventional tubular type of heat exchanger, but preferably the heating and agitation is accomplished in equipment such as a Votator-Swept Wall, Indirect Heater.

The votation treatment is likewise conducted in an inert atmosphere as described above and, in fact, such inert atmosphere or freedom from air is maintained about the banana material beginning with the initial shear treatment in the macerator through this heating and agitation treatment, the holding step, and the cooling step. The heating and agitation, however conducted, bring the banana material up to a temperature in the range from 225° to 300°F. (preferably 255°F.) and it is held at the temperatures recited for 60 to 0.3 seconds (preferably 15 seconds) to control color, flavor and aroma as desired, and to accomplish thermal destruction of the bacteria and enzymes. The raw banana contains 19 to 24% sugar, depending on the degree of ripeness as determined by refractive index, and after the heating and agitation step, the sugar content appears to increase slightly. The heat treatment locks the natural flavor into the product and avoids the development of a hay-like smell in the dehydrated flake.

The puree is cooled to 60° to 120°F., preferably 80°F., and introduced to the surge tank. It is then pumped through an elongated nozzle having a number of small openings delivering the puree to a bight between two rotating, high temperature rolls such as are found on conventional top feed, double drum dryers, and each having, for example, a diameter of about 2 feet. A heating time and temperature are employed to obtain the desired color, flavor, odor and bacterial and enzymatic destruction. The feed of puree to the drum dryer is maintained relatively slow to prevent objectionable accumulations from building up between, i.e., in the bight of the drums. There is used a forced feed to the nozzle or any suitable flow control device which is useful to assure a continuous, steady, uniform feed and flow of a stream to the bight. The puree at this point is exposed to the air and is about 75% moisture and 25% solids.

The puree is deposited on the rotating drums and is dried at a temperature and over a time period on the rotating drums which is found most desirable in the case of the particular pulp to produce the preferred color, odor and flavor. This control is established by film thickness, namely, through maintaining 0.01 to 0.18 inch, preferably 0.015 inch clearance between the drums and by maintaining the rpm of the drums within the range of 4 to 20 rpm. For instance, it can remain on the drums preferably 4 seconds at 8 rpm, and the temperatures of the contact surfaces of the drums is held between 280° and 350°F., preferably 300°F., by steam introduced to the rolls at an appropriate pressure, e.g., 90 psig.

Material dried within the ranges just mentioned and under the preferred conditions recited is removed from the drum surfaces by blades, such as conventional doctor blades. The hot, molten, plastic, glass-like, dried puree films so scraped from the drum surfaces tend to accumulate on the blades, and the films are therefore removed from the blades and stretched to some extent through passing within a relatively cold, dry air blast, the air having a temperature of 40° to 90°F., preferably 65°F. This air blast suspends and cools the continuously moving sheet or web having a thickness in the range of 0.015 to 0.085 inch. Thereby, the continuous films have the sugars therein crystallized and the web itself is made brittle; but the molten web, as removed from the rolls, is smooth, free of perforations and discontinuous breaks, and does not tear under a pulling tension.

The removal of the sheet is maintained at constant speed thereby preventing accumulation of hot, dry product at the blades and with cooling of the product instantly upon removal from the respective drums. The speed of sheet removal is controlled by the drawing of the same between rotating, interiorly air cooled reels and wheels resting freely in suspended contact with the film on the reels. The brittlized sheets extruding from the reels and contact wheels fall into screw conveyors which continuously remove the dried material from the drying area. In this latter travel or conveyor treatment, the brittle sheet forms into particles of miscellaneous sizes and the final particulated material is thereafter further broken up or ground to any desired fineness or size for packaging. This banana powder, or other similarly treated fruit, is useful alone or in combination with other foods.

Sun Drying

L.J. Abalo; U.S. Patent 3,386,838; June 4, 1968 has developed a process for preserving bananas in which peeled bananas are slightly compressed between juice absorbent boards while being exposed to the sun, rested periodically as at night, and the process continued until the preserved product is thus naturally created, free of caramelization, tissue deterioration and the like. The fruit product so dried for from 3 to 4 hours a day for 8 or 9 days is tasteful, having a similar consistency to other sun-ripened-dried products like raisins, dates and figs.

Dried Banana and Plantain Chips

E.J. Sarna; U.S. Patent 3,573,937; April 6, 1971 has developed a process for making banana and plantain products from the mature, unripened fruit. Such products can be prepared completely near their growth source, or intermediate products can be so prepared and then properly packaged and shipped elsewhere for conversion to final products. The process comprises:

(a) heating a section of banana or plantain pulp having a total sugar content up to about 1% by weight and having a thickness from 0.02 to 0.1 inch, to a temperature from 150° to 250°F. for a time interval from 1 to 45 minutes;

(b) cooling the section to a temperature from 35° to 200°F.; and

(c) drying the section to a water content from 5 to 15% by weight.

The dried section of (c) can then be fried. The dried section of (c) can also be reconstituted with water or other edible liquids such as sauces or gravies to serve as a starch-type food. Thus, the dried section can be so used to impart flavor or characterizing texture to numerous foods prepared in a casserole fashion; the dried section can be used in place of or in combination with potatoes, rice or other vegetables, fruits or grains.

In preparing these products the pulp of unripened, mature bananas and plantains is used. The unripened, mature pulp should have a total sugar content of less than about 1, and preferably less than 0.7 percent by weight. Thus, the pulp should have a substantial starch content and little sugar content. Excessive sugar in the pulp is detrimental in that fried chips formed from such a pulp are highly discolored, dark and not as crisp as the desired products. It is advantageous that the fruit be picked when it has attained maximum size but before any appreciable degree of ripening or development of total sugars has occurred.

Since it is difficult to separate the peel and the pulp of the unripened, mature fruit without fracturing the pulp, the fruit is heated sufficiently to loosen the peel from the pulp. This is accomplished by heating the fruit with an atmosphere of steam or by immersing the fruit in boiling water. Temperatures for this treatment range from 150° to 250°F. Depending upon the size of the fruit, type of fruit and temperature, the fruit is so heated for from 1 to 15 minutes, and preferably 5 to 10 minutes at 200° to 212°F. With such a treatment, the peel is loosened from the pulp and the peel bursts. The peel can be stripped or otherwise separated from the pump by hand or by mechanical means.

The pulp is then sliced into thin sections having a thickness from 0.02 to 0.2, and preferably 0.03 to 0.06 inch. The pulp can be sliced transversely against the vertical grain or at a variety of angles with respect to the vertical grain in order to obtain slices of larger size. Mechanical means are employed for slicing the pulp into substantially uniform slices.

At this stage of the process, it is advantageous to treat the sectioned pulp with an antibrowning agent such as sulfur dioxide, sodium bisulfite (a source of SO_2), or citric acid. The agent is an aid to inhibiting or preventing discoloration of the products obtained from the pulp. Temperature of this treatment can range from 35° to 200°F. The quantity of antibrowning agent is from 0.5 to 2% by weight of the sectioned pulp.

The sectioned pulp is next heated (cooked) to a temperature from 150° to 250°F. for a time interval of from 1 to 45, and preferably 15 to 30 minutes, with the lower temperatures being employed for longer time intervals and the higher temperatures being employed for shorter time intervals. The sectioned pulp can be so heated, but is more advantageously heated in water or in live steam. Thus, at the higher temperatures indicated, cooking can be conducted in an autoclave. The heat treatment serves to inactivate natural enzymes present in the pulp, which enzymes appear to cause excessive browning of the pulp. In addition, the heat serves to fix the cell structure of the pulp sections, and partially gelatinizes the raw starch therein.

The cooked, sectioned pulp is promptly cooled to a temperature of from 35° to 200°F. to firm the slices or sections and to reduce breakage in the further handling thereof. That is, the cooked, sectioned pulp is preferably cooled at least about 50°F. below the temperature at which it is cooked. Cooling can be accomplished by flowing cold air, nitrogen or other inert gas over the slices, or by washing the slices with cold water. The water wash is advantageous in removing any excess free starch thereon. If desired, additional firmness of the slices can be gained by contacting the slices while hot or when cooled with a water-soluble calcium salt including the chloride, sulfate and gluconate.

The cooled pulp sections are then dried to a water content ranging from 5 to 15, and preferably 8 to 12% by weight. Higher water contents are to be avoided because mold may form on the product in storage, and because undesirable discoloration and/or flavor degradation may occur. Lower water contents are disadvantageous because insufficient water in the product is available to puff the product when it is fried. Drying can be accomplished with conventional tunnel, tray belt, or still air drying techniques. Temperatures and air flow can be regulated to provide suitable drying rates. For example, temperatures can range from 50° to 300°F. and air rates can range up to several thousand cubic feet per minute.

The resulting, partially dried pulp section has the appearance of a horny chip which is hard, translucent and can be fractured. In these respects, the product is analogous to an uncooked pasta product such as a macaroni. This is an intermediate or first-stage product having a relatively long shelf life as a result of its low moisture content and method of preparation. This chip can be processed further to a final or second-stage product, or can be packaged for shipment to another destination for processing to a final product.

The first-stage chip is converted to a final product by frying in a suitable cooking oil for a short period of time. Expansion or puffing and frying of the first-stage chip occurs with the oil heated to a temperature of from 275° to 400°F. for 5 to 60 seconds, and preferably 360° to 380°F. for 10 to 25 seconds. The first-stage chip expands to approximately the size and volume of the slice or section from which it is prepared, in the initial 5 to 10 seconds of the frying cycle. The balance of the frying cycle serves to develop desired color and flavor of the final product. In view of the relatively short frying operation, only a relatively small quantity of oil or fat is absorbed in and on the chip. It is advantageous to drain any excess fat from the fried product. Generally, the fat content will be less than about 30 and preferably 10 to 25% by weight of the fried chip. Because of the low fat content, the final or second-stage chip has a substantial shelf life, is relatively free of a greasy surface, and is easily packaged. The process is illustrated by the following typical example.

Example: Green, fully mature, unripened bananas (United Fruit Color Chart 1-2) were contacted with boiling water for 10 minutes. As a result, the banana peels puffed and some bursting of the peels occurred. The bananas were then removed from the hot water and were placed in cold water, and were so contacted at 50°F. for about 3 minutes. The bananas were next removed from the cold water and the peels were removed readily from the pulp by hand stripping.

The cooled banana pulp was sliced by cross-cutting with a blade regulated to provide slices from $1/16$ to $1/32$ inch. The slices were then contacted in a container with 1% solution of sodium bisulfite for 2 minutes. The slices were transferred to a container of boiling water and were cooked in the boiling water for 20 to 25 minutes. Physical changes during this cooking operation were observed; the outer portion of the slices became more translucent and radial lines in the slices became more pronounced.

The cooked slices were removed from the boiling water and were placed in cooling water at 50° to 60°F. for 5 to 10 minutes. This cooling treatment served to firm the slices and thus reduced breakage and loss. Excess water was drained from the slices. The slices were then placed on aluminum screening and dried in still air at about 70°F. for periods of time ranging from 16 to 48 hours. The partially dried slices had a water content ranging from

5 to 15% by weight. The slices were then fried in deep fat, comprising mixed vegetable oils (Durkee's Mel-Fry containing small amounts of additives: butylated hydroxy anisole, butylated hydroxy toluene and methyl silicone, the latter serving as an antifoaming agent). Frying was conducted at about 375°F. for 15 to 25 seconds. The slices puffed immediately on contact with the hot oil and browned slightly. The resulting fried product had a crisp, crunchy texture. The fried product was drained of excess oil. A portion was salted. Another portion was dusted with sugar (sucrose).

Treatment with Calcium Hydroxide

A. Damwyk; U.S. Patent 3,728,131; April 17, 1973 has produced an edible dried banana product from peeled ripe bananas which are sliced and dipped in an aqueous solution of calcium hydroxide and thereafter dried at a temperature of 80° to 150°F. for a period of up to about 160 hours, preferably with air circulation. The process involves the use of bananas which are well along in the ripening stage, and the riper the banana, the better will be the final product produced.

The ripe banana to be used in the process is peeled and the tapered ends removed to produce a more uniform product. After peeling, the outer surface of the banana (the fleshy portion) is preferably scraped to produce a smooth surface. Depending upon the size of the banana, it is used whole, sliced in longitudinal sections or sliced in transverse sections. If the banana is smaller than about one inch thick, it can be used whole, sliced in half longitudinally, or sliced in transverse sections. Generally, the sliced banana sections should have a thickness of about one-half an inch since smaller sections are more difficult to handle in the process, and may result in broken pieces. The slicing of the banana in transverse sections is useful when broken bananas or parts thereof are to be used.

The banana, either whole or in sections, is then dipped into an aqueous solution of high calcium hydrated lime for about one-half minute. In 15 minutes, the surface of the banana will harden and become light brown in color. A fragile skin, about 0.004 to 0.008 inch thick, is formed and care is required to prevent damage to the skin, since it acts as an insulator to the inner portion of the banana.

The treatment solution comprises preferably about one-half tablespoon of high calcium hydrated lime powder in one gallon of water. However, the proportion of the lime may vary between one-half and nine tablespoons per gallon of water. The dipping procedure requires care in order to prevent damage to the skin that is forming and also to prevent breaking of the banana into pieces.

After the lime treatment, the banana is dried in any type of air drying system for a period of up to 160 hours at temperatures between 80° and 150°F. Generally, the temperature and time period required are related to the thickness of the banana sections. Thus, for sections one-half to three-quarters of an inch thick, a temperature of about 100°F. and 120 hours drying time is suitable. For thicker sections, drying periods of up to about 160 hours are required. A preferred drying procedure involves placing the banana in a tray having a perforated bottom lined with a material to prevent the bananas from sticking to the bottom. This can be accomplished by lining the bottom with a mat of straw or the commercially available nonstick materials such as polytetrafluoroethylene coatings. Circulation of air through the tray and over the banana is desirable wherein the air temperature is less than 150°F.

Depending upon the condition of surfaces of the sections of the banana being treated, after 60 or 70 hours of drying time it may be necessary to dip the banana sections into the treatment solution again. The second treatment will be required if the surface appears to be sticky, which may be due to the thickness of the section, improper air circulation and drying temperature. The second treatment in the solution is usually for a period of about 5 to 7 seconds. During the drying procedure, the bananas will form the skin and the color of the banana will become light brown. After 100 to 160 hours of drying, the banana will be a darker brown color. The surface will have formed a relatively hard pliant skin with an inner portion that is somewhat softer. The banana will have become about one-third the size of the original banana. It is not easily broken into pieces and has the consistency similar to a licorice stick. After the drying period, the product is cooled at a temperature of 60° to 70°F. for a period of about ten hours. The product can then be packaged. The banana product need not be refrigerated since refrigeration would tend to crystallize the sugar content. The product is very palatable, having a candy-like sweet, banana flavored taste. The taste appears to improve with age.

PUMPKIN

Instant Flakes

A process for preparing a dehydrated pumpkin product has been developed by *M.W. Hoover; U.S. Patent 3,169,875; February 16, 1965; assigned to North Carolina State College of the University of North Carolina.* The process consists essentially of (a) cooking the cut and deseeded pumpkin, (b) pulping the cooked material to remove any undesirable materials and to produce a mashed or pureed pumpkin product, (c) adding controlled amounts of starch and sugar to the puree, and (d) dehydrating the resulting product on a drum dryer to less than 6% moisture.

The conventional procedure generally followed in preparing and processing pumpkin by canning or freezing consists of first cutting the pumpkin in a cutter, and washing the cut pieces in a reel washer to remove seeds and extraneous materials. The pieces are then passed over an inspection belt after which they are cooked with steam by injection or in a heat exchanger. The cooked or partially cooked pumpkin is then run through the pumpkin squeezer to remove a portion of the liquid fraction. The solid fraction remaining is then passed through a series of comminutors and pulpers to give the product a smooth consistency. The pumpkin thus prepared is filled into cans and sterilized or packaged and frozen.

A problem encountered in dehydrating pumpkin is the very high moisture content of the cooked puree, particularly when the squeezing step is avoided in order to retain the soluble solids. Another problem is that straight or unmodified cooked pumpkin puree does not stick properly to the drum dryer during dehydration, thus making it difficult to dehydrate. The drying characteristic of the pumpkin puree was found to be improved by adding controlled amounts of starch and sugar to the pumpkin or pumpkin puree prior to dehydration on a drum dryer.

In the production of instant pumpkin flakes it is important to the quality of the finished product that as much of the condensate be retained as possible during cooking in order to minimize the loss of the soluble solids and flavor components. The cooked pumpkin is then pulped or pureed in a pulper finisher fitted with a screen containing openings ranging between 0.02 and 0.1 inch in diameter. The screened puree is cooled in a heat exchanger to a temperature below 170°F. but generally around 160°F. prior to the addition of starch. Starch and sugar are then added to the pumpkin puree at a rate that will increase the total solids of the puree to above 10%, but generally around 16 or 18% total solids. The amount of starch added to the pumpkin puree should account for no more than 30% of the total dry weight of the pumpkin puree mix. The sugar solids added to the pumpkin should be less than 50% of the total dry weight of the puree mix. Pumpkin solids in the mix should range between 20 and 85% of the total. Generally speaking, the ratio of starch to sugar solids should range between 1:3 and 1:3.5.

In order to facilitate proper mixing of the starch with the other ingredients, it is first necessary to cool the puree to a point below the gelatinization temperature of the starch to be added. A nongelatinized starch is preferred. The gelatinization point may vary slightly with different starches, but a temperature of 160°F. is usually sufficient. If this step is not followed closely, the starch will become lumpy with gelatinization and thus not result in a smoothly mixed product. The pumpkin puree or mix is then dehydrated on a drum dryer that is heated with steam at a pressure ranging from 50 to 90 pounds per square inch.

As one example of the process, a batch of pumpkins was cut into pieces and the seeds removed by washing in a reel washer. The washed pieces were run over an inspection table where the stems and undesirable materials were removed. The cut pumpkin was then fed into a cutter or chopper where it was cut into smaller pieces about one-half inch thick. The pumpkin was conveyed from the cutter to a screw-type steam cooker where it was cooked for 20 minutes with live steam at atmospheric pressure. After the product was cooked, it was pulped in a pulper finisher fitted with a screen containing 0.033 inch openings. The puree was then pumped to a stainless steel mixing tank through a heat exchanger. The puree was cooled in the heat exchanger to 160°F. on its way to the mixing tank.

To a mixing tank containing 1,500 pounds of pumpkin puree, which had a solids content of 7.8%, were added 141 pounds of corn sugar solids, 43 pounds of nongelatinized starch, and 19 ounces of a 10% Tenox IV emulsion. This amount of Tenox IV represented approximately 150 parts per million of butylated hydroxyanisole and butylated hydroxytoluene on a dry weight basis. The contents of the tank were then pumped to a double drum dryer where it was dehydrated to between 2 and 4% moisture. The drum dryer was heated with internal steam at a pressure of 70 pounds. The sheets of dried pumpkin were broken and ground into flakes. The flakes were then packaged in cans under a nitrogen atmosphere.

Pumpkin Pie Mix

F.J. Kane; U.S. Patent 3,597,231; August 3, 1971 describes a unique mixture of dry ingredients suitable for use in the commercial baking field to produce excellent pumpkin pies of uniformly high quality.

In order to produce an acceptable dehydrated pumpkin mix, it was necessary to find the correct formula blend that can make it perform as desired. Experimentation proved that the flavoring ingredients did not affect the performance of the mixture to any significant extent and therefore ingredients such as sugar, salt, spices and corn syrup solids could be employed and utilized as desired without adversely affecting the success of the mix. Other materials such as egg and milk solids were quite critical and percentages of these ingredients had to be quite carefully controlled.

It was also found that the consistency of the dehydrated pumpkin itself was important to the final success of the mix in that normally available shred size dehydrated pumpkin was found to be incompatible with the remainder of the materials and tended to settle out and separate from the mass. In order to overcome this problem, it was necessary to grind the dehydrated pumpkin to a finer consistency. The addition of 12.5 pounds of starch per 100 pounds of dehydrated pumpkin proved to be the optimum mix in order to prevent congealing or lumping during the grinding process.

It was also found necessary to utilize a blend of dehydrated egg albumen and egg yolk in lieu of dehydrated whole eggs in order to produce a whole egg total that will not combat the dehydrated pumpkin for the available water. The optimum mixture was found to be as follows:

	Percent
Dehydrated pumpkin	100
Dehydrated milk	62.5
Dehydrated egg solids	37.5
Sugar	150
Corn syrup solids	37.5
Starch	12.5
Spice mixture	3.5
Algin type gum	–

In order to use the instant dehydrated pumpkin pie mix, all that is required is to add two parts hot water to one part mix and allow to reconstitute for a period of 45 minutes. Following the waiting period, the pie shells can then be filled and baked in accordance with usual practice.

SULFURED FRUITS

Apricots Resembling Those Which Are Sun Dried

In conventional practice, fresh apricots are halved, pitted, sulfured by contact with gaseous sulfur dioxide, then spread on trays and exposed to the sun until dry. The products have certain characteristics which are recognized and desired by consumers, including a rich orange color, a translucent appearance and a gummy texture. Although sun drying yields a product with these desirable characteristics, there are disadvantages to be reckoned with. Notable is the slowness of the process, requiring at least several days even when the fruit is exposed under ideal conditions to the bright sun in the California orchard areas. Another point is that the fruit is exposed to dust, insects, birds, rodents, etc., so that problems of sanitation are encountered. A further problem is the delay and danger of molding in the event of rain during the drying period.

However, if pitted fresh apricot halves are sulfured, then dried by exposing them to a draft of hot air, the products have these characteristics: in color they are pale-yellowish instead of rich orange; in appearance they are opaque instead of translucent; in texture they are pithy rather than gummy.

A process developed by *M.E. Lazar, G.S. Smith and E.O. Chapin; U.S. Patent 2,979,412; April 11, 1961; assigned to the U.S. Secretary of Agriculture* produces apricots which have the desirable attributes of the sun-dried fruit. They are rich orange in color, translucent in appearance and gummy in texture. In addition, the products have an excellent characteristic flavor which is superior to that of sun-dried apricots. In shape, the fruit pieces have the proper curl of the edges toward the cut surface in contrast to the flat slabs often obtained in artificial drying methods. Also, the products reconstitute more rapidly and completely than the sun-dried product.

In applying the process, fresh apricots are first subjected to the conventional preliminary operations of washing, halving, pitting and placing on trays in the cups-up position. The whole fruit may be given a preliminary sulfuring, as by dipping in a solution of sulfur dioxide or alkali-metal sulfite or bisulfite, or by exposure to SO_2 gas, to prevent browning during these preliminary operations, particularly if there is any substantial delay between the successive steps of halving, pitting and traying.

The trayed fresh apricot halves are then sulfured. This may be effected by spraying the fruit with a solution of sulfur dioxide or an alkali-metal sulfite or bisulfite. In the alternative, the trays of fruit may be placed in a chamber where they are exposed to sulfur dioxide produced by burning sulfur or from commercial tanks of liquefied sulfur dioxide. In another alternative, the fresh apricot halves are dipped in a solution of sulfur dioxide or alkali-metal sulfite or bisulfite and then spread on the trays. In any event, the sulfuring should be sufficient that the fruit tissue contains 400 to 1,000 parts per million of SO_2, including that which is introduced by any previous sulfuring step if such is used. A greater proportion of SO_2 than 1,000 ppm may be employed if desired but is not necessary.

The sulfuring step is required to preserve the natural color of the fruit during subsequent processing steps. Also, the sulfuring causes a plasmolysis of the fruit tissue which contributes toward obtaining a final product of desired translucency.

The sulfured fruit is then subjected to partial dehydration. This is preferably effected in any of the usual types of dehydration apparatus which provide a draft of heated air about the fruit pieces to cause rapid evaporation of moisture. Thus forced air dehydrators of the tunnel, tray or continuous belt type may be used. The temperature of the air should be as high as possible to obtain rapid evaporation of moisture yet not so high as to cause damage to color or flavor. Taking account of these factors, the preferred air temperature is 150° to 180°F., most preferably 180°F. The dehydration is continued until the fruit has lost about 50% of its weight.

After partial dehydration, the fruit is contacted with steam. This treatment provides several useful effects. One is that air within the fruit tissue is expelled. This is important to achieve a final product of bright orange color and translucent character; where air remains in the tissue, the final product is yellowish and opaque due to the presence of the minute air pockets in the tissue which by a reflectance phenomenon act like a white pigment.

Secondly, the steaming causes a shrinking and densification of the fruit tissue. This is desirable to obtain a final product with a gummy texture rather than a pithy texture as common to conventional artificially dehydrated apricots. Also the steam treatment brings out the characteristic flavor associated with sun-dried apricots. Another point is that the steam treatment inactivates the enzymes in the fruit tissue, thus ensuring that the flavor and color of the fruit will be preserved against enzymatic action during further processing and storage of the final product. The proper time for the steam treatment may be determined by observing the apricots from time to time and continuing the treatment until the apricots develop a translucent appearance. In general, to obtain deaeration and other desired results without overcooking, the fruit is contacted with steam for a period from 2 to 6 minutes, depending on the size and maturity of the fruit.

Following steam treatment, the apricots are subjected to a second dehydration. This operation may be conducted in any of the types of dryers and under the conditions previously mentioned. However, during this dehydration the maximum air temperature is controlled to avoid attaining a piece temperature above 160°F., thus to prevent darkening and flavor deterioration. The term "piece temperature" is used herein as referring to the temperature of the fruit tissue as opposed to temperature of the surrounding atmosphere. Generally, the second dehydration is continued until the moisture content of the apricots is 15 to 20%.

As evident from the foregoing description, this process is particularly adapted to the treatment of apricots. However, it may be applied to other fruit with similar benefits. Notable among these are pears, peaches and nectarines. It is obvious that in the treatment of these fruits some adjustments of the conditions and procedures may be necessitated by the different properties of these fruits.

Example: Ripe apricots were washed, dipped in 2% aqueous solution of SO_2 for five minutes, then halved and pitted. The fruit was then dipped in 1% aqueous SO_2 solution for one minute, then spread in a single layer cups-up on trays at a loading of about 2 lbs./sq.ft. The SO_2 content of the fruit was 418 ppm. The fruit as placed on the trays was dehydrated to about 50% of initial fresh weight using cross-flow air heated to 180°F. which required about 2¼ hours. The partially dried fruit was then contacted with live steam (212°F.) for four minutes without disturbing the position of the fruit on the trays. The trays of partially dried, steamed fruit were returned to the dryer and dehydration was continued at an air temperature of 160°F. until the moisture content of the fruit was about 17%. This second dehydration required about 7 hours.

It was observed that the product had a rich orange color, translucent appearance, and a gummy texture. Also, the product had the characteristic shape with inturned edges typical of sun-dried apricots. In flavor, the products were superior to the best quality sun-dried apricots. Moreover, the products were more uniform in color and shape and rehydrated faster on contact with water.

Pretreatment to Shorten Sulfuring Time

The dehydration of apricots, plums, peaches, grapes, pears, etc. has been developed by

A. Dolev, C.H. Mannheim and M. Schimmel; U.S. Patent 3,692,546; September 19, 1972; assigned to Centre for Industrial Research, The National Council for Research and Development, Israel.

The conventional dehydration of various kinds of fruit, e.g. of the genus Prunus, comprises two main operations: sulfuring the fresh fruit and drying the sulfured fruit. The sulfuring has the purpose, inter alia, of destroying microorganisms apt to cause fermentation or similar deterioration of the fruit, and in order that this aim can be achieved, the sulfuring agent should deeply penetrate into the fruit and should be retained there as much as possible during the subsequent drying operation. The sulfuring agent used almost exclusively is gaseous sulfur dioxide produced in situ by the combustion of sulfur.

Attempts have also been made to soak the fruit in aqueous SO_2 solutions, but it has been found that the penetration is too slow for practical purposes and gives rise to the loss of valuable flavor constituents. Even the sulfuring operation by means of gaseous SO_2 takes many hours, depending on the drying method used, since the fruit must retain a sufficient amount of sulfuring agent even after the dehydration: if the sulfured fruit is dried in the open air at temperatures below about 33°C., the fresh fruit should be exposed to the SO_2 gas for 8 to 9 hours. If forced-draft-drying is used, especially at somewhat elevated temperature, the sulfuring still requires 5 to 6 hours. The long duration of the sulfuring operation is a waste not only of time, but also of working area. In the case of apricots, the drawbacks of conventional drying methods have prevented the development of the mainly small-scale, rural by-production of dried fruit, with its inherent lack of uniformity of quality of the product, into a large-scale industrial operation yielding products of standardized quality.

The described process avoids the time-consuming conventional sulfuring and includes the steps of (a) fissuring the skin by treating the fruit with a dilute alkaline aqueous liquid, preferably at elevated temperatures; (b) neutralizing the fruit with a dilute aqueous edible acid; (c) impregnating the fruit in vacuo with an aqueous sulfuring solution and (d) forced-draft-drying the fruit.

The step (a) is preferably effected by means of an alkali metal hydroxide solution, but other alkaline substances may be used, e.g., alkali metal carbonates or ammonia. With sodium hydroxide, for example, preferred concentrations of the solution are of the order of 1 to 5% by weight, and the temperature may be as high as 90°C. Thus, with a 2% solution at 90°C., the treatment achieves its purpose within as short a time as 30 to 60 seconds. The operation may be effected in batches in suitable vats, or continuously.

As an optional operation preferably interposed between steps (a) and (b), the fruit may be rinsed with water, e.g. a stream of water, for removing a large part of the alkaline matter before neutralizing the remainder thereof. The acid used in step (b) may be, for example, citric, tartaric, acetic or phosphoric acid, and indeed any edible acid the salt of which does not unfavorably affect the flavor and taste of the fruit. The term neutralizing does not necessarily mean that the pH of the fruit has to be adjusted precisely to 7.0.

Step (c) is an essential feature in this method. The fact that the impregnation of the fruit is effected under subatmospheric pressure and by means of a liquid sulfuring agent makes for the removal of air from the fruit and the deep penetration of the sulfuring agent into the fruit and its satisfactory retention therein during the subsequent dehydration. The magnitude of the pressure, i.e. the degree of evacuation, can be chosen within wide limits. It will determine to some extent the duration of the impregnating operation; within limits the time may be shorter, the lower the pressure. With pressures of 50 to 70 mm. Hg the impregnating time may be reduced to a few minutes. Of course, there are also other determining factors, e.g. a satisfactory fissuring of the skin in step (a), and the concentration of the sulfuring solution which is an aqueous solution of SO_2 or of a water-soluble bisulfite or meta-bisulfite. For example, at a pressure of 35 to 85 mm. Hg, the impregnation with a 2% by weight sodium bisulfite solution takes about 10 minutes, and with a 4% solution about 5 minutes.

The forced-draft-drying operation may be carried out in any suitable manner and by means of any suitable conventional equipment. The use of a tunnel oven is mostly preferred since it enables continuous operation. Suitable temperatures and drying periods will be selected empirically in accordance with the nature of the fruit to be dried, the relative humidity of the outer air, and other variable factors.

Apricots dehydrated by this method have a pleasant orange-to-deep-orange color and an attractive glossy appearance, and a sufficiently high SO_2 content for keeping for a more than adequate period of time. If the process is carried out in the same manner but the forced-draft-drying is replaced by open-air sun-drying, the fruit turns brown and is not glossy. The process is illustrated by the following examples.

Example 1: Two kilograms of fresh apricots (of the Israel-grown Raanana variety) were dipped in a 2% by weight aqueous solution of NaOH at 90°C. for 30 seconds, rinsed with fresh water and a 3% solution of citric acid, pitted, halved, and immersed in a 4% by weight aqueous solution of sodium bisulfite in a vacuum vessel for ten minutes at a pressure of 60 mm. Hg. The fruit was dried in a forced-air tunnel oven for 12 hours at 70°C. whereafter it contained 4,960 ppm of SO_2 and 10% water.

Example 2: Two kilograms of fresh peaches (of the Red Haven variety) were dipped into a 2% by weight aqueous solution of NaOH at 90°C. for 30 seconds, rinsed with fresh water and a 3% solution of citric acid, pitted, halved, and immersed in a 4% by weight aqueous solution of sodium bisulfite in a vacuum vessel for 5 minutes at a pressure of 60 mm. Hg. The fruit was dried in a forced-air tunnel oven for 12 hours at 70°C., whereafter it had a moisture content of 12% and contained 4,540 ppm of SO_2.

DRIED FRUITS FOR BAKING MIXES

Fruit Fragments Having Easily Hydrated, Gel-Like Form

J.H. Forkner; U.S. Patent 3,020,164; February 6, 1962; assigned to The Pillsbury Company has developed a process for making fruity fragments for incorporation in a bakery product which absorb moisture at a controlled rate so that they completely hydrate during baking.

Products made in accordance with the process are in the form of small masses that are generally spherical in shape. Each mass has a porous sponge-like interior, and a substantially unbroken and relatively impervious exterior surface or skin. The ingredients from which the masses are formed are such that when introduced into a moist bakery or confection mix, controlled penetration of moisture occurs through the outer surfaces and into the pores of the masses, with transformation of the masses into solid gel-like masses of the same volume. In other words, moisture absorption and transformation into solid gel occurs without any substantial amount of contraction or expansion, and such transformation occurs over a substantial period of time comparable to that required to prepare a wet cake mix, and to subject the same to conventional baking. The following examples illustrate the process.

Example 1: 45 pounds of fresh frozen blueberries were defrosted and comminuted in a suitable hammermill, such as one of the Reitz type. The comminuted material or pulp was placed in a suitable high speed mixer at a temperature of about 170°F., and 25 ounces (by weight) of low methoxyl pectin (Exchange Pectin #466) introduced and dissolved therein. This material was then introduced into a vacuum evaporator and intermixed with added corn syrup and citric acid. 69 pounds of commercial corn syrup was employed (refractometer reading about 80) and 2.5 ounces (by volume) of citric acid. In the evaporating operation the mix was heated to a temperature level of about 150°F. and the vacuum applied ranged from 26 to 28 (mercury column). Concentration was continued until the material contained about 85% solids.

This concentrate was then subjected to sheeting with application of starch to prevent sticking to the surface of the rolls. The sheets produced measured ¼ inch in thickness. They were chilled to a temperature level of about 50°F. by placing them in a refrigerated atmosphere. Thereafter the sheets were supplied to a mill of the Fitzpatrick type which reduced the sheets to fragments, whereafter the fragments were subjected to sizing with removal of a small amount of undersized material passing through a No. 10 screen, and with removal and recrushing or recutting of oversized fragments remaining on a No. 3 screen. The fragments were then intermixed with dry powdered starch, in equal proportions (by weight). This mix was then introduced into trays to produce layers about 1½ inches in thickness. The trays were then placed in a vacuum dehydrator, where the trays were heated to a temperature level of 180°F., and a vacuum applied corresponded to 28 inches mercury column. After remaining in the vacuum dehydrator for about two hours, the trays were removed and the expanded product separated from the starch by screening.

The product had a bulk density of about 415 grams per liter. The moisture content was about 3%. It had desirable characteristics including sufficient inherent strength to enable mechanical intermixing with various wet mixes, such as cookie, pancake and dough mixes without mechanical disintegration. With certain mixes like foam cake and yeast raised goods, the product may be incorporated into the batter just prior to the final mixing stage. In the finished baked products made by use of this product, islands of gel-like masses were formed, directly identifiable with the original dry masses.

Example 2: Apricots were predried by conventional dehydration to produce dehydrated fruit containing about 14% moisture. They were then subdivided to fragments whereby the bulk of the material remained upon a No. 12 screen, but passed through a No. 4 screen. Seventy-five pounds of commercial corn syrup, together with 23 ounces of low methoxyl pectin, were evaporated to produce a concentrate containing about 90% solids. Twenty-five pounds of the dehydrated fruit fragments were introduced and intermixed with this concentrate, and the concentrate then sheeted to produce sheets measuring ¼ inch in thickness. Thereafter the sheets were cooled and reduced to fragments and the fragments processed all as in Example 1.

The final product obtained in this manner was generally in the form of a small spherical mass. Each of the masses contained dehydrated fruit fragments which were readily identifiable as such.

Simulative Fruit Granule

Many persons enjoy eating baked goods which contain edible fruit, such as fruitcake, blueberry muffins and blueberry pancakes. For reasons of convenience, cost and the prevention of variation in ingredients, it is clearly advantageous to both the manufacturer of the prepared mix and to the consumer that the desired type of fruit be intimately mixed with the dry ingredients of the prepared mix rather than being added after preparation of the batter by the consumer. It is not commercially feasible to intimately mix fresh fruit with the normal prepared mix, both because the substantial quantity of moisture present in fresh fruit initiates the leavening reaction during the normal marketing and culinary storage period, and because at the termination of such period both the fruit and mix are not fit for human consumption.

Although the use of dried fruit obviates the difficulties created by moisture, such use gives rise to various other disadvantages. The rate of rehydration of dried fruit is not sufficiently rapid to permit the fruit to rehydrate to the required degree during the baking cycle. Consequently, upon completion of the baking cycle the fruit does not possess the desired fresh fruit flavor and is tough. Moreover, it is difficult and expensive to dehydrate fruit commercially so that it can be readily rehydrated in the form of whole pieces.

S. Barton; U.S. Patent 3,102,820; September 3, 1963; assigned to The Procter & Gamble Company describes a simulative fruit granule which, when intimately combined with the dry ingredients of prepared mixes, made into a batter, and baked, demonstrates properties which are comparable to a high degree with those properties demonstrated by uncombined fresh fruit. The simulative fruit granule comprises an intimate mixture of from 18 to 71%

of dried, pulverulent, bland-flavored fruit selected from the group consisting of apple, white grape and mixtures thereof; from 15 to 70% of sugar selected from the group consisting of sucrose, glucose, dextrose, lactose and mixtures thereof, at least about 50% of this sugar being sucrose; from 3 to 30% of edible filler selected from the group consisting of starch, cereal flour and mixtures thereof; fruit flavor; and from 0 to 6% water by weight based on the weight of the granule.

The fruits employed in the process are limited to those having a bland flavor, as, for example, apple or white grape. Apple is preferred. The fruit should be in a dry and pulverulent form. If less than 18% of dried, pulverulent fruit is used, the finished granule is too soluble and fails to retain a definite solid form during baking. The use of more than 71% of such fruit gives a granule which fails to rehydrate sufficiently during the baking cycle. The fruit may be dried by any known means, including vacuum drying, freeze drying, air drying and foam drying. For convenience, the fruit may be ground subsequent to the drying step. Two examples will illustrate both the preferred composition of the simulative fruit granule and of the preferred method of making it.

Example 1:

	Parts by Weight
Dehydrated applesauce (pulverulent, will pass through a 30 mesh screen, colored blue)	36
Sucrose (baker's special sugar)	36
Cornstarch	15
Blueberry flavor (imitation)	2

The above ingredients, in the proportions specified, were placed in a Sunbeam Mixmaster set at speed setting No. 1, and were dry blended. Then 11 parts (solids basis) of a 60% sucrose syrup were slowly added while simultaneously mixing the ingredients in the mixer at speed No. 1, thereby forming granules. The granules were dried overnight at 100°F., which drying was sufficient to give a granule containing 1.5 parts by weight of water.

The dried granules were then screened and 50 grams of the screened granules (20 grams of which would not pass through a standard 10 mesh screen, 20 grams of which would not pass through a standard 8 mesh screen, and 10 grams of which would not pass through a standard 6 mesh screen) were mixed in a basic yellow cake batter. (All of the granules had a size within the range of $\frac{1}{16}$ to $\frac{1}{2}$ inch.) Then the batter and granules were baked. When the finished cake was eaten, it was found that the granules imparted a distinct and delicious blueberry taste sensation.

Utilizing the above method, simulated fruit granules comparable to the above granules were also prepared employing 15 parts of cake flour rather than 15 parts of cornstarch.

Example 2:

	Parts by Weight
Dehydrated applesauce (pulverulent, will pass through a standard 30 mesh screen)	18
Sucrose (baker's special sugar)	55
Cornstarch (containing a minor amount of orange color)	7
Orange flavor (natural)	5

The above ingredients, in the proportions specified, were dry blended in the same manner as described in Example 1. Then 15 parts (solids basis) of a 60% sucrose syrup were slowly added while simultaneously mixing the ingredients in a Sunbeam Mixmaster at speed No. 1, thereby forming similar granules. The granules were then dried in order to reduce the water content to 6%, based on the total weight of the granule.

Protectively Coated Fruit Pieces

R.A. Shea; U.S. Patent 3,516,836; June 23, 1970; assigned to The Pillsbury Company has developed a process for adding dehydrated fruit pieces to baking mixes in which the partially dehydrated fruit pieces are encapsulated with an edible protective coating, the protective coating being further characterized by providing a moisture rise rate of less than 1.4 and a melting point ranging from 95° to 185°F.

The dry, culinary composition containing a chemical leavening base and the fruit particles provide unexpectedly, superior fruit containing baked products. The dry, culinary compositions can be reconstituted with an aqueous medium to provide a batter or dough in which the fruit pieces are uniformly distributed. Consequently, the resultant baked product contains a more uniform distribution of the fruit pieces therein. The encapsulating coatings for the fruit pieces remain intact under normal storage conditions and are not removed from the fruit pieces until the dry mix is reconstituted and baked. In general, upon baking the coating is melted and the fruit pieces then absorb sufficient moisture so that the resulting fruit pieces have a moisture content similar to that of baked goods containing freshly cut fruit pieces.

For convenience, whole fruits are advantageously reduced to about the size desired in the culinary composition prior to their dehydration. Advantageously, the fruit pieces are partially dehydrated to a moisture content of 15 to 25% by weight with a water content of about 20% by weight being the preferred level.

Exemplary dehydrated fruit pieces include acid-containing dehydrated fruit pieces such as apples, blackberries, strawberries, raspberries, blueberries, cranberries, orange peelings, oranges, raisins, peaches, apricots, grapes, figs, pineapples, tangerines, lemons, lemon peelings, boysenberries, prunes, etc. Particle size of the dehydrated fruit may vary broadly (e.g., 50 to 3,000 cubic millimeters); however, it is advantageous to employ dehydrated fruit pieces which have a particle size of 100 to 500 cubic millimeters with an optimum size being between 200 and 350 cubic millimeters.

An essential feature of the process is to encapsulate or coat the partially dehydrated fruit pieces with an edible coating which possesses the following characteristics:

(a)   an effective moisture barrier in respect to absorption and retention of moisture by the fruit piece as evidenced by a moisture rise rate of less than 1.4; and

(b)   normally solid at room temperature with a melting point in the range of at least 95° to 185°F.

A variety of coating compositions can be employed. One general class of materials are the plastic edible shortening materials which are normally solid at room temperature. Many such materials are the vegetable shortenings derived from animals, acetylated monoglycerides and others. A preferred shortening material is distilled acetylated monoglyceride which is a nonfracturing, flexible and normally solid waxy material. An additional suitable shortening type coating is confectioners coating butter (generally containing up to 30% by weight cocoa butter) obtained from hydrogenated vegetable oils.

The fruit particles are preferably homogenously mixed with the desired bakery mix ingredients in a weight ratio ranging from 1 to 12 parts by weight fruit particles to 40 parts by weight of the bakery mix ingredients. From an economic and taste viewpoint, the bakery mixes usually should contain from 1 to 3 parts by weight fruit particles to 20 parts by weight of remaining mix ingredients.

Sweetening and Flavor Enzymes Added

W.J. Motzel and F.J. Baur; U.S. Patent 3,224,886; December 21, 1965; assigned to The

Procter & Gamble Company have developed a process for the preservation of fruit to be incorporated into prepared culinary mixes of various types which is achieved by the process of separating the edible fleshy portion of the fruit from the waste materials; treating the fleshy portion of the fruit in a suitable form with a sweetening material selected from the group consisting of monosaccharides, disaccharides and hexitols, so that the sweetening material infiltrates the fleshy portion; dehydrating the treated fleshy portion to a moisture content of not substantially more than 15% by weight; extracting flavorese enzyme from the fruit waste materials; and drying the enzymes so obtained to a moisture content of below 5% by weight. When water and the enzymes are added to the suitably dehydrated fruit, the reconstituted dried fruit will be found to have essentially the flavor of the original fresh fruit.

A wide variety of fruits can be treated according to the process. Examples of fruits are the berries (such as strawberries and raspberries), pomes (such as apples), drupes (such as peaches) and various tropical and subtropical fruits (such as pineapples and bananas).

The fruit is prepared by removing the waste materials, such as skin, rind, core and leafy petals, from the edible flesh of the fruit. The next step in the process is the treatment of the edible flesh of the fruit with sweetening material so that the flesh is essentially entirely infiltrated. For illustrative purposes, sucrose has been selected as a specific sweetening material in the following detailed description. In general, the fruit can be soaked in a 40 to 70% sucrose solution or it can be covered with 20 to 25% by weight of sucrose in a suitably comminuted form. The fruit must be subjected to the action of sweetening material for at least a sufficient time to permit substantial infiltration of the fruit. A period of 20 minutes has been found to be sufficient. A period of 30 minutes at sucrose concentrations of 60 to 70% is about optimum for the process.

In treating bananas, they can be handled in the form of slices 1/8 inch thick; for pineapple 3/8 inch cubes have been satisfactory. The fruit flesh can also be very highly comminuted or it can be pulped so as to reduce it to a puree. Indeed, for some applications of the process, the pureed form is preferred. When the fruit flesh is reduced to a pulp or puree, the sweetening material can more readily be incorporated. The requisite amount of sweetening material can be added to the fruit flesh after the waste materials have been removed and the fruit can be reduced to a pulp in the presence of the sweetening materials.

It is always helpful, and sometimes necessary, to incorporate a protective substance, such as ascorbic acid, into the fruit. For example, about 0.5% by weight ascorbic acid added to bananas prevents deterioration before drying. This can be conveniently done by adding the ascorbic acid to a solution of sweetening material in which the fruit is soaked, or by adding it to solid sweetening material before the fruit flesh is pulped. If general, about 0.1 to 1.0% ascorbic acid will give satisfactory results.

The next step in the process is dehydration of the fruit flesh so that when dried it contains substantially not more than 15% water by weight. For best results, a water content of 2 to 7% is preferred. The drying operation should be so carried out that substantially all enzymatic activity in the fruit flesh is terminated.

The method of dehydrating the fruit flesh after it has been treated with sweetening material is not critical. Desirably, the temperature of the drying medium should be kept as low as possible, and it is preferable that it not exceed about 170°F. However, some fruits are less sensitive than others, and for these the preferred temperature limit could be increased somewhat beyond 170°F. One method which has been used very successfully is drying of the treated fruit flesh in a forced draft oven. Air temperatures in the range of from 115° to 170°F. have been used. The more economical dehydration range appears to be from 140° to 155°F. because optimum product quality is combined with a relatively short time.

While the flavorese enzyme extraction step of the process is described subsequent to dehydration of the fruit flesh, it may be, and generally is, desirable that the enzyme materials be processed contemporaneously with the processing of the fruit flesh. Briefly, the enzymes are extracted by comminuting the waste materials, extracting them with water, and precipitating the enzymes. The techniques of enzyme chemistry are used for the extraction.

The process involves the addition of sufficient water to rehydrate the dried fruit flesh, addition of enzyme obtained from the waste product to the mixture, and permitting the enzyme to act upon the substrate, whereupon the natural flavor of the dehydrated fruit is obtained. The addition of water to the dehydrated fruit pulp is not critical. It is more conveniently accomplished with water at room temperature.

The enzyme may be added to the dried fruit flesh before the water of rehydration is added, the enzyme may be dissolved in the water of rehydration or a portion of it, or the enzyme may be added while the fruit flesh is rehydrating or after it has rehydrated. It is generally convenient to add the enzyme to the dried fruit flesh just before rehydration or just after the water of rehydration has been added, so that the enzyme flavor development and rehydration of the fruit flesh take place simultaneously. Since the enzyme acts upon the substrate in the presence of water, it is generally desirable to store the enzyme separately from the dehydrated fruit flesh until reconstitution.

The amount of water used to reconstitute the dried fruit flesh will generally be equal to the amount of water removed, where the physical characteristics of the original fruit are desired. For instance, pineapple fruit comprises about 85% water. On reconstitution of the dried fruit flesh containing 5 to 10% moisture, water in the amount of 5.5 times the weight of the dehydrated fruit should be added properly to reconstitute the pineapple properly. When the fruit flesh is reconstituted by incorporating it into a batter, it may or may not be desirable to add extra water to the batter. Naturally, less water may be added to the dehydrated fruit pulp if the effect of a candied fruit is desirable. Even if insufficient water is added so that the forecited candied-fruit texture is obtained, the fruit flesh treated with enzyme will nevertheless have the full flavor of the fresh fruit.

Example: 650 grams of bananas are peeled and sliced transversely to the long axis of the banana into ⅛ inch slices. The 480 grams of slices so obtained are soaked for 30 minutes in a 70% sucrose solution containing 0.5% of ascorbic acid. The slices are drained for 30 minutes after removal from the sucrose solution. The drained slices are placed in a forced draft oven at a temperature of 155°F. and dried for 24 hours.

The peels are comminuted in a blender and 210 grams of water are added. The mixture is permitted to stand for 1 hour and then the supernatant liquid is separated from the solids by centrifugation. To the supernatant liquid is added 250 grams of acetone. The solution is agitated and then permitted to stand until a precipitate has formed. The precipitate is separated from the fluid by centrifugation and is dried under a pressure of 75 cm. of mercury. Upon rehydration of 20 grams of the dried slices with 60 grams of water and the addition of 10 mg. of the dried enzyme, the resulting slices are found to have a flavor which is essentially that of fresh banana slices.

FRUIT FOR BREAKFAST CEREALS

A process for the preparation of a fruit product particularly suited for incorporation in a breakfast cereal has been developed by *J.L. Holahan, W.L. McKown, R. Moen and V.E. Weiss; U.S. Patent 3,134,683; May 26, 1964; assigned to General Mills, Inc.* It comprises preparing a thoroughly mixed composition comprising fruit, fat, sugar and starch, extruding this composition, partially drying the extrudate, heating the dried extrudate, and flattening the heated dried extrudate. Virtually all fruits are useful in the process. Specific examples of particularly useful fruits include apples, apricots, bananas, cherries, dates, figs, raisins, loganberries, nectarines, peaches, pears, prunes, oranges, pineapple, lemons, raspberries, strawberries, blueberries and blackberries.

The fruit should be dried to such a degree that the total composition has the desired moisture content. For amounts of fruit generally employed, the moisture content of the fruit should be in the range of 8 to 40%. Since fresh fruit is generally damaged by high temperatures, it is preferable to carry out the drying operation at relatively low temperatures. In order to obtain optimum drying rates, it is generally necessary to carry out the drying step at reduced pressures. Thus, the dryer is normally the type referred to as a vacuum dryer.

The process will be more clearly understood by reference to some examples, in which all percentages are by weight.

Example 1: Into a Simpson Porto-Muller mix-muller was charged 40 lbs. of raisins having a moisture content of 10%, 20 lbs. dry sucrose, 21 lbs. of dextrose containing 10% by weight of moisture, 12 lbs. of refined wheat starch, containing 10% moisture, 5 lbs. of rearranged lard containing 7.5% glyceryl lactyl palmitate, 3.75% soft monoglyceride, and 2 lbs. of five-fold grape juice concentrate. The material was mulled for 15 minutes.

The mixture, which had dough-like properties, was charged to a 4 inch cereal-dough extruder having a grooved barrel and equipped with a multihole die having 195 orifices of $\frac{1}{8}$ inch diameter lined with Teflon fluorocarbon resin. A rotating blade scraped the face of the die to produce pellets of approximately $\frac{1}{8}$ inch. The pellets dropped onto a moving bed of wheat starch, thereby enrobing the pellets with starch. The excess was removed by sifting so that 5% of starch remained with the pellets. The coated pellets were spread 1 inch thick on drying trays and placed in a vacuum dryer having heated shelves. The pellets were dried for 6 hours at a shelf temperature of 150°F., at a pressure of 15 mm. Hg absolute.

The dried pellets, having a moisture content of 3%, were fed to a vibrating sifter equipped with wire mesh screens having openings of $\frac{13}{64}$ inch and $\frac{7}{64}$ inch. Pellets which passed through the larger screen and were retained on the smaller screen were conveyed under an infrared heating unit by means of a vibrating conveyor. After the pellets had been heated for 15 seconds to a temperature of 140°F., they were immediately fed to a pair of cereal flaking rolls spaced to give 15 mil flakes. The flakes were immediately mixed with a flaked wheat cereal at a fruit flake-to-cereal ratio of 1:2. This provided a nutritious, appetizing dish when served with milk. Storage tests of the cereal-fruit flake product showed that the fruit portion of the product retained its flavor and texture longer than the cereal.

Example 2: Using substantially the same procedure and equipment as in Example 1, a dough was prepared from 15 lbs. of dried strawberries having a moisture content of 10%, 33 lbs. dry sucrose, 32 lbs. dextrose having a moisture content of 10%, 16 lbs. of wheat starch having a moisture content of 10%, 4 lbs. of the lard of Example 1, and 6 lbs. of water. After processing the dough as in Example 1, there was obtained a strawberry flavored fruit flake.

FRUIT-CEREAL BABY FOOD

F.W. Billerbeck, F.S. Hing and V.J. Kelly; U.S. Patent 3,506,447; April 14, 1970; assigned to Gerber Products Company have developed a process for obtaining a dehydrated reconstitutable fruit-cereal by incorporating an ester-containing organic releasing agent in a fruit-cereal slurry prior to dehydration of the slurry on a drying surface.

Precooked, dried products, which lend themselves to easy reconstitution or rehydration to yield a smooth, fluffy textured, edible mass when mixed with a liquid such as milk or water, have found increasing use in the feeding of infants and adults who require geriatric or postoperative care. Dehydrated products of this general type are available to the consumer in the form of cereals and fruits. These dehydrated products can be produced by preparing a paste or thick suspension, i.e. a slurry or puree, obtained by heating the raw materials to form a pulp which may be strained or sieved to ensure uniform particle size. The slurry or puree is then applied to the surface of conventional drying equipment such as a drum dryer, where substantially all (90 to 98%) of the water contained therein is removed. The dried product is conveniently flaked and packaged for use. When the dehydrated puree is prepared from a cereal, it is easily removed from the drying surface by scraping with doctor blades. However, when the sugar content of the slurry is appreciably higher, for example, when a puree is prepared from fruit, the puree dehydrates to a sticky, plastic sheet which is difficult to remove from the drying surface. In the case of a fruit puree, it is necessary not only to employ scrapers to remove the dehydrated product from the dryer, but also to utilize draw-off rollers or tension devices to prevent product build-up at the scraper blades, to ensure that the dehydrated product will not be retained on the drying surface where its flavor and ability to reconstitute would be diminished by excessive drying or heating. Equipment of this nature is exemplified by U.S. Patent 2,352,195. Similar difficulties, which result in loss of production, have been encountered even when only a portion of the puree is prepared from fruit, e.g., when up to half the puree is formed from a cereal composition.

It has been found that when certain ester-containing additives are included in the fruit-cereal puree compositions, conventional dehydration techniques produce a dried product which is neither sticky nor difficult to remove from the drying surface. Simple drum dryers, which merely employ standard scraping devices without tension rollers or the like for sheet removal, can be utilized for the drying of fruit-cereal purees and the desired continuous sheet obtained.

The organic additives, referred to as releasing agents, which have been found to minimize, in some unknown manner, the previous detrimental effect that a high sugar content has on certain physical properties of fruit-cereal slurries, are those which contain at least one ester linkage resulting from the condensation reaction of an organic acid and an organic alcohol. The esters are derived from fatty acids or alcohols and therefore contain a long hydrocarbon chain. In addition, the esters are commonly identified as emulsifiers and have the requisite chemical structure although they do not necessarily function as emulsifiers in the present environment. Examples of preferred monobasic acids as starting materials for the esters are linoleic, arachidic, palmitic, stearic, myristic and oleic acids. Examples of dibasic acids include malonic acid, succinic acid, glutaric acid, adipic acid, pimelic acid and the like.

The types of alcohols which, when reacted with the abovementioned acids, result in edible esters effective for the purposes of this process, include saturated and unsaturated monohydric alcohols of from 12 to 26 carbon atoms, preferably 14 to 18 carbon atoms; and saturated and unsaturated polyhydric alcohols of from 2 to 10 carbon atoms, preferably of from 4 to 8 carbon atoms. Illustrative of monohydric alcohols suitable as starting materials for the esters of this process are myristyl alcohol, cetyl alcohol, stearyl alcohol and the like. Dihydric alcohols advantageous for this process include ethylene glycol and the like, while trihydric alcohols include glycerol, sorbitol and the like.

In addition to the fatty hydrocarbon chain which is obtained with the above compounds, the ester molecule will also elsewhere contain an oxygenated polar group such as a hydroxyl group, a carboxylic acid salt radical, or a phosphatide linkage by reason of which the ester exhibits emulsification properties under other circumstances. More specifically, phosphatides, a group of complex lipids containing a phosphoric acid grouping and a nitrogenous base, in addition to other acid components, have been found to be especially desirable. A preferred ester of this group has been found to be lecithin, a mixture of the diglycerides of stearic, palmitic and oleic acids linked to the choline ester of phosphoric acid.

The ester-containing releasing agents may be employed in quantities up to 5% by weight, based on the initial fruit-cereal composition, but have been found to be advantageous in amounts as small as 0.01% by weight. When employed in concentrations from 0.1 to 1.0% by weight, these releasing agents have been found to be especially effective in assisting in the production of a dehydrated puree which, not only releases easily from the drying surface, but also results in a reconstitutable product which has a true and desirable fruit flavor.

This reconstitutable fruit-cereal puree can be prepared by any of the conventional methods.

For example, a slurry can be prepared by mixing cereal grain flour, fresh ripened fruit, granulated white sugar, milk (or milk concentrate), vegetable oil and the ester-containing releasing agent along with any other ingredients, such as minerals or vitamins which may be deemed desirable, and sufficient water so that the solids content of the resulting slurry will be from 15 to 30% by weight and preferably 25%. It will be recognized that the solid concentration of the slurry will control the characteristics and properties of the resulting dehydrated product, as well as controlling the amount of water which necessarily must be removed in the drying operation.

The slurry is then passed through a screen to produce a puree of uniform consistency. Conventional screening devices with openings of about 0.033 inch are frequently employed. Subsequent heating of the slurry to 140° to 230°F., and preferably 190° to 205°F., in conventional equipment such as atmospheric, vacuum, pressure-cooking tanks or in-line agitating heaters, produces a gelatinized puree suitable for dehydration.

Next, drying is accomplished not only to remove a substantial portion of the water contained in the puree, but to complete the cooking process. Drying means include, e.g., any of the conventional equipment readily available, such as single or double drum dryers well known in this art. The dehydrated puree is most easily removed from the drying surface as a continuous sheet, preferably having a thickness of 4 to 7 mils, and without sticking, balling or lumping with, e.g., a conventional doctor blade.

When using such drum dryers, the film thickness will obviously be dependent on the spacing between the drums, the drum speed, the type of puree being treated and the steam pressure in the drums. The steam pressure can be advantageously employed in the range of 30 to 90 psig. It has been found that an optimum product is obtained when the drying means, especially double drum dryers, are run at high drum temperature and drum pressure. However, none of these factors will deter the effectiveness with which the ester-containing releasing agents of this process enhance the release of the dried sheet from the drying surface.

The dried fruit-cereal sheet may be prepared for packaging and ultimate consumption by flaking into various sizes depending, of course, on the rehydration characteristics of the particular fruit-cereal puree being treated. Here again, the flaking is accomplished in standard equipment such as a 10 mesh U.S. sieve series screen (0.030" wire). However, when the particle size of the finished product is too fine, the particles may tend to lump and therefore be difficult to rehydrate.

Example: A fruit-cereal slurry was prepared in approximately the following proportions.

Ingredients	Percent by Weight
Banana puree* (total solids, 23%)	50
Oat flour	31
Granulated white sugar	10
Skim milk powder	6.4
Lecitreme 40**	0.6
Vitamins, seasoning, acid, etc.	2.0

*The banana puree was prepared from fully ripened, frozen bananas by riving frozen bananas through a riving machine and finishing through a 0.033 inch screen.
**Lecitreme 40 is a powdered product containing 40% crude double bleached lecithin.

Sufficient water was added to reduce the overall solids content to about 22.6%. The slurry, having a pH of about 5.0, was pumped through a line strainer having a screen size of 0.060 inch to a holding tank at a temperature of 160° to 180°F., from where it was transferred to an agitating heater at about 205°F. The heated slurry was then fed to a conventional double drum dryer, each drum operating at an internal pressure of 80 psig. The drums were rotated at 3 rpm and the dried sheet removed with doctor blades. The resulting sheet was nonplastic, continuous and had a film thickness of about 5 mils. The doctor blades were loosely held to the drums, yet the sheet was readily removed. Once equilibrium of operation was established, no losses of dried puree were incurred on the dryer surface.

The resulting dried fruit-cereal sheet was continuously conveyed to a flaker employing a No. 5 (U.S. Standard) screen size. The resulting flaked product had a moisture content of 2%. The dried product was stored overnight, then reconstituted with about five times its weight in water and a smooth textured fruit cereal with a distinctive natural banana flavor was obtained.

WITH ADDED SUGAR

Fibrous Fruits and Peels

H.E. Swisher; U.S. Patent 2,976,159; March 21, 1961; assigned to Sunkist Growers, Inc. describes a process for dehydration of firm, fibrous fruit products including citrus peel, pomes, cranberries and pineapples. The process will be described as relating specifically to the treatment of citrus peel.

The citrus peel constituting the basic preferred raw material of the process, which is available in the form of cups or half shells as a result of the juice extracting operations of a citrus plant, may be utilized in this form or diced, shaved or otherwise subdivided with a minimum degree of oil gland rupture to the size desired. It is important for many product applications that the oil pores of the peel remain intact to as great an extent as possible. If they are ruptured by subjecting the peel to grinding or some equally harsh process, the volatile essential oil, which is primarily responsible for the taste of the peel, is lost to the product. A dehydrated citrus peel product having variable and controllable levels of peel oil flavor can be produced by selective rupturing or destruction of the peel oil cells prior to the dehydration and syrup impregnation process. This purpose is accomplished most effectively by pin-point pricking the peel on the flavedo side, thereby rupturing a portion of the oil cells and liberating the oil therefrom.

The corn syrup solids-glycerol solution, which is utilized as a heat exchange medium and as an impregnating agent, consists of not more than 40% nor less than 15% by weight of corn syrup solids and not more than 85% nor less than 30% of glycerol. In addition, dextrose monohydrate may be incorporated in the solution in amounts not exceeding 40% by weight of the solution.

The first step of the process after adding citrus peel to the hot corn syrup solids-glycerol solution is the vacuum hydrodistillation, at a temperature of from 40° to 130°C., of the resulting peel-glycerol mixture. The distillation is carried on a sufficient length of time to reduce the moisture content of the peel to a maximum of 16% and preferably to a moisture content of not over 8%. After the distillation step has been completed, it is necessary to separate the dehydrated, impregnated citrus peel from the corn syrup solids-glycerol solution. This may be accomplished by simply draining the distilland through a screen, by centrifugation, by vibratory screening or by any other suitable means. The peel may then be cut to the final size and shape desired.

The product resulting from the foregoing treatment will be composed of not less than 10% corn syrup solids, not less than 22% glycerol, and will not contain more than 16% moisture. It will be of any particle size desired and may be used as an article of commerce as soft preserved peel in the preparation and flavoring of sherbets, ice cream, cakes, marmalades, candy fillings, candied or glacéed peel and other products.

If desired, the product may be prepared in the form of a free-flowing dry preserved peel. This is accomplished simply by adding to the corn syrup solids-glycerol-containing dehydrated peel sufficient starch or other tasteless water-soluble polysaccharide powder coating agent exemplified by the vegetable gums (locust bean, acacia and guar gum for example), soluble dextrine, pectin and carboxy methyl cellulose, to substantially coat the individual peel particles. The treated peel and coating agent are mixed and classified until individual peel pieces are completely coated.

For some purposes it is desirable to utilize a two-step process. Instead of initiating treatment with a glycerol-corn syrup solution, glycerol alone is used in the dehydration of the peel. Upon completion of moisture removal, excess glycerol is quickly drawn off while boiling under vacuum is continued. Since glycerol has a substantially lower viscosity than the glycerol-corn syrup solids solution utilized in the one-step process, moisture removal is greatly accelerated, thus reducing the time-temperature factor so important in flavor deterioration.

Just as soon as the glycerol is removed, a warm corn syrup-solids-glycerol solution is added, while the vacuum is maintained. If a vacuum is not utilized, the peel will collapse as a result of the moisture having been removed from the cellular pores. After sufficient corn syrup solids-glycerol solution has been introduced to completely cover the dehydrated peel, the vacuum is disrupted intermittently in order to impregnate the peel with the warm preserving solution. The following examples illustrate the process.

Example 1: A quantity of 400 grams of raw ¼ inch size diced Valencia orange peel was added to 900 grams of a syrup consisting of 40% corn syrup solids and 60% glycerol at a temperature of 115°C. in a round bottom flask. The syrup temperature dropped to approximately 70°C. and with heating gradually increased to 95°C. after 12 minutes under a vacuum of 22 inches of mercury. Heating was discontinued and the vacuum was released several times to permit impregnation of the peel with corn syrup solids and glycerol and to produce a semitranslucent peel. The contents of the flask were cooled under a vacuum of 27 inches of mercury until the temperature dropped to 80°C.

The syrup-peel mixture was then removed from the flask and passed through a coarse screen to separate excess syrup. A yield of 331 grams of soft preserved peel, having a composition of 30.8% corn syrup solids, 45.9% glycerol and 23.3% orange peel solids was obtained. A quantity of 45 grams of pregelatinized cold water swelling starch was then thoroughly mixed with 165 grams of the soft preserved peel obtained above to yield 210 grams of dry peel having a composition of 24.2% corn syrup solids, 36.1% glycerol, 21.4% pregelatinized starch and 18.3% original orange peel solids.

Both the soft preserved and the dry preserved peel of the foregoing example were stored at room temperature in a closed jar for a period of six months, at which time they were rehydrated by immersing in water. Both samples had a fresh peel color, a fresh peel flavor, and a fresh peel firmness. On the other hand, a sample of untreated peel taken from the same batch deteriorated to an inedible extent within two weeks under identical storage conditions. This sample was brown, soft and mushy and was covered with mold growths.

Example 2: Raw whole cranberries were treated in accordance with the two-step method by introducing 300 grams of the cranberries and 700 grams of USP glycerol into a round bottom flask and heating the flask under vacuum for 14 minutes until the temperature rose to 120°C. At this temperature the glycerol was removed under vacuum from the flask and a warm syrup of 60% glycerol-40% corn syrup was substituted. Heat was again applied to a temperature of 85°C. and the vacuum was interrupted six separate times to effect impregnation of the cranberries while under the surface of the solution. Finally, the excess warm glycerol-corn syrup solids solution was withdrawn to yield 151 grams of semitranslucent dehydrated cranberries, a corn syrup solids content of 13.7%, a glycerol content of 69% and an original cranberry solids content of 17.3%. Total moisture content of this product was 2.3%.

The cranberries retained their flavor and prospective long storage life after six months' storage at room temperature. In connection with the processing of cranberries in accordance with this method, it has been found that it is desirable to puncture or perforate the skins of the cranberries prior to treatment in order to avoid rupture of their outer skin. This treatment is not, however, necessary to production of a satisfactory cranberry product, since the cranberries retain their flavor and preservability irrespective of loss of shape.

Expanded Medium Process

The drying of fruits by the sun or by heat seriously affects various heat sensitive constituents of the fruit. This impairment occurs largely in the final stages of drying. Better quality products are obtained by vacuum dehydration, but even with this process there are disadvantages, resulting primarily from poor heat transfer.

J.H. Forkner; U.S. Patent 3,057,739; October 9, 1962; assigned to The Pillsbury Company has developed a process which is characterized by the use of a special expandable medium which performs certain important functions during evaporation of moisture from food product masses. More specifically, the expandable medium is associated with the food product masses from the beginning of the dehydrating operation and during dehydration, during which time the food product is subjected to reduced pressure and elevated temperature conditions to effect evaporations of moisture. Concomitantly the medium is caused to expand to provide a greatly increased area of contact with the exterior surfaces of the masses, whereby during a substantial or major part of the drying operation, the expanded medium forms cellular means for effectively transferring heat by conduction to the masses.

With respect to the expandable medium, good results have been secured by using a medium consisting mainly of a reducing sugar, such as glucose, dextrin and maltose. Varying amounts of maltose and other reducing sugars are supplied by using commercial corn sugar, corn syrup, malt syrup, glucose, or mixtures of the same. Commercial corn syrup contains from 18 to 22% moisture, which largely evaporates during treatment in partial vacuum to effect expansion of medium and increased area of contact with the product and subsequent dehydration of the product and solidification of the medium. Generally it is desirable to concentrate the syrup by preliminary evaporation. Minor amounts of other sugars can be used, such as sucrose, although the major portion of the sugar employed is of the type mentioned above.

While reducing sugar syrup can be used by itself as an expandable medium, it is desirable to provide small amounts of additives, particularly materials such as soluble casein or low methoxyl pectin which stabilize the medium by affording better control over the extent of expansion and time of solidification or setting. The casein can be added as a soluble caseinate (e.g., sodium caseinate) or as an alkaline solution of casein in water. Skim milk solids, with or without removal of a portion of its lactose content, can be used as a source of casein.

Example: Fresh peaches were cut into halves, pitted and peeled. By the use of a conventional dehydrator, they were dried to a moisture content of 40% by use of a drying atmosphere at 140°F. over a period of five hours. The partially dehydrated peach halves were then cut into quarters and placed on a vacuum drying tray, together with an amount of corn syrup sufficient to provide a syrup layer about ¼ inch in depth. The peach quarters were arranged as a single layer. The trays were placed in a vacuum dehydrator on shelves heated to a temperature fo 170°F. The vacuum within the dehydrator was rapidly increased over a period of ten minutes to a value corresponding to 28 inches mercury column, and kept at this value for the remainder of the dehydrating period. The entire vacuum dehydration period was about ten hours.

Upon removing the trays from the vacuum dehydrator at the end of the dehydration period, the corn sugar solids were in the form of a solid, porous, sponge-like medium extending between the peaches and the bottom of the tray and enveloping and contacting the side surfaces of the peaches. The dried peaches had a moisture content of about 5%. The peaches, together with the solidified corn sugar solids, were removed from the trays and subjected to tumbling to disintegrate about 50% of the sugar. Thereafter the disintegrated sugar was removed from the peaches by screening. The final dehydrated product was of excellent quality, without metal contact burned areas, and with a natural color and flavor. The volume was substantially the same as after preliminary drying. It was porous and readily hydrated upon soaking in water.

Dried Fruits with Fine, Porous Structure

M. Pader and C.G. Richberg; U.S. Patent 3,365,309; January 23, 1968; assigned to Lever Brothers Company have as their objectives a process for preparing a particular type of dehydrated fruit, namely, one which has the color, texture, appearance and the aromatic and flavor notes of natural fruit after it has been reconstituted; has a fine, porous structure in the dehydrated state; has a relatively undamaged cell structure; is opaque and nonsticky; has a good rate of reconstitution; is edible after limited rehydration; and has a relatively low moisture content, e.g., 5%. They have accomplished these aims by employing a process containing several critical steps.

Thus, in accordance with this process, a fruit, e.g. an apple, is prepared for dehydration by washing, coring or pitting, slicing and/or treating with browning inhibitors. All of these operations are usually conducted without applying any amount of heat, such as blanching, which will raise the temperature of the fruit substantially. This prepared fruit is immersed in a sugar solution until the sugar uptake value as defined below is at least about 0.4 at conditions of time, temperature and concentration which provide after dehydration an opaque, rather than a translucent, dehydrated fruit. After the fruit is removed from the solution, it is dried, for example, in a hot air oven.

Any fruit compatible with sucrose is applicable to this process, such as apples, pears, peaches, apricots, cherries, strawberries, grapes, pineapples, plums, bananas, watermelons, raspberries, oranges and the like. As defined herein, a fruit is a fresh ripe or green fruit or a frozen fruit in which the cellular structure is not broken down.

After the fruit has been prepared by washing, slicing, coring or pitting, etc., the prepared unblanched fruit is then immersed into an aqueous sugar solution. It is necessary to employ a crystallizing sugar, such as sucrose. Noncrystallizing sugars, such as corn syrup, are inoperative since the resulting dehydrated products are tacky, rubbery and excessively hygroscopic. The temperature and concentration of the sugar solution and the time for immersion will vary according to the fruit being treated. However, it is critical to use time, temperature and concentration conditions to provide a sugar uptake value (SUV) at least of about 0.4. The sugar uptake value, which is the measure of sugar uptake by the fruit, is determined as follows:

$$SUV = \frac{(\text{dry weight of treated fruit}) - (\text{dry weight of starting fruit})}{(\text{dry weight of starting fruit})}$$

In other words, this value is determined by measuring the weight of the fruit solids before the sugar treatment and by measuring the weight of the fruit solids and sugar in the fruit after sugar treatment and dehydration. The sugar in the fruit includes the sugar on the surface of the fruit besides the sugar inside the fruit. An SUV under about 0.4, e.g., 0.1, is unsatisfactory since there is not enough sugar in the fruit to provide the benefits of the treatment with respect to flavor and texture. The preferred value is between 0.5 and 2. An SUV above about 2 may be impractical and unecessary but it is satisfactory if achieved without harm to the fruit. However, the rate at which sugar is taken up by the fruit increases markedly as the temperature of the sugar-impregnation operation is increased. Although at high temperatures it is generally very easy to obtain a sugar uptake value substantially over 2, this is unsatisfactory if the temperature used causes the dehydrated product to be nonopaque.

It is also critical that the conditions of time, temperature and concentration be selected to provide a fruit which on dehydration is opaque. If conditions are chosen which provide a dehydrated fruit which is translucent, the desirable qualities obtained by this process are lost. The opacity of the dehydrated fruit undoubtedly is at least partly the result of its porous character. Some fruits, such as strawberries, are fairly translucent in their fresh state. When properly treated and dehydrated according to this process, however, they provide an opaque, porous structure. When improperly treated, on the other hand, they may give a dry product which reconstitutes to fruit with reduced flavor and soft, undesirable texture.

The most important factor in preventing loss of porosity and opacity is temperature. Apples or other hard fruits may be impregnated with sugar by using a solution with a temperature as high as 80°C. without substantial loss in opacity. Pears or other fruits of intermediate hardness, on the other hand, become translucent and lose flavor if the temperature of the sugar solution is over about 50°C. and the dehydrated fruit is translucent. Even a lower temperature is required for the soft fruits; if peaches are treated, room temperature is approximately the maximum temperature for the sugar solution. The time for the treatment will depend on the temperature, concentration which generally is over 30°Brix, e.g., 50° to 60°Brix, area of exposed surface and amount of agitation.

After treating, the fruit is removed from the sugar solution and dehydrated to a relatively low moisture content. A dehydrated fruit is defined herein as a product having enough water removed to cause improved stability with respect to flavor, texture and bacterial decomposition. Therefore, it may have little water content, e.g., 5%, or an appreciable amount of water, e.g., 25%. The dehydration can be performed by any artificial means known in the art so long as excessive conditions which will harm the fruit are not used. The fruit may even be dehydrated in hot air without losing its desirable properties. Vacuum drying is also applicable here but it is less economical.

The final, dehydrated product is sponge-like with small, tightly packed pores. It is only when the sponge-like characteristics are achieved that the fruit reconstitutes to a product with full body and good, full flavor rather than a washed-out flavor, and only then is the dehydrated product firm and nontacky. The texture of the product is tough and relatively nonsticky which facilitates handling. Despite its high sugar content, the dehydrated fruit does not reconstitute unusually quickly. It rehydrates better than conventionally dried fruit in starch-containing water, undoubtedly due to osmotic effects, but at about the same rate as conventional low moisture fruits in boiling water; in contrast, prior art sugar treated fruits reconstitute very rapidly.

Cooking behavior of the dehydrated fruit of this process is important. Dehydrated fruits, in the past, usually were considered done and edible only after thorough cooking even though the raw, fresh fruit was edible. The products prepared here, however, are quite edible after only limited rehydration; they are also similar to fresh apples or other fresh fruits since they are edible when not completely cooked and during cooking their flavor changes from that of the fresh apple to that of the fresh apple cooked.

The following examples illustrate the process. Unless otherwise indicated, all parts and percentages are based upon weight.

Example 1: Fresh New York State McIntosh apples, which had been held in controlled atmosphere storage for approximately four months, were prepared for a sugar treatment by paring, coring, slicing to 10 to 12 mm. diameter segments, dipping for one minute in an aqueous 0.1% $NaHSO_3$ solution to retard browning of the exposed surfaces and draining. The prepared slices were placed in a stainless steel mesh basket and dipped in 67°Brix sucrose solution with 0.1% $NaHSO_3$ contained in a stainless steel steam jacketed kettle (10 gallon capacity). During a 45 minute dipping period, the temperature of the syrup was maintained at 80°C. and a centrifugal pump (flow rate, 7 gallons per minute) circulated the syrup constantly. After the dipping was completed, the basket was removed from the kettle and the sugar treated slices therein were allowed to drain for 20 minutes. The sugar treated slices were subsequently placed on perforated trays (at a tray load of 2 to 3 lbs./sq.ft.) and dried in a Proctor and Schwartz cabinet hot air dryer at 150°F. for

16 hours by using maximum air velocity and by directing the air through the bed of apple slices. The dried slices were then removed from the trays, placed in moistureproof bags and stored at room temperature. The following results were noted during this procedure:

Initial weight of fresh apple slices	10,000 grams
Initial moisture of fresh apple slices	88 percent
Initial Brix of syrup	67°
Initial weight of syrup	25,000 grams
Brix of syrup after dip	59°
Final weight of dehydrated slices	2,600 grams
Final moisture of dehydrated slices	5.8 percent
Final sulfite level	250 ppm
Sugar uptake value	1.0

The sugar treated dehydrated fruit of this process, after being cooked in water, had a much stronger apple flavor and aroma than comparable fruit that was air dried by conventional means.

Example 2: Fresh Bartlett pears were peeled and cored in a conventional manner, treated as described in Example 1 with sugar syrup except that part of them were treated at room temperature of 50°C. for 16 hours and the remaining part of them were treated at 80°C. for one or three-quarters hour and air dried. The sugar uptake value was 1.0. The dehydrated fruit treated at 50°C. was opaque and it reconstituted in only a few minutes in boiling water to a product with exceptionally good texture, flavor and aroma. Conversely, the dehydrated fruit treated at 80°C. was translucent and it reconstituted to a relatively soft product that was somewhat washed out in flavor.

Dried Fruit in Form of a Bar

A process has been developed by *G.A. Fitzgerald; U.S. Patent 3,006,773; October 31, 1961* which is directed to the preparation of fruit, especially dried fruit, in a convenient and attractive form, and in a manner that will preserve the original natural flavor. Essentially, the fruit preparation consists of the natural fruit pulp in the uncooked condition with the inedible parts of the fruit removed, the pulp being reduced by comminution to a uniform, finely divided condition thereby forming a mass having a consistency suitable for molding into the form of chunks, sticks or bars of convenient size which may be wrapped and distributed for eating.

It has been found that if a minor proportion of sugar is added to the fruit while controlling the moisture content, and comminution of the mixture is carried out at freezing temperatures, for example, 0° to 25°F., a product may be obtained which has desirable characteristics of natural flavor, low sugar content, homogeneous blending, and comparatively dry, nonsticky handling qualities, and a firm texture.

The fruit forming the basis of the process is advantageously in a partially dehydrated condition. Such fruit which may contain about 18% moisture is reduced to a suitable degree of subdivision as, for example, by means of cutting into fine shreds. With this dried and subdivided fruit there is incorporated a mixture composed essentially of minor proportions of sugar and water, these proportions being based on the amount of the fruit. The temperature of the resultant mixture is reduced below the freezing point of water. With a fruit of naturally high sugar content a lesser amount of sugar may be used, and the amount of water will depend on the degree of dryness of the fruit.

The mixture of dried fruit, sugar and water, reduced to and maintained at a temperature below freezing, is subjected to a comminuting and homogenizing treatment in a precooled high speed blending machine, such as a hammer mill. This blending treatment may be completed in a matter of minutes. Thereafter the temperature of the blended product may be raised. In order to stabilize the moisture content, a humectant may be incorporated with the product, such as glycerin, propylene glycol, and crystalline D-sorbitol. Further, a fungicide such as potassium sorbate may be included. In some cases the product may be subjected to a subsequent caramelization treatment. Thereafter, the product may be molded by pressure into suitable forms and shapes and packaged for distribution.

It has been further found that whereas certain sugars in the product, particularly sucrose, have a tendency over a period of time to take up water of crystallization from retained moisture and thereby give rise to the development of an undesirable grainy or granular texture in the product, and whereas either a humectant such as sorbitol, glycerol or propylene glycol, or of a sugar that is soluble under the low moisture conditions existing in the product, such as fructose, commercial glucose or beta lactose, or a mixture of such sugars will when incorporated in the product to a considerable degree act as a crystallization inhibitor, the combination of both a humectant and a noncrystalline sugar overcomes crystal formation to a greater degree than would be expected from the results of using either separately.

Example 1: Dried apricots, cooled to a temperature of 20° to 25°F., cut into shredded form, are admixed with beta lactose, sucrose and water in the following proportions.

Dried apricots	28 parts by weight
Beta lactose	12 parts by weight
Sucrose	8 parts by weight
Water	10 parts by weight
Potassium sorbate	0.1 percent based on the total weight

The above ingredients while being cooled to 20° to 25°F. are placed in a precooled hammer mill and comminuted and mixed for three minutes until the larger particles in the mixture will pass through a 50 mesh sieve. The product now is molded into sticks and wrapped separately in waxed paper.

Example 2: The above procedure is carried out using the following components.

Prunes	25 parts by weight
Crystalline D-sorbitol	5 parts by weight
Beet sugar	10 parts by weight
Water	4 parts by weight

Fruit Bar with Added Lecithin

Compacted, dehydrated food bars have been utilized by the Armed Forces in field rations, by astronauts during space explorations, by earth explorers, hikers, and other who must carry their food supplies along with them. An outstanding advantage of rations in this form is that they provide highly concentrated nutritional values in compact and convenient forms. They also may be stored for considerable periods of time without spoilage, especially when the moisture content is sufficiently low to prevent growth of microorganisms in the compacted food bars.

One of the outstanding problems encountered with fruit bars is that when they are compacted by application of pressure in the formation of the bars, if the moisture content is as low as is desirable for stability, the fruit bars become so hard that they are extremely difficult to eat directly without rehydration. In some cases they have been known to become so hard that when an attempt is made to eat them directly, they have caused breakage of teeth. On the other hand, if enough moisture is left in the compacted fruit bars to permit direct eating of the bars without danger of damage to teeth, the fruit bars may be unstable in long-term storage. Further, they may be very difficult to form because of the extrusion of the fruit pulp from the mold during compression in forming the bars.

A directly edible, compacted and dehydrated fruit bar has been developed by *A.R. Rahman and G.R. Schafer; U.S. Patent 3,705,814; December 12, 1972; assigned to the U.S. Secre-*

tary of the Army. The bar is comprised of one or a plurality of fruits suitably subdivided and coated with a lecithin or modified lecithin containing composition and dried to a moisture content in the range of from 7 to 14% on a weight basis, then compressed at a pressure of from 200 to 3,000 psi into bar form of such dimensions as to facilitate the direct eating of the bar without prior rehydration thereof.

A relatively small quantity of lecithin or a modified lecithin, usually less than 5% by weight, when well distributed through a fruit bar on the surfaces of the subdivided particles that are compressed together to form the fruit bar was found to produce a change in the texture of the fruit bar such that the resulting compacted fruit bar can be easily bitten through and chewed without damage to the teeth of the consumer and with enjoyment, in contrast to the difficulty with which a similar fruit bar lacking the lecithin or modified lecithin can be eaten directly. The lecithin or modified lecithin used in making the compacted, dehydrated fruit bar is an edible lecithin or modified lecithin. It may be derived from various natural sources, but the lecithins or modified lecithins produced from the natural lecithin of soybeans have been found to be particularly effective for this purpose.

The lecithin or modified lecithin may be applied to the particles of subdivided fruit as a solution, such as soybean oil solution, or as a spray, such as an aerosol spray employing a nontoxic propellant. An aerosol spray of lecithin under the trade name PAM pure vegetable food release contains about 12% pure lecithin and is useful for spray coating the particles of fruit prior to compression thereof to form compacted fruit bars. If applied to the food particles in the form of a soybean oil solution, it is preferred to use a solution having a concentration of about 15% lecithin on a weight basis, such as Centromix C.

Generally speaking, the fruit is subdivided to form particles of from ⅛ to ½ inch in the longest dimension and the particles are dried to from 7 to 14% moisture content by weight prior to the application of the lecithin or modified lecithin containing composition thereto. A sufficient amount of the solution or suspension of the lecithin or modified lecithin is applied to the particles of fruit to obtain a reasonably uniform coating on the particles of from 1 to 5% of lecithin or modified lecithin on a weight basis. About 2% lecithin or modified lecithin has been found to be particularly effective.

The pressure required in compressing the subdivided fruit coated with lecithin or modified lecithin will, in general, depend on the amount of moisture in the fruit. The lower the moisture content of the fruit down to as low as 7%, the higher the pressure required, up to 3,000 psi, to obtain a fruit bar of the proper degree of adhesion. The higher the moisture content of the fruit, up to as high as 14%, the lower the pressure required, down to 200 psi, to obtain a fruit bar of the proper degree of adhesion. The type of fruit has some effect on the pressure required, figs and pears in general requiring higher pressures than most other fruits.

This process has been found to be particularly effective with dried dates, raisins, cherries, figs and pears; however, it may also be applied to other fruits. Also, other components, such as edible seeds and nuts, may be added to the fruits in order to impart variety and interest to compressed fruit bars. Also, cereals, proteins, fats, chocolate, spices and various other flavoring or chemical additives may be incorporated in the compressed fruit bars in minor proportions in relation to the fruit components.

VEGETABLES

For purposes of convenience, the processes in this chapter are organized by product. Many of the processes, however, can be utilized for other vegetables than those under which they appear.

BEANS

Processes for Minimizing Butterflying

Heretofore, various types of legumes, including beans, have been cooked and dehydrated, but problems have been encountered in "butterflying" of the beans during dehydration. Precooked beans normally split wide open immediately upon exposure to a dehydrating environment. This splitting, known as butterflying, apparently is caused by a differential rate of drying between the bean skin and cotyledon. The skin dries more rapidly and contracts. The slower-drying cotyledon develops internal vapor pressures of such magnitude that the pressure ruptures the skin and produces the butterflying effect. One method for reducing the butterflying is to freeze the beans prior to dehydration. However, blast freezing equipment is relatively expensive and requires plant area and facilities completely eliminated by this process.

W.R. Dorsey and S.I. Strashun; U.S. Patent 3,290,159 December 6, 1966; assigned to Vacu-Dry Company describe a process to almost completely eliminate butterflying by reducing the initial bean drying rate of the precooked beans until the moisture content is below the 40% level by weight. The beans then are rapidly dried to their final low moisture content. The described process is useful for producing beans having a variable cooked condition so that they are useful as "simmer type" beans for use as an admixture with other dry ingredients in soups or casseroles which upon reconstitution are subsequently cooked slightly or simmered.

The process also is useful in producing an "instant" type bean which upon rehydration instantly reconstitutes and can be eaten without further cooking. The process is applicable both to dry legumes such as navy beans, red beans, lima beans, kidney beans, pinto beans, etc., and to freshly harvested beans such as fresh lima beans.

In one embodiment of the process for instant type beans, fresh lima beans, after passing the customary inspection and prewashing procedures, feed directly to an atmospheric steam blancher for approximately 15 minutes at 212°F. During residence in the blancher atmospheric steam thoroughly cooks the beans. For simmer type beans for use in soups, casseroles and other uses, the residence time is reduced to 10 minutes at atmospheric conditions in the steam blancher.

During the described blanching step the beans are treated with a ¼% water solution of monosodium phosphate applied over the bean exterior. The gentle corrosive action of this alkali on the skin tenderizes the skin. It also preserves the chlorophyll in lima beans so that they retain their green color after dehydration. The tenderization of the skin results in a normal texture balance between skin and cotyledon after dehydration. The beans then are quenched or cooled and dewatered as they pass from the cooking facility to dehydrating equipment. Enroute a sulfite spray contacts each bean. The sulfite concentration is adjusted to yield 200 to 800 ppm sulfur dioxide in the final product.

Dehydration takes place next in conventional drying equipment such as the air dryer described in U.S. Patent 2,541,859. To eliminate butterflying of the beans, dehydration must be in two stages. A first stage of dehydration at a slow moisture removal rate in the order of one pound of moisture per pound of dry bean solids per hour reduces the bean moisture content from an initial value normally around 55% to 35 to 40%. This is followed by a second stage of more rapid dehydration down to the approximate final moisture content. By way of example, lima beans are exposed to a drying temeprature of 130°F. for about 2 hours to reduce the moisture content to 35 to 40%. This is followed by subsequent rapid drying at 160°F. for about 1 hour to bring the final moisture down to 6 to 10% by weight on a dry basis.

The beans then may be finish dried in a finishing bin by exposure to a stream of dry air passing through the beans at 130°F. This reduces the bean moisture to less than 10% in 4 to 6 hours. The beans then are given a final inspection, treated with antioxidant, such as butylated hydroxyanisole and packaged. Similar processes are used for dehydrating navy, red or kidney beans, with variations in times of blanching and dehydration.

Another process for preparing beans to prevent butterflying has been developed by *J.J. Thompson and W.F. Allen; U.S. Patent 3,291,615; December 13, 1966; assigned to Kellogg Company.* The process is carried out without the deliberate employment of means to rupture the skin, and on the contrary is directed to prevention of such rupture, splitting or bursting and to such end the process is characterized by control of the redrying conditions from a stage where the legumes which have been soaked to moisture content of from 50% or more are dried down to about 20% or less moisture content and whereby in this drying stage of the process there is very little difference in drying rate on all sides of each unit.

Further, the rate of removal of moisture is held in check so that the outside of the skin and cotyledon do not dry too rapidly and whereby the difference in moisture content between the interior of the legume and the outside areas thereof is held below the point of causing sufficient strain to result in splitting or bursting. A specific example will illustrate the process.

Example: A body of white navy beans are spread over a #5 USS screen, washed in tap water, using 1 part of water to 2 parts of beans by weight, with constant stirring for about 2 minutes, after which they are spread out onto a 6 or 8 mesh screen to drain. A second wash of 1 part of water to 4 parts of beans is given by sprinkling the beans on the screen and then allowing them to drain for about 2 minutes.

The washed beans on the screen-bottom tray spread to a depth of ½ to 1 inch are immediately steamed in an autoclave for about 15 minutes under a steam pressure of 0.5 psig equivalent to a temperature of about 212°F. The blanched beans are then quickly transferred to a receptacle in hot water at a temperature of from 120° to 150°F., and preferably a temperature of from 130° to 145°F., in the proportion of about 2 parts of water to about 1 part of dried beans by weight.

The beans are soaked in the hot water for a period of from 3 to 6 hours and suitably will be maintained at the preferred temperature for 4 to 5 hours with the beans submerged during the entire period. Any small, hard beans that are resistant to soaking in water and which will not be completely cooked are here removed by screening through a #3 USS screen. The screening and removal of undesirable units may be done near the end of the soaking period and the beans placed back in the same soak water for an additional short period, following which the liquid is drained off.

The soaked beans which have a moisture content of approximately 52 to 56% are immediately spread onto a 6 to 8 mesh screen bottomed tray to a depth of 1 to 2 inches, and cooked in an autoclave with steam for approximately 1¾ hours at steam pressure of 5 psig or approximately 220°F. At the end of this period the steam pressure is released very slowly and the door of the autoclave opened 1 or 2 inches for release of water vapor and to allow as much water as possible to vaporize from the beans while they are hot and the humidity in the autoclave remains high. This occupies about 20 minutes and during this time the beans are spread out as thin as possible on the foraminous tray or trays within the autoclave.

Thereafter, the beans are removed from the autoclave and placed into a receptacle covered by means for example with a cloth, and then transferred to a drying room or cabinet having automatic temperature and humidity control. They are here spread out in thin layers of from ¼ to ½ inch deep on 6 to 8 mesh screen bottomed trays with air circulation all around the trays. The temeprature at the start for about 3½ hours is maintained at 100° to 120°F., and at 62 to 65% relative humidity, and then the heat is gradually raised to approximately 130° to 140°F., with slightly lower relative humidity such as 58 to 62%. The constant temperature in this area is accomplished by radiant heat units or infrared lamps.

After a period of 2½ to 4 hours at the higher temperature, or when the beans contain less than 20% moisture they are transferred to a vat or dryer and agitated therein with gentle hot air current at a temperature starting at 130° to 140°F., in about 1½ hours until the moisture content of the beans is reduced to 8 to 12%, after which the beans are allowed to cool in shallow trays to room temperature.

The resulting product is stable and will keep for many months or indefinitely provided the beans are kept in a container that allows slight breathing of outside air and not exposed to dampness to the extent of unduly raising the moisture content of the beans. These thoroughly cooked beans are easily made ready to serve by merely boiling in water for approximately 10 minutes.

The entire process can be accomplished by the use of a series of horizontally mounted drums rotating at 2 to 10 rpm. Perforated or 5 to 8 mesh drums can be employed for the cleaning and presteaming and drying operations and a solid metal drum for the soaking operation. In such continuous process a calculated amount of hot water can be added for the soaking step that will completely soak into the beans to give the optimum moisture content of from 52 to 58% for the required time period. In the drying operation infrared lamps or radiant heat units can be evenly spaced or disposed around the revolving screen drum for uniform application of heat to the beans.

In order to shorten the soaking and cooking time a weak alkali such as sodium bicarbonate can be added to the soaking water in an amount that will provide a pH of about 6.9 at the end of the soaking period. This can also be accomplished by addition of a weak acid such as acetic or citric acid to the soaking water such that the pH at the end of the soaking period is about 5.5. Although the foregoing example is specific to navy beans, the process is equally applicable to other dried beans and peas of commerce.

Yet another process for producing quick-cooking beans without butterflying is described by *L.B. Rockland, R.J. Hayes, E. Metzler and L. Binder; U.S. Patent 3,318,708; May 9, 1967; assigned to the U.S. Secretary of Agriculture.* It involves the following steps: (1) Dry beans are hydrated, using (a) a special operational technique and (b) a special aqueous hydrating medium. (2) The hydrated beans are then dried. If desired, the hydrated beans may be washed with water prior to this drying step.

The basic feature of the process concerns the operational technique by which the dry beans are hydrated. The dry beans are placed in a vessel together with an excess of the hydrating medium, for example, 2.5 to 5 times the weight of the beans. The vessel is closed and a vacuum is applied, then released. This cycle involving application of vacuum and release of vacuum is repeated several times, usually 2 to 20 times, depending on the hydration resistance of the beans, over a period of 5 to 60 minutes.

Such vacuum impregnation or hydration yields results entirely different from those obtained with ordinary soaking. Thus where dry beans are simply soaked in water, the penetration of water is strictly from the outer surface of the beans toward the inside. This type of penetration is very slow and, moreover, it causes the establishment of stresses and strains in the seeds so that when they are subsequently dried they crack and split. In this procedure the entry of the water follows an entirely different course.

As a first step, the withdrawal of air from the bean tissues by the vacuum, followed by release of the vacuum causes the liquid to rush into the space between the skin and the cotyledon, the liquid entering this space through minute openings in the skin and/or through the hilum or the micropyle. Thus as a first step the water is actually placed within the structure of the bean. Once the water is thus inside it can diffuse readily into the various structures of the bean. For example, it diffuses inwardly into the cotyledon and outwardly through the various layers of the skin. It is of interest to note that the skin is composed of several layers and the outermost is the most impervious to liquids.

As a consequence moisture can penetrate more easily from the innermost to the outermost layer of the skin than it can be in the reverse direction. Another important factor is that when the water enters into the bean by the vacuum treatment it contacts the innermost (water-permeable) layer of the skin and quickly makes the skin plastic so that it can expand without developing any tendency to rupture. As a matter of fact, after applying the several cycles of vacuum and release of vacuum, it is noted that each bean is in a swollen distended condition. This is caused by the swelling of the skin and the establishment of a pool of liquid inside the skin and surrounding the cotyledon. When this state is reached the beans are simply allowed to remain in contact with the hydrating medium to complete their hydration. The time required for completing the hydration will vary with such factors as the size, variety, moisture content, postharvest age, and time-temperature history of the beans.

In general, a time of 2 to 24 hours is used. In any particular case, the time of hydration is readily determined by simply feeling the beans from time to time as they are soaking. Thus when the cotyledons have taken up so much water they have swelled to fit the swollen skin and the skin no longer feels loose, the procedure is complete. It is to be noted that the hydration of the beans is conducted at ambient (room) temeprature or below. Increased temperature will increase the rate of hydration but cause a mottling of the beans and is hence avoided.

After the beans have been hydrated as above described, they may be washed with water to remove the hydrating medium from the surface of the beans. This washing is conveniently carried out by placing the hydrated beans on a screen and spraying them with water for a short time. Additives may be incorporated into the wash water to accomplish certain ends. As an example, glycerin may be added in a concentration of 5 to 10% whereby to plasticize the skins, that is, keep them in a soft or elastic condition. Also, a minute amount of ascorbic acid may be incorporated in the wash water as a means of minimizing oxidation of oxidizable components in the skin when the products are stored before use.

Following washing, or directly after hydration, the beans are dried. This is conveniently effected by placing the beans on trays and exposing them to heated air. Generally, the air temperature and conditions of treatment are selected so that the drying takes place relatively slowly. Normally an air temperature of 100° to 180°F., preferably 100° to 140°F., is used and circulation of the air is restricted to avoid too rapid drying which could cause checking or cracking of the beans. In any event, the drying is continued until the beans contain from 9.5 to 10.5% moisture. This moisture range is preferred as providing an optimum level for retention of quality during storage since moisture levels above the preferred tend to encourage reversion (loss of quick-cooking character) and moisture levels below the preferred one tend to encourage flavor changes associated with oxidative changes of labile components. Following drying, the products are packaged, preferably under vacuum or with replacement of air by nitrogen. Although the medium used in the hydration step is largely water, it contains certain additives which provide advantageous results which could not be obtained with water alone.

These results are attained by having present in the hydration medium, sodium chloride and a chelating agent. The sodium chloride has the principal effect of softening the skins. Usually, the sodium chloride is present in the hydrating medium in a concentration of from 1 to 3%. The chelating agent exerts a variety of useful effects, including the following:

(1) It softens the pellicle or skin.
(2) It aids in the solubilization of proteins, such as those in the cotyledons.
(3) It acts as a buffer to maintain pH.
(4) It facilitates uniform penetration of additives into the centers of the cotyledons, so that the final products have a uniformly smooth texture.
(5) It tends to lighten the color of the product.

Various conventional chelating agents may be used such as the alkali metal salts of ethylenediamine tetraacetic acid (hereinafter referred to as EDTA), alkali metal pyrophosphates or tripolyphosphates, citric acid or its alkali metal salts, etc. Generally, the chelating agent is added to the hydrating medium in a concentration of from 0.25 to 5%. Particularly preferred as the chelating agent is a single EDTA salt, or a mixture of EDTA salts, which contain both sodium plus calcium or magnesium or all three of these cations.

To assist in the tenderizing effect, it is preferred that the hydrating medium additionally contain a trace, i.e., 0.01 to 0.05%, of a reducing agent, preferably one having a sulfhydryl group. Typical of the components which may be used are alkali metal sulfides, mercaptoacetic acid or its alkali metal salts, mercaptopropionic acid or its alkali metal salts, thioglycollic acid or its alkali metal salts, cysteine, etc.

It is also hypothesized that the reducing agent may reduce the flatulent effect of the products by reducing oxidized derivatives (i.e., sulfides, or disulfides) of sulfur-containing amino acids or sulfur-containing peptides to the corresponding thiol forms. During cooking of the products, it is, of course, essential that moisture penetrates throughout the bean tissue. For this reason the preferred forms of the hydrating medium will contain a minor proportion, for example, 0.05 to 0.5%, of a conventional edible-grade surface agent.

Also, it has been found that addition of a surface-active agent improves the final appearance of the treated, dried beans. Typical examples of surface-active compounds are sorbitan monolaurate, sorbitan monomyristate, sorbitan monopalmitate, sorbitan monostearate, sorbitan monooleate, and sorbitan monolinoleate. The corresponding fatty acid esters of mannitan may also be used.

Examples 1 through 12: A solution was prepared containing water and the following ingredients:

	Percent
NaCl	1.5
EDTA, Na_4 salt	1.0
EDTA, Na_2Mg salt	1.0
EDTA, Na_2Ca salt	0.25
$NaHCO_3$	0.25
$Na_2S \cdot 9H_2O$	0.025
Polyoxyethylene sorbitan stearate	0.25
Butylated hydroxyanisole	0.003
Butylated hydroxytoluene	0.003

The pH of the solution was 9.0. Several different lots of dried beans and dried peas were treated with the above solution, using the following technique in each case. The dried beans (or peas) were placed in a vessel together with enough of the solution to cover them. The vessel was closed and connected to a source of vacuum. The vacuum was maintained for about a minute, then the vacuum was released. After about a half minute the vacuum was again applied, held for a minute, then released, and so on. In all 10 cycles of vacuum application and release were used. The system was then allowed to stand, at room temperature and at atmospheric pressure, with the beans (or peas) in contact with the solution until the cotyledons swelled to fill the swollen skins.

The hydration time applied in each case is given in the following table. (It is to be noted that the time of hydration varies, depending on many factors, as hereinabove explained, and it is not maintained that the particular times used are necessarily applicable to other lots of legume seeds.) Following residence in the hydration medium, the products were dried in a current of air at 140°F. until their moisture content was 10%. The products were then ready for packaging or use. It was observed that in all cases the seeds retained their structure intact with no noticeable cracking or splitting.

To test the cooking quality of the products, each lot thereof was added to boiling water and simmered until the products reached a standard tenderness, typical of properly cooked beans. In all cases the cooked product had an excellent flavor and the seeds were essentially intact with no significant mushing or sloughing. The results are given in the following table.

Example		Time of Residence in Hydrating Medium, Prior to Drying, hr.	Time Required to Cook Product, min.
1	Lentil	6	13
2	Lima, large white	4	25
3	Lima, baby	6.5	25
4	Beans, Pinto	6.5	30
5	Peas, whole	18	35
6	Beans, red kidney	6	35
7	Beans, red	24	35
8	Beans, pink	24	35
9	Beans, small white, Calif.	6	35
10	Beans, Great Northern	18	35
11	Blackeye	18	45
12	Soybean	24	50

J.J. Mancuso; U.S. Patent 3,340,068; September 5, 1967; assigned to General Foods Corp. has found that butterflying can be prevented by removing moisture from the cooked legume in a dehydrating medium. The method of this process is applicable to legumes in general and is particularly useful for beans, e.g., navy beans, pinto beans, red kidney beans, dark red kidney beans and red beans.

After being cooked, the legumes are quenched, thoroughly soaked in cool water, drained and placed in a dehydrating medium. The dehydrating medium may consist of an organic dehydrating agent selected from the group consisting of water-soluble alcohol and ketones, e.g., methanol, ethanol, isopropyl alcohol, tertiary butyl alcohol and acetone, or mixtures thereof. The minimum amount of dehydrating agent employed is that amount which permits equilibration to take place between the moisture being extracted from the beans and the alcohol replacing the moisture. This amount is assured if the dehydrating agent is present in a quantity sufficient to completely immerse the beans.

As the dehydrating agent becomes diluted with water, its efficiency diminishes. Accordingly, it may be desirable to use a large excess of dehydrating agent to overcome the decrease in efficiency due to dilution. On the other hand, smaller volumes of dehydrating agent may be used by employing two or more sequential treatments with dehydrating agent. A preferred dehydrating agent is 95% ethanol. The length of time required for dehydration depends on the size of the legume and the ease of water extraction. In general, in the case of beans, small beans, such as navy and red beans, will dehydrate faster than beans such as kidney and pinto beans. The dehydration time can be shortened if the skins of the cooked beans are pierced. Generally, the extraction time will require a minimum of about 1 hour at ambient temperature. The following example illustrates the process.

Example: Raw navy beans were soaked in tap water and held at room temperature for about 16 hours. The beans were then drained, placed on screens and autoclaved at 15 pounds per square inch gauge for 25 minutes. They were then quenched in cool water (50° to 60°F.) for 30 minutes. Two pounds of the quenched beans were thoroughly drained and placed in an excess of 95% ethyl alcohol. After 1 hour, the alcohol was drained off and the beans were placed in an excess of a fresh quantity of 95% ethyl alcohol. After 30 minutes, the beans were drained of the second quantity of alcohol, placed on screens to permit residual alcohol to evaporate and then subjected to mild drying at about 140°F. for 1 hour. The final dried beans exhibited a marked reduction in butterflying.

When placed in boiling water, the dried beans rehydrated in about 5 minutes. The rehydrated beans were similar in appearance and texture to beans prepared by the lengthy soaking and cooking time required to prepare raw beans and were almost completely free of butterflying.

The process developed by *K.H. Steinkraus, D.B. Hand, J.P. Van Buren and R.L. La Belle; U.S. Patent 3,510,313; May 5, 1970; assigned to Cornell Research Foundation, Inc.* involves hydrating the dry beans by soaking in water, precooking in steam, coating with sugar, and dehydrating. The process is applicable to dry beans of various kinds. Significant among the advantages of the process are the following:

(1) Butterflying of the beans during cooking and dehydration which ordinarily results in loss of bean structure and identity is economically controlled.

(2) Baked bean flavor and color are produced without a baking step.

(3) The amount of brown color produced can be varied by varying the concentration of dextrose in the coating mixture.

(4) Storage beans with poor inherent cookability can be processed to yield a quick-cooking product with uniform and characteristically smooth texture when rehydrated.

The initial step of the basic process is to hydrate the dry beans. In the case of pea beans this may be effectively accomplished by soaking the dry beans in water at 80°F. for 5 hours. In the case of marrow beans the necessary absorption of water is much improved by a 15 minute presteaming of the beans followed by soaking in water at room temperature (80°F.) for 8 or 9 hours. The presteaming of marrow beans may be omitted and complete hydration may be attained by soaking for 15 hours at room temperature.

To secure maximum hydration in a minimum length of time it has been found that hydration is best carried on at elevated temperatures. For instance, in water at 180°F. hydration of pea beans immersed for time periods varying from 15 to 90 minutes produced hydration factors ranging from 1.69 to 2.0. The hydration factor is the ratio obtained by dividing the weight after hydration by the original starting weight of the dry beans. In the process the beans must be well hydrated before cooking and the hydration factor should be in the neighborhood of 2.0. Hydration of pea beans for 5 hours at 80°F. will satisfy this requirement. In other words, the beans take up an amount of water approximately equal to their starting weight.

The next step in the process is precooking the hydrated beans to a degree which eliminates the grainy texture of the beans while retaining the bean morphology. The cooking time and temperature in the precooking step is believed to be of great significance. By way of example, this may be accomplished by first draining the beans and then cooking the drained beans in steam at about 245° or 250°F. for approximately 90 minutes. Following precooking as described previously, the beans are immersed in a sugar solution for example, 20° Brix, at 160°F. for 5 minutes. The use of a hot sugar solution keeps the beans, which come to the sugar solution from the precooker, hot until they go to the dryer. The sugar coating step may also be practiced by using powdered sugar in place of a separate sugar solution since the moisture on the surface of the beans will form a layer of dissolved sugar. The sugar coating is provided for the purpose of protecting the bean structure during subsequent processing.

The principal difficulty which is obviated by means of the sugar coating is the opening up of the bean during drying. This is variously referred to in the art as bursting, butterflying or puffing. While the above specified time, temperature and sugar concentration are highly effective, various adjustments in these relationships may be made. Sucrose, dextrose and lactose, the three most economical sugars, appear to give about equal control of butterflying. However, dextrose caramelizes more readily and provides a greater degree of browning in the final cooking. By selecting suitable proportions of sucrose and dextrose the exact degree of browning which is desired may be achieved.

The final step in the process is dehydration. A final moisture content in the range of 8 to 10% is preferred for the precooked beans. It has been determined that in the case of beans which have been quickly hydrated in water at elevated temperatures a longer dehydration is required to reduce the moisture level to the required dehydration range.

Where the beans have been subjected to slow hydration at room temperature dehydration to a 9% moisture content can be accomplished in 50 minutes of drying time whereas rapidly hydrated beans required about 80 minutes at the same drying temperature to reduce the moisture content to the same level.

The drying times referred to here were effected in an atmosphere having drying and wet bulb temperatures of 200° and 110°F., respectively (relative humidity 7.5%) and an air velocity of 200 to 300 ft./min. down through a 1" deep bed of beans. Beans prepared according to the foregoing method have been stored for 1 year without detectable deterioration. No special rehydration is required. The precooked beans are merely placed in water and cooked for approximately 30 minutes.

Treatment with Phosphate to Reduce Cooking Time

As is well-known, dried beans require a long cooking time. Normally this time is on the order of 2½ to 3 hours. *J.P. Nielsen; U.S. Patent 3,108,884; October 29, 1963; assigned to Idaho Bean Commission* has developed a process which reduces their cooking time to about 35 minutes. The process also results in beans which have low flatulence properties so that the gastric distress and social embarrassment which frequently accompanies the eating of beans is largely or wholly eliminated. Thus, the process may be employed in situations where quick-cooking is not important as in the preparation of canned beans.

Generally speaking, the process is carried out by soaking the beans in a phosphate solution of a specified concentration for a certain time, and dehydration. If desired, the beans may be presoaked and blanched prior to treatment with the phosphate solution although these steps are not generally necessary.

The phosphate solutions which are used can be prepared from any sodium, potassium or lithium phosphate salts but it is normally preferred to use the sodium salts of phosphoric acid since they are readily available, of low cost and have no adverse effect upon the taste of cooked beans. It is important that the phosphate solution have the correct pH and a pH in the range of 6.6 to 10.7 and preferably from 8 to 10.7 is employed. The concentration of the phosphate ion (PO_4) can be from 0.01 to 0.2, and is preferably 0.03 to 0.09 mol/l. or, even better, 0.03 to 0.07 mol/l.

The phosphate soaking solution is prepared to have the desired pH and phosphate ion concentration and 3 to 5 parts by weight of the soaking solution are used for each part of beans treated, although the exact amount of solution used is not critical. At least as much solution should be used as will keep the beans covered throughout the soaking period. The soaking time is preferably from 2 to 5 hours at a temperature of 45° to 50°C., although the temperature can vary from as low as 40°C. to as high as 55°C. With some beans having delicate skins, such as small white beans, the soaking time can be reduced somewhat. It has been found that the temperaturre of soaking is quite critical. If the temperature is below 40°C., the soaking time is unduly long and the beans will sour before the desired results are achieved. If the temperature is too high, undesirable results will be produced.

After soaking, the beans are drained and dried. Drying is preferably accomplished in two steps. First, the beans are subjected to a temperature of from 80° to 110°C. for from 45 to 90 minutes to reduce the moisture content of the beans to between 18 and 40%, and then the temperature is lowered to between 40° and 75°C. and drying is completed at this lower temperature for a period of from 12 to 18 hours whereby the moisture content of the beans is reduced to between 6 and 15%.

If the temperature of the beans is allowed to exceed a certain temperature, for example, over 65°C., the beans will have a poor color and flavor and the quick cooking properties may be impaired. The higher tememprature is permitted when the beans are wet since the evaporating water keeps the beans themselves at a temperature lower than the ambient air.

However, as the beans approach dryness, it is preferable to lower the temperature in order to obtain maximum quality of color and flavor. Although it is usually most convenient to use two fixed temperatures, it is feasible to gradually lower the temperature while the beans are drying. Further, it is possible to do all the drying at a temperature of say 40°C., although a longer period of time is required.

Generally speaking, the beans are dried to a dryness of 18 to 25% moisture at the higher temperature and drying is continued to a moisture content of from 6 to 15 or preferably 6 to 10% at the lower temperature. When drying is complete, the beans will have a moisture content normal for dry beans and will be substantially identical in appearance to untreated beans. Various types of dryers may be used, such as fixed tray, moving tray, continuous belt, cylinder, dielectric, infrared and other dryers. The following example illustrates the process.

Example: A quantity of 100 grams of pinto beans was soaked in water at 50°C. for one-half hour. Sufficient water was used to keep the beans covered at all times. The beans were then removed from the water, and blanched in steam at 90°C. for 2 minutes. Immediately thereafter, the beans were placed in 300 cc of a phosphate soaking solution which was maintained at a temperature of 45°C. The phosphate soaking solution had a concentration of 0.034 M trisodium phosphate adjusted to a pH of 9 with phosphoric acid.

The beans were soaked in this phosphate solution for a period of 3 hours and then were drained and placed in a dryer maintained at a temperature of 80°C. for about 50 minutes, whereupon the moisture content of the beans was 20%. The temperature of the dryer was then reduced to 40°C. and drying was continued for about 15 hours at which time the beans had a moisture content of 8%. The beans appear similar to beans which have not been treated but upon cooking them in the normal manner, it was found that they had a good color, flavor and texture with only 45 minutes of cooking. Similar beans, but untreated, required a cooking time of 3 hours to achieve the same degree of softness. This process may be applied to other legumes such as lentils, peas, etc.

Perforating Procedure for Making Beans Quick Cooking

A process has been developed by *J.J. Thompson and C.A. Doan; U.S. Patent 3,203,808; August 31, 1965; assigned to Kellogg Company* to prepare quick cooking, storage-stable, dry beans and peas which substantially retain their natural and individual shapes even though they have been pretreated to enable them to be rapidly prepared for eating.

In general, the method comprises preparing quick cooking, dried whole beans or peas, by subjecting dried, raw, whole beans or peas to contact with moisture to at least soften their surfaces and adjacent interior, followed by perforating or puncturing the skin or seed coat and the body or cotyledon at a number of points and from opposed surface areas. The thus perforated legumes are then subjected to cooking to a desired extent while still substantially retaining their integrity.

Thereafter, the precooked particles are subjected to heat drying to reduce their moisture content to substantially equal to that of or below that of the original dried particle employed. For example, the dried peas and beans of commerce, to be storage stable, have a moisture content of from 10 to 12% by weight, and this process redries them to at least this moisture content, or preferably to 5 to 8% of moisture by weight. This perforating procedure permits the dried beans and peas to be cooked and then dried to a low moisture content without causing them to lose their individuality. The perforations also allow the moisture to escape readily from the particle during the redrying operation.

Softening of the dried particle preceding the perforating step may be carried out by steeping in water overnight or in a more rapid manner by, for example, placing in boiling water and cooking for several minutes, and then draining. As an aid in moistening and surface softening chemical adjuvants can be employed such as a mild alkaline solution, for example, a 1% sodium bicarbonate solution.

The raw, dried whole legumes may be perforated either batchwise or continuously after the softening treatment. Thus, in the batch treatment they can be impaled on steel pins mounted on a block on 0.1" centers. For example, a 3" x 3" grid of 10 x 10 to the inch was laid out on a ⅜" thick steel block and then 0.035" diameter holes were drilled through the plate at the intersections of the grid lines. 2" steel pins were then inserted into these holes and a backup plate was then attached to the block to hold the pins in place. One or an opposed pair of such blocks may be employed for perforating the skins and adjacent interior of soaked beans or peas so as to provide, for example, 5 to 10 small holes, more or less, on each side of each bean, depending on its size. In the alternative, the particles may be perforated by being pressed onto a fine wire brush. In continuous perforating processes, the particles may be fed between pairs of brushes or between pairs of rollers having radially extending pins. An example will illustrate the process.

Example: A batch of raw, dried navy beans of commerce were soaked overnight in a 1% sodium bicarbonate solution, 4 parts of solution being employed to 1 part of beans. The soaked beans were then removed from the solution and pierced all the way through using steel pins that were mounted on a block as aforesaid. After being pierced the beans were removed from the pins and placed on a stainless steel tray and cooked in an autoclave for 1 hour and 30 minutes at 5 psig steam pressure. The cooked beans were then dried for 2 hours at a temperature of 170° to 175°F. in a forced hot air dryer to a moisture content of approximately 7%. These beans were then storage stable and could be cooked rapidly for eating in a much shorter time than is conventional for dried beans which were not treated in accordance with the process.

Quick Cooking Beans Produced by Heat and Pressure

A process was developed by *A.K. Ozai-Durrani; U.S. Patent 3,388,998; June 18, 1968* for treating a large variety of edible beans, including soybeans. The treatment of the original, raw beans to convert them into the form requiring little or no cooking preferably involves the steps of first soaking the beans to cause them to absorb moisture to such an extent that the moisture content is increased to 50 to 70% of the total weight of the soaked beans. This soaking step may be carried out in any suitable way to insure the desired increase in the moisture content of the beans. This may be accomplished, for example, by introducing the beans and the water into a revolving drum which is revolved at a relatively low speed, say about 1 rpm, during the soaking period. If matured in the pod, garden beans are used as the starting material, the soaking step may be eliminated or greatly reduced.

When the beans have been soaked to bring their moisture content to the desired point, they are subjected to a heating step to bring about a mealy and homogenized condition of the interior structure of the beans. This heating step may be carried out in any of a variety of different ways. For example, the beans may be placed in a revolving drum or tank and then subjected to the action of steam that is introduced into the tank under a suitable pressure, which will raise the pressure within the tank up to between say 1 to 50 psig. The steam may be continuously introduced into and withdrawn from the tank, or it may be simply introduced and allowed to remain within the tank for the period of time required, provision being made for the removal of any excess condensate.

Heat may also be supplied, by electrical or other means, to maintain the steam at the desired temperature within the tank. The duration of the steam treatment will vary somewhat in accordance with the nature of the beans being treated. In general it may be said that the steam treatment may be carried out for from 2 to 45 minutes depending upon the character of the beans. In applying the process to Michigan Navy beans it has been found desirable to subject the saturated beans to an initial treatment with steam under a pressure of 10 lbs. gauge for 15 to 25 minutes and then to retain the beans at the elevated temperature, with just enough introduction of further steam as may be needed to maintain the desired temperature, for another 10 minutes.

Before the heating step mentioned above, it has been found quite desirable and effective to first subject the beans to a tempering action at a temperature below 100°C., preferably between 90° and 95°C., for a period of 10 to 150 minutes and to then raise the temperature within the heating vessel or drum to 100°C., or somewhat higher, to insure plasticizing or cooking or homogenization of the beans. This second heating step may be carried out for a period of 2 to 40 minutes. Such two stage heating may be effected by the use of steam or through the supply of heat in any other suitable way.

It has been found highly desirable in connection with the heating and cooking of the beans to perform it in such a way as to maintain a relatively high moisture content in the beans commensurate with the content upon soaking, and preferably slightly higher. Following the cooking treatment, the beans are subjected to drying under controlled conditions to provide a final product suitable for packaging, which has substantially the same appearance as the original beans, except for a slight change in color and the development of hardly noticeable fissures in the skins of the beans. Thus the final product, which may be prepared for serving within the period of only minutes, has a wholesome aroma, a very attractive appearance, and has almost all of the nutritive components of the original beans. Through the use of certain special steps, in one or more of the stages of the process, the final product may be rendered edible without cooking. Therefore the beans may be used as a very appealing cold snack.

An important feature of the process which has been found to bring about the final product in substantially its original whole bean form, is the subjection of the beans, in certain stages of the process and particularly the cooking and drying stages, to a certain amount of pressure or compression which serves to retain the beans in their original shape. Various means may be employed for imparting the desired pressure, but it has been found particularly effective to impart it through the treatment of the beans in a rotating cylinder, which is rotated at a relatively slow rate, sufficient to cause the beans to pass successively from a zone of substantially no pressure through a zone of increasing pressure up to a certain maximum but not too high a pressure, and then back to a zone of no pressure. In this way all of the beans of a large quantity being treated may be subjected to the same changes in pressure with the overall result that the beans retain their original shape and wholeness and enlarged volume.

Drawings and a detailed description of the apparatus in which the beans may be treated, are given in the patent. An example will illustrate the process.

Example: 100 lbs. of Michigan Navy beans having a specific gravity of 0.8678 and a moisture content of 14.30% were soaked in cold water, having a temperature of 22°C., for 12 hours. This was done in a stationary container in which the beans were all immersed in the soaking water. After soaking, the beans were found to be at a temperature of 25°C. and were found to have a moisture content of 55.5% and a specific gravity of 0.7555. These soaked beans were placed into a tank or container and this container was placed within an outer container in the drawings. The beans were steamed for 15 minutes by introducing substantially dry steam, through injection tubes of the character indicated in the drawings, into the inner container under a pressure of 10 psig. During this period the inner container was rotated at 0.75 rpm.

At the end of the 15-minute period, the beans were retained within the container at the elevated temeprature therein for another 10 minutes, and the condensate and remaining steam was withdrawn. The beans were then removed from the container and found to have a moisture content of 55.7%. In this condition the beans were placed in trays with perforated bottoms having an overall length of 10 feet and a width of 20 inches. Air at a temperature of 110°C. was then forced through the beans on the trays at the rate of 300 cubic feet per minute, the air being distributed evenly over the surface of the beans on the trays and being forced through the beans and outwardly through the perforations in the bottoms of the trays. After 10 minutes of this air blowing procedure, the beans were found to be at a temperature of 70°C. and to have a moisture content of 44%. Substantially all of the beans were found to have ruptured and opened up to a certain extent but none had broken apart.

The partially dried beans were then placed in the revolving container or cylinder in which the steam treatment had been carried out. At this time the bean containing cylinder was rotated at higher speeds than during the steaming treatment, and hot air having a temperature of 110°C. was blown through the beans at the rate of three hundred cubic feet per minute. It was found that in about thirty minutes the moisture content of the beans could be reduced to the desired point of 12.5 percent and the beans were found to have a specific gravity of 0.6825.

During this drying stage of the beans, the cylinder containing them was rotated 3.9 rpm for about 15 minutes, and this was found to bring about substantially complete closing of the gaps which had developed in the beans in the course of drying them in the trays. The speed of rotation of the cylinder was then reduced to 2 rpm and the beans were subjected to the further drying action of the hot air for the balance of the time required to reduce the moisture content to 12.5%.

To determine the cooking quality of the beans treated in the manner described, one cup of the beans was introduced into two cups of water in a beaker and the mixture was boiled for 5 minutes. The beaker containing the beans and water was then covered and allowed to stand for another 5 minutes. At the end of this period the beans were found to be an excellent edible product.

BRUSSELS SPROUTS

Brussels sprouts consist of a more or less spherical part with a diameter of from 1.5 to 4 cm., hereinafter called the head, which tapers at one side to a more or less cylindrical part which is the cut residue of the stalk by which the sprout was originally attached to the main stem of the plant; this residue is hereinafter called the butt. That part of the head diametrically opposite the butt is known as the apex and the center and diameter of the head are, for simplicity, called the center and diameter of the sprout. The center of the sprout will be taken to be a spherical volume with a radius of 0.2 cm., about the true geometrical center.

Although the dehydration of vegetables in general is well-known and is practiced on a very large scale, it is a matter of difficulty to dehydrate brussels sprouts satisfactorily, other, perhaps, than very small ones, in the whole state. Dehydration can be carried out on brussels sprouts halved by a diametrical cut through the butt, but the resultant product, after dehydration, has an unattractive appearance; the consumer normally likes to deal with sprouts that are whole, or substantially so.

N.B. Hazeldine and M.K. Withers; U.S. Patent 3,190,761; June 22, 1965; assigned to Lever Brothers Company have found that brussels sprouts can be satisfactorily dehydrated in the substantially whole state by making in them, before dehydration, at least two cuts in a special manner. These cuts extend from the periphery of the sprout to the center, thus lying substantially in the planes of great circles, and at least one of these cuts passes within 45° of the axis of the butt and at least one other cut passes within 45° of the apex. That is, at least one of these cuts intersects the base of a cone having:

(a)   an axis coincident with the axis of the butt,
(b)   a vertex coincident with the geometrical center of the sprout,
(c)   a generatrix forming an angle of 45° with the axis of the butt, and
(d)   an intersection with the surface of the sprout encircling the butt,

and at least one other of these cuts intersects the base of a cone having:

(a)   an axis coincident with an extension of the axis of the butt,
(b)   a vertex coincident with the geometrical center of the sprout,
(c)   a generatrix forming an angle of 45° with the extension, and
(d)   an intersection with the surface of the sprout encircling the
       apex of the sprout.

The cuts are so arranged that they do not divide the sprout into two or more separate pieces; and if none of the cuts passes through the butt at least one additional cut is made to pass therethrough. It is desirable that any such additional cuts which are made in the butt should extend across its width and should not be in the form of a round or slit-shaped hole. At least one cut extending from the region of the apex to the center must be an extensive cut, that is, one which at the level of the center of the sprout has a length of not less than 75% of the diameter of the sprout, and preferably has a length of 100% of the diameter of the sprout, that is, it passes completely through the sprout at any given level.

The other cut or cuts may be shorter than the extensive cut, although it is preferred that one cut from the butt region to the center is also an extensive cut in the sense given above. Where there is only one cut that is extensive, one other cut preferably passes completely through the sprout from butt to apex.

The cuts will normally be plane cuts, but cuts curved in one or two directions may also be used, provided that the radii of curvature are large in comparison with the radius of the head of the sprout. The cuts through the butt region and through the apex region may meet or overlap at the center provided that the integrity of the sprout is preserved. Thus, for example, in a preferred case, where there are only two plane cuts, the first of which passes through the butt and the second of which passes substantially through the apex, it may be arranged either that the depths of the cuts are such that they do not quite meet at the center or preferably, that they do intersect at the center (each passing a short distance, say, up to one-quarter of the radius of the sprout, beyond the center), in which case, to preserve the integrity of the sprout, it is necessary that the cuts meet at an angle, suitably an angle greater than 30° and preferably at an angle of about 90°.

In addition to the cuts already mentioned, extra minor cuts may be made in the sprout if this is desired for any reason. For example, it has been found that additional minor cuts (of the order of 0.5 cm. deep) in the butt region of the sprout may improve its appearance on rehydration. The cuts may be made by hand or by machinery.

The sprouts cut in accordance with the process are subsequently subjected to normal dehydrating procedures (preferably drying at elevated temperatures, as by circulation of hot air to contact the sprouts) and dehydrate uniformly and well, with no extensive discoloration. On rehydration, they readily take up water and regain substantially their original appearance, the cuts being barely noticeable. After cooking, their flavor and consistency are found to be satisfactory.

Example: Brussels sprouts with a diameter of between 1.5 and 3.2 centimeters were hand-trimmed and in each were made two extensive diametrical plane cuts, one from the butt to the center and the second at right angles to the first, from the apex to the center, the cuts intersecting at the center to the extent of about 0.3 centimeter. The width of the cut is, for clarity, somewhat exaggerated. The cut sprouts were held for 10 minutes at 45°C. in a solution containing 2.5% of common salt, 1.25% of sodium sulfite and 1.5% of sodium carbonate.

After straining, they were blanched in steam for 2 minutes and dried by the forced circulation of air at 52°C. through them for 3 hours, the air velocity being 178 cm. per second. This was followed by bin drying overnight (approximately 16 hours) at 52°C. but with an air velocity of 61 cm. per second. The dried sprouts had a water content of 5.7%, the ratio of the initial weight being 7.2:1. They were cooked by being put in cold water, brought to the boil and simmered for 6 minutes. Rehydration was satisfactory and the sprouts had a good color and appearance, the cuts being barely noticeable. The ratio of the rehydrated weight to the dried weight was 6.6:1, this being a 92% rehydration.

CARROTS

In accordance with a process developed by *W.M. Bright, M. Pader and W. Wiesner; U.S. Patent 3,295,995; January 3, 1967; assigned to Lever Brothers Company* carrots and other root vegetables are prepared for dehydration by cutting to the desired size and shape and then heated to case harden the exterior surface. During case hardening, a differential moisture content is formed between the exterior surface and interior portion of the prepared vegetable. It is also important for the prepared, case hardened vegetable to have an average moisture content of at least about 40%. The prepared, case hardened vegetable which is also partially dehydrated is then treated, e.g., crushed, to mash the interior portion without damaging substantially the exterior surface except for at least one perforation therein. The resulting piece is dried further to provide a hollow, dehydrated vegetable piece, such as dehydrated carrot dice.

It is critical to heat the prepared vegetable, e.g., carrot dice, to case harden the exterior surface, to remove a required amount of average moisture and to provide a different moisture level in the exterior surface and the interior portion. In this process the exterior surface is considered to be of finite thickness. The dice are case hardened when the exterior surfaces are tough, but pliable, but the interior portions are still relatively soft. This tough surface will resist subsequent crushing pressure and will be substantially undamaged except for perforations caused by the pressure of the interior portion being squeezed outwards. In contrast, the interior portions of the case hardened dice will be mashed during the crushing operation.

During the case hardening of the prepared vegetable, some of the original moisture is removed therefrom. For example, carrot dice containing about 90% original moisture should have from 40 to 85%, preferably 50 to 75%, moisture after the heat treatment. It is important to dehydrate the exterior surface quickly and to dehydrate the interior portion to a relatively small extent to provide the distinct contrast in toughness. To accomplish this result, the air temperature is usually between 180° and 290°F., depending upon time, air velocity and type of drying apparatus. The dice are preferably moved frequently during this step to case harden uniformly all the surfaces.

The partially dehydrated, case hardened vegetable piece, e.g., dice or julienne strips, are then treated to impart certain desired characteristics. The preferred treatment is crushing and this step should follow the case hardening with a minimum amount of delay to maintain the substantial differential in moisture content between the interior portions and exterior surfaces of the case hardened dice; otherwise, the moisture within the dice may be in equilibrium and uniform throughout. During the crushing, the exterior surfaces of the dice remain intact; however, the pressure creates a definite perforation at one or more weak points in a surface of each of the dice.

To accomplish this, the crushing may take place in any suitable equipment. The conditions for the crushing vary according to the size of the dice, their moisture content, the extent to which the dice were blanched and the like. However, rollers set 0.09 to 0.14 inch apart are preferred. The crushed vegetable piece, e.g., carrot dice, is generally flattened to a roughly rectangular shape; the flat surfaces show little or no damage while the ends appear to be ruptured.

During the final dehydration, the exterior case hardened surface is relatively impervious to water evaporation and free water vapor is liberated in the mashed interior portion. This water vapor is channeled out through the one or several perforations in the exterior surface. Therefore, the material in the center of the vegetable piece dehydrates before the material around the periphery, i.e., below the case hardened exterior surface. Consequently, shrinkage occurs from the inside toward the outside, and a hollow product is formed with one or more perforations in the exterior surface.

This is the reverse of the shrinkage during normal hot air dehydration where the water from the inside goes through the surface. At the same time, water is evaporated off the surface at a faster rate and the surface becomes partially dry. The shrinkage, therefore, occurs from the outside towards the inside to form a caved-in product. The vegetable may also be coated with starch or sugar to retard changes in color and flavor during drying.

The dehydrated piece, e.g., carrot dice, is hollow and shrunk from the inside to the outside. The instant dehydrated vegetable, if boiled in water, reconstitutes in a relatively short time, i.e., about 15 minutes or less. Furthermore, the size of the reconstituted product is about the same size as the original piece with the hollow inside being filled. The reconstituted product has good flavor and is tender and plump. The following example illustrates the process.

Example: 50 lbs. of long, California-grown Imperator carrots were prepared for dehydration by peeling, washing, dicing them into ⅜ inch cubes, blanching for 7 minutes at 200°F. using flowing steam, and dipping in a 2.5% starch solution. The prepared dice were case hardened and partially dehydrated to a moisture content of 70% by heating at 195°F. for 35 minutes with the air at high velocity through a 1 inch-layer bed. The case hardened dice were then crushed by passing them between two rolls which were separated by 0.13 inch and heated at 140°F. The crushed dice were further dehydrated for ¾ hour at 160°F. with medium air velocity through a ¾ inch drying bed and subsequently dehydrated for 16 hours in an air oven at 120°F. The moisture content of the dehydrated carrot dice was 5%.

70 grams of these dehydrated carrot dice were added to 2 cups of water containing ½ teaspoon salt; brought to a boil; simmered for 15 minutes; and drained. The reconstituted vegetable had approximately the same size as the vegetable in the initial raw state. Texture, color, and flavor of the reconstituted product were good. Conventional dehydrated carrot dice having the same initial size were similarly cooked. However, the reconstituted dice were hard and rubbery. The sides were concave rather than flat, which indicated incomplete hydration.

CORN

A.I. Nelson; U.S. Patent 2,989,404; June 20, 1961; assigned to the U.S. Secretary of the Army has developed a process for preparing a precooked dehydrated food, such as cream-style sweet corn, which can be almost instantly rehydrated and yet be wholly acceptable in color, flavor, consistency and texture. The process consists essentially in segregating two components of the grain, namely, a cream component and a kernel component, and thereafter dehydrating them, and then combining the two dehydrated products.

The cream component was prepared from mature deep-cut sweet corn kernels that were steam-blanched for 7½ minutes, comminuted in a Fitzmill with the cutting blades running at approximately 5,200 rpm and a #5 screen, and finished through an 0.045" screen. When the grain was immature, waxy pregelatinized starch, 3% by weight, was added to the cream to alleviate stickiness, and the cream was then dehydrated on a double-drum dryer operated under atmospheric conditions with a roll steam pressure of 35 psig, roll spacing of 0.0015", and roll retention time of about 14 seconds. The final moisture content of the dry product was less than 5%.

In this preferred method, the kernel component was likewise prepared from mature, deep-cut kernels that were blanched and comminuted. This material was spooned onto a stainless tray in cookie-like gobs and frozen. The trays were placed in a freeze-dryer with the temperature of dehydration at 75°F. for the first 4 hours and then at 130°F. for an additional 4 hours. The dry gobs having a moisture content of less than 5%, were then crumbled, and the particles passing through a $^{12}/_{64}$" screen but retained on a $^1/_{12}$" screen were selected to serve as the simulated kernel component.

The precooked dehydrated cream-style sweet corn consists of a mixture of the cream component and the kernel component to which sugar and salt are added and the proportions of cream component and simulated kernel component may vary between wide limits. For instance, the cream component may vary between 80 and 20% of the dehydrated mixture exclusive of the sugar and salt and the simulated kernel component may vary between 20 and 80%. One acceptable formula may be proportioned about as follows:

	Percent by Weight
Drum-dried cream component	68.5
Simulated corn kernel component	23.3
Waxy maize starch	1.2
Sugar	4.9
Salt	2.1

This formula was rehydrated by adding about 75 cc of boiling water to 25 grams of the dehydrated product, stirred thoroughly, and allowed to stand for 5 to 10 minutes before serving. The resulting product was found to be quite acceptable in all respects and to be the equal in color, flavor, consistency and texture to conventional creamed sweet corn.

ONIONS, GARLIC AND THE LIKE

Fluidized Bed Process

In conventional dehydration of such materials as onion and garlic, customary practice is to slice or chop the fresh or partially dried material and then subject it to continuous drying by slow movement through a tunnel dryer. Drying may be accomplished in stages, a first stage involving concurrent flow of relatively hot gases (e.g., 160°F.) adjacent the wet stock and later stages involving somewhat cooler gases (e.g., 120°F.) moving countercurrent to the partially dried material. Such drying may remove about 99% of the moisture content of the treated material. Final drying is then accomplished in large bins of the material through which warm gases (about 120°F.) circulate for prolonged periods of time.

Tunnel drying is not only wasteful of space, time and manpower, but also is quite wasteful of the material being treated. A principal difficulty is that the material tends to stick to the trays or conveyor belts on which it is being transported. Loss of material due to such sticking is substantial, frequently amounting to as much as 10% of the weight of the material being treated. Losses may arise in part due to mechanical damage, for example, in attempts to rake or otherwise free the stuck material, or scorching may occur. The sticking also necessitates that the trays or belts be washed or scrubbed prior to each reuse so that the useful life of equipment is shortened. When toasting of the material is also accomplished, the higher temperatures involved only serve to intensify the problems mentioned.

A process has been developed by *A. Van Gelder; U.S. Patent 3,063,848; November 13, 1962; assigned to Basic Vegetable Products, Inc.* to provide a method and apparatus useful in dehydration processing by which space requirements can be reduced to a fraction of the requirements and by which drying rates can be increased as much as 10 to 20 times the rates of drying. The process also eliminates the sticking problem. The process is predicated upon the discovery that the effect and rate of gaseous treatment of materials is greatly enhanced and increased when the material is subjected in the treating zone simultaneously to:

(1) A high velocity gas flow passing upwards through the material and in particular a high velocity gas flow sufficient to cause "flotation" or fluidization of the material within the treating zone.

(2) A vibratory motion of such a nature and extent that the material in the treating zone maintains a constant level, exposes fresh surfaces of the material to the flow of treating fluids or gases and prevents sticking to the surfaces with which the material may come in contact, and in particular, a vibratory motion having primarily a vertical component with little or no horizontal component.

(3) Variable gravitational forces sufficient to provide an efficient, effective control over material movement through the treating zone.

According to this process, treatment can be with heated, chilled, dehydrated and/or humidified gases, or any combination of these. As a further variation, treatment may be accompanied by exposure of the material to electromagnetic radiation and in particular to exposure to radiation of the high frequency range (e.g., microwaves or radiations in the infrared range), or by exposure of the material to ultrasonic sound waves, or to a combination of these. The treated material can also be suspended in liquid in the treating zone.

It is contemplated that the process will have specific application to the drying, freezing or toasting of various foods, including the separation of various impurities such as chaff and skins of such foods. By way of illustration, the process is particularly effective in the dehydration of such foods as onions and garlic with removal of from 70 to 99% of the total moisture present in such materials being easily accomplished in one or two hours or less. The dry outer skins of such foods, which have no inherent food value, are simultaneously removed with the exhaust treating gases. Similar dehydration results can be had with other foods such as potatoes, sliced pears, apricots, tomatoes, etc. Drawings and a complete description of the apparatus are given. An example is given using the described apparatus in two separate six-section stages.

Example: Sliced raw onions having an approximate total solids on a weight basis of 14% are fed to the first section of the first stage at a rate of 500 lbs./min. Each dehydrating section of the two-stage system is 20 feet in length and 6 feet in width providing a total dryer length of 240 feet and a total screen area of 1,440 square feet.

The first stage is vibrated at an approximate rate of 12 oscillations per second. The sections of the second stage are not vibrated. The air inlet of each section is through a twill cloth wire screen of 20 x 200 mesh. Heated air is supplied to each section at a rate to produce an air velocity through this screen of 500 ft./min. The moisture present in the air admitted is approximately 52 grains of moisture per pound of dry air.

The temperature of the air supplied to the first two stages of section one is 240°F., to the third section, 200°F., and to the remaining stages of sections one and two, 140°F. The total volume of air used in the processing is 720,000 cubic feet per minute. Such processing produces dehydrated onion slices (moisture content 7%) at a discharge rate of 70 lbs.

per minute, and acts also to effectively separate the lighter onion skins from the denser portions of the onion.

If desired, the produce can be binned as it is removed from the third section of the second drying stage, with final drying being accomplished in large bins through which warm gases are passed at a temperature of the order of 120°F. The approximate moisture content of the material removed at this point in the drying is 20%.

Toasted Onion Products

Toasted onion products are made from the already dehydrated onions by subjecting the dehydrated pieces to heat for long periods of time. The products are highly desirable as food seasonings because of their unique aromatic caramelized flavor and odor of sautéed onions. Toasted onions are regularly added to soups and liver sausage, especially Braunschweiger, liver pudding, and liverwurst.

It has been found that certain varieties of onions, for example, White Sweet Spanish, Grano and Southport White Globe, when dehydrated in pieces or slices in the usual way, may be subsequently readily toasted to give well flavored and uniformly brown toasted onion products. On the other hand, certain varieties of onions, such as, for example, W-45 and Creoles, while highly desirable for making the commercially desired light-colored dehydrated onions, may be readily toasted to yield onion products of uniform flavor or color, and require, even for a nonuniform toasted product, a longer time for toasting.

Since these varieties (producing inferior toasted products) are highly desirable for making the plain dehydrated onion products, because of their light color and high solids content, an unsolved problem has existed on how to bring about uniform toasting of dehydrated onions made from these varieties. The distribution of the reducing sugars and other ingredients which produce the toasting, within the structure of many varieties of onions, is not uniform, and results in a nonuniform toasted product.

A.N. Prater and T.M. Lukes; U.S. Patent 3,098,750; July 23, 1963; assigned to Consolidated Foods Corporation have discovered that toasted onion products of improved seasoning quality and of more uniform appearance may be made by partly or wholly rehydrating the dehydrated onion pieces prior to toasting, and then subjecting them to toasting temperatures for the time necessary to produce the desired color and flavor. It has been further found that the seasoning quality and appearance may be further improved by soaking the hard-to-toast dehydrated onion products in either hot or cold aqueous suspensions of finely divided onion powder, or in aqueous extracts of raw or dried onion, or in raw or blanched onion juice, or in aqueous solutions of reducing sugars, amino acids, or hydrolyzed protein, separately or in combinations.

The toasting effect in every case is a combined function of temperature and time of toasting. The toasting to a desired degree may be hastened by raising the temperature or by lengthening the time. Practically, the toasting temperatures are held in the range from 160° to 350°F. and the time may vary from a few minutes to 30 hours or more.

The amount of the aqueous treating liquid absorbed also influences the toasting results, and may vary from complete or maximum rehydration to a moistening of the surfaces. Dehydrated onions normally have about 4% moisture content. When the raw juice is heated or the extracts of dehydrated onion are made with hot water, the toasted product is a little different in flavor, perhaps described as less harsh. It was also observed that filtered extracts of dehydrated onion powder gave toasted onion products a more uniform appearance, free from spots, as compared to the results using suspended onion powder or slurries.

The preferred toasting agent for addition to hard-to-toast dehydrated onion pieces is raw onion juice derived from varieties of onions (for example Southport White Globe) which are relatively easy to toast. Substantially all of the hard-to-toast varieties are improved as to color uniformity, time of treatment and flavor by moistening the dehydrated pieces with onion juice prior to toasting. The improved rate of toasting when a toasting agent is employed, has also resulted when much higher temperatures (and much shorter times) are used, as in a continuous toaster, where toasting at 300° to 320°F. may be accomplished in minutes.

When toasting those varieties of onions which are difficult to toast to a uniform color, the addition of the toasting agents as above described produced uniformity of appearance and of flavoring quality. The quality of the flavor is also improved because of the shortening of the time of exposure for toasting.

Onion Powder by Spray Drying

A process of preparing dehydrated onion powder by spray drying is described by *P.P. Noznick, R.H. Bundus and A.J. Luksas; U.S. Patent 3,183,103; May 11, 1965; assigned to Beatrice Foods Co.* The process utilizes the following steps.

The raw onions of the desired pungency and of any suitable variety are washed to remove any dirt or debris, rinsed and ground to the desired fineness or size in any suitable reduction equipment which will result in a puree containing particles of the size of approximately 0.040" or less. For best results the particles should pass through a 50 mesh sieve, i.e., have a size not over 0.012".

The comminution should be fine enough that the viscosity of the puree is reduced to a level to permit formation of small spray droplets. Thus at a solids content with 9 to 25% by weight the viscosity of the puree should be between 10,000 and 20,000 cp. The onions can be peeled, e.g., brush peeled to remove skin, or the peeling can be omitted. In the event the skins are retained the initial water washing should be more thorough.

The onion puree at room temperature or below is then mixed with 30 to 40% by weight of dextrin (as dry solids). The addition of the dextrin has been found to improve the drying characteristics of the puree, e.g., to make it easier to dry. The puree is then spray dried in a conventional spray dryer at a temperature between room temperature, e.g., 70°F. or lower, and a temperature of 155°F. The total time of treatment of the puree at elevated temperatures up to 150° to 155°F. should not be over 4 minutes.

It has been found that with Southport White Globe onions it is critical to heat the onion puree containing the dextrin at a temperature of 150° to 155°F. for 3 to 4 minutes, e.g., by use of a suitable heat exchanger of the tubular type. The puree is then spray dried, preferably at room temperature although as indicated higher temperatures can be employed providing the total time at temperatures of 150° to 155°F. is not over 4 minutes. If Southport White Globe onions do not receive the treatment at 150° to 155°F. they develop an undesirable red color.

With onions other than Southport White Globe onions, e.g., Yellow Globe onions, there is no problem of color formation and the heating of the puree containing dextrin to 150° to 155°F. can be omitted. The powder produced by this method is at least 20% stronger in its flavoring effect than powder produced by conventional dehydration of the same onions. When the heat treatment is employed it destroys all pathogenic bacteria and greatly reduces the bacterial population.

Example: Southport White Globe onions were washed and brush peeled, rinsed and ground to below 0.04". The onion puree at room temperature was mixed with 35% by weight of dextrin (as dry solids) and heated at 153°F. for 4 minutes. The puree was then cooled to room temperature (70°F.) and spray dried.

Onions Impregnated with Salt Before Drying

H. Truckenbrodt and G.M. Sapers; U.S. Patent 3,493,400; February 3, 1970; assigned to

Corn Products Company have developed a process for dehydrating precooked toasted onions which reconstitute in 2 minutes or less in boiling water and which upon reconstitution possess a desirable dark brown color and meat-like flavor, free from bitter notes. Dehydrated food mixtures, such as dehydrated soups, stews and casseroles, composed of a powdered base mass and one or more dehydrated precooked foodstuffs can also be prepared by the process. The dehydrated precooked foodstuffs are prepared by: (1) precooking and impregnating 100 parts of the foodstuff with from 2 to 50 parts by weight of sodium chloride and (2) dehydrating the impregnated precooked foodstuff to a moisture content of from 1 to 10% based on the total weight thereof.

The foodstuffs are prepared for cooking in a conventional manner and precooked by any method customarily used in the art. The impregnation step is preferably accomplished by thoroughly mixing the foodstuff pieces with a suitable quantity of dry granular or crystalline sodium chloride for whatever period of time is required to achieve impregnation of the food with the desired amount of sodium chloride.

The impregnated precooked foodstuff is then dehydrated by conventional procedures, typically air convection drying for vegetables, so that the final moisture content of the final dehydrated impregnated precooked food product is from 1 to 10% by weight, preferably from 1 to 5% by weight so that the dehydrated product is compatible with any other moisture-sensitive dehydrated foodstuffs with which it may be mixed.

Toasted onions produced by this process possess a richer brown color, a meatier flavor free of bitterness, and a softer texture than were heretofore available. These advantages are obtained by precooking and impregnating 100 parts by weight of subdivided onions with 2 to 5 parts of sodium chloride and about 1 to 3 parts of monosodium glutamate, for example, by mixing 5 parts of NaCl, 2 parts of MSG and 100 parts of precooked subdivided onions in a rotating drum; partially dehydrating the mixture to a moisture content of from 5 to 20%, say in an air dryer, then toasting the onions at a high temperature for a few minutes. If necessary, the toasted onions may be further dehydrated to a final moisture content of from 1 to 5% based on the total weight thereof, say by bin drying. The flavor and color developed by this process is believed to be the result of a Maillard-type reaction between the amino groups of the MSG and the carbonyl groups of the reducing sugars in onions.

Example 1: Fresh leek was washed, trimmed and cut into approximately ½ x ½ inch dice using an Urschel Model RA Dicer. The leek was precooked for 5 minutes in a steam blancher, cooled, and dipped in an 0.1% sodium sulfite solution for 3 minutes. After sulfiting, 20 parts of sodium chloride were mixed with 100 parts of leek for 5 minutes in a rotating drum. The salted leek and an unsalted precooked control were dehydrated in a cross draft air dryer for 8 hours at 140°F. The dehydrated salted product had a moisture content of 2.9% and a sodium chloride content of 65.4% equivalent to a retained salt level of 19.4 parts of sodium chloride per 100 parts of fresh cooked leek.

Example 2: Spanish onions were peeled, trimmed and diced with an Urschel Dicer Model RA, yielding approximately ¼ x ½ inch pieces. The onion dice were precooked for 9 minutes in a steam blancher. After cooling, 5 parts of sodium chloride and 2 parts of monosodium glutamate were mixed with 100 parts of onion dice for 5 minutes in a rotating drum. The salted onions and an unsalted precooked control were partially dehydrated in a cross draft air dryer for 1½ hours at 140°F. by which time the moisture content was reduced to 10.4%. The partially dehydrated onions were toasted in a cross draft air dryer at 300°F. for 7 minutes and were then bin dried at 120°F. for 16 hours.

The final dehydrated product had a moisture content of 3.92% and a sodium chloride content of 28.2%, equivalent to a retained salt level of 2.9 parts of sodium chloride per 100 parts of fresh cooked onions. After 2 minutes reconstitution in boiling water, the salted onions were found to be superior to the unsalted control (reconstituted in salted water), having a richer brown color, a meatier flavor free from bitterness and a softer texture.

Free-Flowing Onion, Garlic or Horseradish

When heat sensitive products are dehydrated, the finished quality is greatly influenced by time-temperature conditions to which they are exposed. In general, low temperatures, low humidity air and small particle size are employed in an attempt to minimize the damage which occurs during the drying procedure. Low temperatures result in extended drying times; low humidity air is often obtainable only at a great cost in dehumidification equipment; and finally, particle size cannot be reduced to a point where drying is rapid, without excessive damage to the product during the reducing operation.

Normal slicing or dicing can be conducted commercially, but if one attempts to reduce a raw or cooked vegetable, for instance, to very fine pieces, cells are torn, juices are liberated, and a mush of no commercial value results. Volatile constituents are often liberated; the particles cannot be uniformly exposed to drying gases; and bruising and oxidation are inevitable.

In the production of relatively small particles of heat-sensitive dehydrated products, it has been the practice to reduce the product to normal sized dehydratable pieces in the form of thin strips or slices or dice as small as one-quarter inch cubes. These pieces are then compeltely dehydrated while in piece form and after the desired end moisture is reached, the normal sized pieces (strips, slices, cubes, etc.) are reduced by milling to the desired size.

Normal sized strips and slices which are customarily dehydrated are made as thin as practical since this exposes more surface and speeds the drying process. If this is carried to an extreme, the pieces become so thin that they lose their rigidity and mat together defeating the purpose of thinner cut. Also, as more cuts are made on the raw or cooked products, more cells are progressively damaged. This causes poor quality and poor rehydration characteristics.

The bruising, which is normal even with the most efficient cutting or dicing to form these pieces, has an adverse effect on quality, appearance, and color. In addition, there is a particular reason in the case of onion, garlic and horseradish why a minimum number of cut cells is desired. In this specialized group of products, flavor is formed by an enzyme action which is initiated by severing cell walls which liberates enzymes which then react to form the flavoring oils. In such cases, the volatile flavors formed are lost in any type of subsequent dehydration. If the products are cut into thick sections, economical dehydration becomes an impossibility.

J.S. Yamamoto and R.M. Stephenson; U.S. Patent 3,378,380; April 16, 1968; assigned to Basic Vegetable Products, Inc. have developed a process in which onion, garlic or horseradish is first partially dehydrated (until the product is no longer susceptible to bruising, pinking or flavor formation), the dehydration is stopped and the material reduced in a mill to the desired size, and the product is then dried to completion.

The process starts with garlic which has been conventionally sliced to produce normal sized dehydratable pieces about $3/32''$ thick and predried by any drying procedure such as continuous belt, tunnel, vacuum dryers, or the like, to a moisture content of 12 to 14%. With optimum temperatures and humidity, this requires about 12 hours in conventional tunnels.

As much clove skin is removed from the garlic flesh as possible at this point. The material is then fed from the first drying step into a breaker which breaks attached garlic clove skin free from the partly dried slices but not vigorously enough to break the garlic flesh to produce fines. The mixture of loose clove skin and broken garlic slices is fed to a clipper which separates the flow into five streams according to size.

Each stream from the clipper goes to an individual aspirator where pure garlic skin is separated by air and discarded. Several types of aspirators are commonly used, but a spinning disk is preferred which discharges the material in a thin uniform layer into a rising air stream sufficient to carry the skins away and to allow the heavier garlic particles to fall to a collection belt. The skins are collected in a cyclone separation unit and discarded. The cleaned garlic particles are reunited and passed through an inspection station where any defective parts are removed by hand. Electronic sorting equipment can be substituted for removal of any discolored garlic particles.   .

The inspected garlic particles are then fed uniformly into a mill which is equipped with a screen adjacent to the grinding chamber so that all particles must be finer than the screen perforations before they can pass to the next operation. The mill is operated at a speed and the screen opening is so chosen that a minimum of –100 mesh parts are produced and at the same time a maximum percentage of –40 +100 mesh material is formed. A Fitz mill operating at 5,000 rpm and equipped with a $\frac{3}{32}$" round hole perforated screen is preferred.

The mill discharge is preferably air conveyed by an air lift (using the air which cools the drying chamber) to a separation cyclone. A dust collector may be used at this point, if needed. The collected garlic then passes through a star valve in the bottom of the cyclone and is taken immediately to a predryer (if screening is done immediately after milling each stream would go to a separate single dryer).

The predrying equipment in this process serves several functions; the air entrains and removes all residual skin particles and almost all –100 mesh powder, the conditons are adjusted so that entrained material is completely dried by the time it is collected, the predryer dries all pieces below the moisture point where there would be any tendency to be sticky even in hot storage conditions, the predryer removes at least 50% of the total moisture which must be removed in a very short period without raising product temperature appreciably, and the predryer discharges garlic at a moisture where screening is efficient and rapid. The moisture content of material discharged from the predryer varies with the particle size. Conditions are preferably adjusted so that entrained material is approximately 6.25% moisture (a 6.5% maximum is prescribed by the American Dehydrated Onion and Garlic Association). To obtain this, a final fluidized bed dryer is used.

The entrained material from the predryer is caught in a collection cyclone and bag room. Analyses of this fraction show that 50% of the entrained material is blown off in the first 30 seconds, 35% in the next 30 seconds, and the balance in the final 30 seconds. The +100 mesh fraction is largely garlic skin and fibrous fractions, and represents less than 20% of entrained material (2% of total product). This is discarded. The –100 fraction can be combined with the more coarse –100 material which was not entrained. This mixture is a powdered garlic of superior quality.

The discharge from the predryer goes either to final dryers or can be screened at this point and each screen fraction sent to a separate final dryer. Screening at this point eliminates any fine material which could be entrained from the final dryers thereby eliminating the necessity of hoods and dust collection equipment. A further advantage is that particles of like size and like moisture levels are dried together to give a uniform end product with no wet or overdried material.

The final dryers can be fluidized bed, suspension, cyclone, continuous or any other conventional type. The discharge from the final dryers is fed by gravity into a cooler which is similar in construction to the dryers except that room temperature air is forced through the dry product. This is a precaution against "stack burn" which can occur when warm heat-sensitive products are filled into containers under conditions which do not allow for rapid cooling to ambient conditions.

Modifications of the basic process are given, including several processes for converting –100 mesh powdered products into stable larger particles by agglomeration.

Large Onion Pieces

A process for producing large dehydrated onion pieces by submerging large cut onion pieces in a liquid, repeatedly drawing a vacuum over the liquid in order to remove at least one epidermal membrane and subsequently drying has been developed by *J.H. Hume; U.S. Patent 3,607,316; September 21, 1971.*

The process relates to a different means of treating the layers of onion flesh, also called sheaths or fleshy scales, to render them permeable, thereby making them readily available for further processing, for example, rapid dehydration, rehydration, and other treatment with additive ingredients.

The unique layer structure of the onion bulb has heretofore largely determined the size and shape of the piece which is commercially dehydratable and rehydratable. Each layer of the onion is protected on both sides by an epidermis which is substantially one cell in thickness and is a significant barrier to further treatment, processing and water vapor removal during dehydration, and to liquid reentry during rehydration. This unique structure of the onion bulb has forced dehydrators to slice onions into slices about one-eighth inch thick, or to dice or kibble, prior to dehydration.

During dehydration almost all drying, i.e., removal of water, takes place through the cut surfaces. If onions are sliced more than one-eighth inch thick, the rate at which the onions can be dehydrated is reduced sharply and the quality of the pieces being dehydrated is impaired. The slices, rings, or dice produced by this practice are too thin vertically from cut-surface to cut-surface to be used for many normal applications of onions, e.g., in stews, casseroles, salads, snack items, oriental-type and other dishes. For these latter applications, layer or sheath sections of onion produced by quartering rather than slicing, are desired.

Onion rings having a vertical dimension greater than one-eighth inch are also in demand for preparing the popular french fried onion rings. Most restaurants have had to prepare their own from whole fresh onions, since no factory-prepared dehydrated rings of the requisite dimension are available and no commercially practicable process for producing them is known.

It was found that if the epidermis is removed from either or both sides of a piece of onion layer, the piece can be readily processed, for example, rapidly dehydrated or impregnated regardless of the very small cut surface that it may have. The epidermis on the inner or concave side of an onion piece is simpler to remove and comes away without breaking the adjacent flesh cells and hence there is no loss of flavor. This epidermis can be removed by a process of immersion in an aqueous bath which is subjected to repeated applications of vacuum with moderate agitation, permitting simultaneous impregnation with additives, as desired.

The vapor barrier supplied by nature by means of the epidermis on each side of an onion layer or sheath is easily demonstrated by attempting to dry a section of such layer or "shell". Drying starts at the periphery of the shell and slowly progresses toward the center. If high temperatures are employed in an attempt to speed the drying process, the center of the shell rapidly reaches the temperature of the applied gas because there is no adiabatic cooling where no moisture is being released. As a result, the shell "stews" instead of dries, and the product develops a pink or brown discoloration and distorted flavor.

It has been discovered that it is practical to dehydrate such shells or sections of onion layers of substantial vertical dimensions if one or both epidermal membranes have been removed or have been rendered porous enough to allow passage of water vapor during dehydration. Microscopic examination of an onion shell reveals a tightly adhering epidermal membrane on the convex side and a more loosely adhering epidermal membrane on the concave side. Both epidermal membranes are like tight films covering the fleshy onion cells. Removal of the convex epidermal membrane normally ruptures cells. The epidermis on the convex side can be removed mechanically by inserting a point under the epidermis and peeling it back as one would peel a peach. Pieces peeled in this manner dry quickly. The concave epidermis can be removed mechanically in an easier fashion. The inner or concave epidermis appears to be held by a weak cementing substance and when pulled away, the cells beneath remain intact. Pieces of onion sheath with uncut surface dimensions greater than three-sixteenths inch with both epidermal membranes intact dry so slowly that serious quality degradation occurs. On the other hand, pieces with uncut surface dimensions greater than three-sixteenths inch from which either or both epidermises are removed dry as rapidly or more rapidly than onion slices of one-eighth inch or less in thickness. These results indicate that in an untreated piece of onion: (a) both epidermal membranes are practically impervious to moisture vapor, and (b) removal of either or both membranes permits rapid commercial drying of pieces which cannot be achieved satisfactorily with both epidermal membranes intact.

Submerged sections of shell up to 1 inch square were successfully treated to loosen the concave epidermis by the repeated application and release of vacuum. The loosened epidermis was separated from the layer oxygen shell section by agitation. Larger pieces were successfully treated by applying rapidly repeated vacuums of 7" to 20" Hg, followed by release under liquid.

In one test, a whole onion was peeled and top and root ends removed. When this whole onion was subjected to five short vacuum applications of 15 inches and released under water, the concave epidermis of each sheath was loosened. When the sheaths were separated, this epidermis, in many instances, was lightly attached to the convex surface of the sheath. The loosened epidermis was easily removed by stirring in water.

When onion layers and removed epidermal membranes are together in a liquid, care should be taken to isolate one from the other to prevent accidental clinging of epidermal membrane to a surface from which the epidermis has been removed. Such accidental reapplication of epidermis prevents effective dehydration substantially to the same degree as when the epidermis is intact.

With the partial vacuum described above, removal of intercellular air was avoided, and the dried products had the same desirable appearance as those from which the epidermis had been removed by hand. Addition of an appropriate amount of an approved sulfite to the water produces an end product of lighter color. Other approved additives can also be introduced into the onion by being added to the water to realize desired changes in various product characteristics, such as flavor and texture.

Reducing Bacteria Level of Dehydrated Onions with H_2O_2

Attempts have heretofore been made to produce a dehydrated onion powder of satisfactorily low bacteria level from onion pieces of low economic value such as aspirations (dried husks), roots and screenings from other onion processing operations. However, such materials frequently have bacteria levels which are well above the desired maximum of 100,000 (standard, total plate count) for a powdered onion product of satisfactory quality.

J. Gilmore and J.A. Scarlett; U.S. Patent 3,728,134; April 17, 1973; assigned to Rogers Brothers Company have discovered that dehydrated onion pieces, or those of other members of the allium family, which have an unduly high bacteria level can be converted to the the corresponding products of an acceptably low bacteria level by the practice of a method whereby the pieces are first sprayed with a dilute aqueous solution containing 1.3 to 15% by weight of H_2O_2, and then dried to the desired, moisture level. This treatment, while highly effective in reducing the bacteria content of the treated pieces and of powders or other products prepared therefrom, has little if any effect on the flavor and appearance of the product.

While the method can be employed with all members of the allium family, including onions, garlic, leeks, chives and shallots, the principal raw materials with which this process can most usefully be practiced are dehydrated onion aspirations, roots and screenings as separated and recovered in the processing of dehydrated chopped onions and the like. Many of these materials have been in direct contact with the soil, and the bacteria level thereof is, accordingly, well above the acceptable level of approximately 100,000 for the dehydrated onion product or the powder. In this connection, as well as elsewhere herein, the bacteria levels expressed have been determined by the standard, total plate count method.

In carrying out the process, dehydrated onion pieces of the type referred to above are sprayed with an aqueous hydrogen peroxide solution containing from 1.3 to 15% by weight H_2O_2. In general, the lower the bacteria level of the starting material, the lower the concentration of the peroxide solution to be employed.

Peroxide concentrations up to about 15% can be employed with onion pieces of high bacteria content. The amount of peroxide solution which is sprayed onto the onion pieces should be such as to effectively wet the latter without producing an excessive runoff, and good results can be obtained using 1.0 to 3 gallons of the aqueous peroxide solution per 100 lbs. of the dehydrated onion starting material.

Following the step of spraying the onion pieces with the peroxide solution, the wet materials are dried to a moisture content of about 4.25% by weight, or below. This drying can be effected either in a continuous fashion, as by passing the product on a travelling belt through a heated oven or the like, or it can be accomplished by a batch drying procedure. In the latter case a stream of air heated to 110° to 140°F. is passed upwardly through the moist onion pieces, care being taken to stir the mass from time to time to facilitate even drying.

It has been discovered that particularly good results can be obtained by placing the high bacteria dehydrated onion pieces in a suitable feeder bin such as a Syntron feeder, belt feeder, screw feeder, vibro screw or vibra belt unit, and then feeding the material from the bin onto a downwardly inclined shaking table or trough which can be either electrically or mechanically vibrated. The shaking action of the trough distributes the oven pieces evenly across the bottom surface as they are brought to a discharge point from which they fall into a receptical for drying. The peroxide solution can be sprayed onto the onion pieces as they pass along the trough, the shaking action having the effect of turning the pieces to permit of a relatively even application of the spray.

However, in a preferred practice, the spray is directed against the onion pieces as they are in free fall from the trough to the receiving bin. By operating in this fashion, one avoids the difficulties such as particle sticking and the like which are encountered when the particles are sprayed as they pass along the trough. Further, the free fall spraying method insures that all surfaces of each onion piece receive an even application of the peroxide solution.

However the spray is applied, the wet, falling onion pieces can be recovered in a bin having a false screen or other regularly perforated bottom against which the onion pieces come to rest. When the bin has been filled to the desired level, hot air at temperatures of from 110° to 140°F. can then be introduced into the chamber below the false bottom for uniform introduction into the overlying onion mass. The latter is turned from time to time as by a fork or the like as the drying continues to insure that all of the pieces are dried at a uniform rate.

Samples can be withdrawn from the bin from time to time for moisture determination, and when the desired level of about 4.25% or less by weight is reached, the drying operation is discontinued and the pieces can then be ground in a conventional onion mill. If desired, the material can be mixed with other dehydrated onion pieces either before or after the peroxide treating step, with the mixture then being ground to a powder.

When practicing the process in connection with onion materials of low economic value such as aspirations, screenings and the like, it is found that a powdered product having a bacterial count below 100,000, a good optical index and a maximum hot water-insoluble solid content of 30%, can readily be produced. However, when it is desired to provide a powdered product of fancy grade, with a maximum hot water-insoluble solid content of 20%, the aspirations and screenings are blended with from 30 to 70% of regular dehydrated onion pieces, usually before spraying with peroxide, though the mixing can take place after the aspirations and screenings have been subjected to the peroxide treatment.

Example: Aspirations (450 lbs.) having a bacteria count of 1,010,000, as measured by the standard plate count method, are fed through a Syntron vibrating feeder onto a stainless steel mechanically vibrating trough and sprayed as they pass along the trough with 8.5% H_2O_2 from two nozzles at a pressure of 45 psig, resulting in the feeding of 0.208 gallon per minute of hydrogen peroxide to 7.35 lbs. of aspirations per minute. This represents a ratio of 24.2 lbs. of 8.5% peroxide to 100 lbs. of onion. The aspirations drop from the inclined vibrating trough into a bin with a perforated screen in the bottom. The bin is then connected to the hot air (120°F.) supply, where the onions are dried down to 4% moisture and then ground to powder in the onion mill. The ground product has a bacteria level of 64,500 and has a good appearance, as well as a good taste and odor.

Fried Onion Rings

There are many disadvantages to encased, deep fat fried onion rings. Natural onion rings vary in size, thickness, solids, and flavor. Because only the outer rings can be utilized, high wastage results. Coatings are broken loose from the ring and are often tough and greasy. The interiors are soggy and after frying, the exteriors rapidly become soggy.

M.A. Shatila; U.S. Patent 3,761,282; September 25, 1973; assigned to American Potato Co. has developed an encased onion for deep fat frying in which two separate formulations are prepared and then uniquely combined. One formulation comprises a dough of starchy foodstuffs, such as dehydrated instant mashed potato and potato starch. Although the ratio of dehydrated instant mashed potato and starch can be varied widely, it is preferred that from 85 to 95% by weight of dehydrated potato be employed, the remainder being starch. The second formulation comprises film, high solids, discrete onion pieces in a mixture of modified nongluey starch or equivalent filler with additives.

The shell component was satisfactory when the percentage of raw starch admixed with dehydrated instant mashed potato product ranged from 5 to 15%. Starch percentages below and above this range resulted either in shells lacking in cohesiveness or in excessively greasy end products. Although several types of raw starches worked satisfactorily in the above range, potato starch was preferred because of its mild flavor.

All types of dehydrated instant potato products such as potato flakes, potato agglomerates, or Potato Buds gave satisfactory results but potato granules were preferred. A binder, such as raw starch, must be incorporated with such products because they consist essentially of intact potato cells which have very little adhesive quality when formed into a dough with hot or cold water. Various core mixture components were found to give satisfactory results. For example, both Amylaise VII and Textaid gave excellent results.

The percentage of salt and sugar, in the core mixture, can be varied to meet any taste requirement. In general, a sugar level in the range of 1 to 5% by weight and salt in the range of 0.3 to 1.0% by weight of the core mixture were satisfactory. The percentage of onion in the core mixture component can be varied to provide the level of onion flavor desired. Satisfactory products were made varying this component from 10 to 80% by weight of the core mixture, 50 to 80% being the preferred range. The amount of onion used would also be dependent upon the strength of the dehydrated product, as well as the level of onion flavor desired. Test mixtures employing reconstituted domestic dehydrated onions made from special pungent white varieties were preferred.

When dehydrated onion pieces are employed, the amount of water used in reconstituting can be varied to give the desired finished texture. In commercially available frozen onion rings, the onion component is about 90% moisture prior to frying. This level is far above that of any satisfactory coating or breading. In contrast, when employing dehydrated onions in producing the product of this process, the preferred moisture level of the reconstituted onions is reached by using 1 to 1½ parts by weight water to each part by weight dehydrated onion. This results in a moisture content of the refreshened onion of 50 to 60%, and onion solids of 40 to 50%. This level of onion solids is 4 to 5 times the solid content of the average market onion used in making commercial onion rings and allows the product of this process to be fried to a crispy interior without overfrying the shell. The core mixture has a preferred moisture range of 40 to 60% by weight. This factor is undoubtedly largely responsible for the improved texture and mouthfeel of this product as contrasted to commercial onion rings.

Example: Component A (Shell) — A mixture of 90% by weight commercial potato granules and 10% raw potato starch was mixed with hot water, approximately 32 parts by weight potato starch mix and 68 parts by weight hot water at about 190°F. to form a cohesive dough of about 70% moisture content.

Component B (Core Mixture) — Dehydrated ¼ inch diced onion pieces were reconstituted with water using a ratio of 1 part by weight onion to 1.4 parts by weight water. To the reconstituted onion was added a water mixture of Textaid, a modified starch, sucrose, and salt to make a core mixture of the following composition:

	Percent
Textaid	7.0
Sucrose	3.0
Salt	0.3
Reconstituted onion	71.8
Water	17.9

By calculation the above core mixture has a moisture content of about 60%. The shell Component A was extruded in hollow tubular form with a shell thickness of approximately ³⁄₆₄ to ⁵⁄₆₄ inch and with a central void of about ¼ inch in diameter. Simultaneously, by coextrusion, Component B was added to fill the central void of Component A completely with no distortion. The coextrusion was cut in lengths of 5 to 10 inches, looped, and the ends "welded" to form a continuous circular ring. Other lengths were left in stick form. The formed pieces were then parfried in oil at 300° to 350°F. for 40 to 60 seconds, frozen and packaged.

After frozen storage, the parfried product was either finish fried in deep fat at 350°F. for 1 to 1½ minutes or baked in an oven at 450°F. for 10 minutes. The finished product was crisp throughout, onion-like and free from sogginess or gumminess. The fried product was held under a standard restaurant heat lamp for 15 to 30 minutes, after which the product was still crisp and excellent in flavor and mouthfeel. The finished product of this example was uniformly golden brown in color, had a moisture content of 40 to 50% and a fat content of 5 to 20%. In contrast thereto, commercially available frozen onion rings purchased at the local supermarket were found to be greasy. The fat content when prepared according to package directions was 15 to 50%.

PEAS

Prevention of Shrinkage

It is a well-known phenomenon in the field of dehydrated foodstuffs that, on reconstitution, the foodstuff does not swell to its original volume. Thus, for example, a proportion of dehydrated peas tend to remain wrinkled. This fact is believed to be due to permanent damage to the capillary system of the foodstuff during the dehydration operation.

A simple and inexpensive process for dehydrating vegetables, particularly peas, was developed by *R.A.S. Templeton; U.S. Patent 3,281,251; October 25, 1966* which provides results comparable to the use of the expensive freeze drying process. The dehydration is carried out in a conventional way, except that the foodstuff is immersed either before or during the drying process in a heated solution of a crystalline substance, preferably a sugar, for a sufficient length of time to achieve appreciable takeup of the crystalline substance into the foodstuff.

The crystalline substance must be a substance which is readily soluble in water, nontoxic and of acceptable taste and preferably colorless. Sugars such as sucrose, dextrose and maltose are particularly useful but in some cases may be replaced by salts, such as sodium chloride or certain other salts of inorganic acids, which satisfy the above criteria, and also in some cases by the alkali or ammonium salts of organic acids.

The solution of crystalline substance is preferably held at a temperature above 140°F. and preferably about 200°F. (about 93°C.). The foodstuff can however be immersed in a boiling solution provided that the immersion time is sufficiently short to avoid much cooking and structural change of the foodstuff and this will, of course, depend on the nature of the foodstuff under treatment.

One way of carrying the process into effect is to treat a raw vegetable, for example whole peas or carrots cut into the form of dice, in a 5 to 15% sucrose solution at a temperature of 150° to 212°F. for 2 to 5 minutes prior to the commencement of dehydration in a conventional manner in warm air. It has been found with this example that the immersion of the vegetable in a 5% sucrose solution for as short a period as four minutes at 200°F. had a considerable effect on the ultimate product.

A significant improvement in the process can be achieved if it is carried out under alkaline medium at the time of any initial heat treatment, such as blanching in hot water or steam or between that treatment and immersion in sugar or other crystalline solution. The alkaline treatment leads to a significant improvement in the color retained by the final dried product.

In certain vegetables, such as peas, there is present a wax-like layer at or just inside the surface of the peas. By the use of alkaline conditions it is possible to render the foodstuff much more permeable for release of interior pressure due to blanching and to penetration by sugar solution and this is believed to be due to the leeching away of wax-like substances by alkali.

This treatment has been found to produce a product, which, when reconstituted, more nearly resumes the shape of the original material than does a similar product prepared from similar raw material and using otherwise the same dehydration conditions. The reconstituted product is of more regular texture and of relatively larger size and is freer of objectionable hard particles of imperfectly rehydrated material.

The use of an alkali softening step employing a mild alkali, such as sodium carbonate, is found to reduce very substantially the time required for cooking dried peas, as compared with peas dehydrated by conventional methods. The takeup of sugar (sucrose) into the product during dehydration in the case of many fruits and vegetables, such as peas, carrots or apples, results in a product, which on reconstitution, is more palatable than a corresponding product which had not been subjected to the treatment during its production.

It is also found that in many instances the product shows better color retention and may be reconstituted more rapidly. It has been found possible to add additional flavor, such as mint, to the product, simultaneously with the treatment with sugar solution. Furthermore, the color of the product may be improved by the inclusion of natural vegetable colors, such as chlorophyll, anthocyanins and carotenoids, to impart green, red and yellow colors respectively.

Example: Peas were blanched in a sodium carbonate solution having a pH value of 9.5 at a temperature of 205°F. for 3 minutes. This treatment rendered the peas more permeable for the escape of vapor and intake of sugar solution than they would have been if blanched under normal conditions. The peas were then immersed in a 10% sugar solution at a temperature of 150°F. for 5 minutes and dried to a low residual moisture value in warm air. The reconstitution in water after drying was facilitated by the treatment.

Debittering Dried Peas

Raw dry peas (*Pisum sativum* L.) taste bitter and are impalatable without debittering. Pea flour is too bitter to use in food preparations that do not require long moist cooking. The preparation of split pea soup is an example. Here the peas are boiled in water for a substantial time.

G.T. Austin and H.F.Austin; U.S. Patent 3,317,324; May 2, 1967; assigned to Washington State University Research Foundation have described a process for treating dried peas to make them palatable, and to provide a method whereby raw dry peas may be debittered with a minimum of protein denaturation and minimum addition of water so that the treated peas may be ground into flour without drying.

The preferred method that has been employed is to support the peas in an autoclave in the presence of water but out of contact therewith. The temperature in the autoclave is then raised to create steam and this steam is exhausted through a vent until the air in the autoclave is purged. The vent is then closed. This takes about 10 minutes. Then pressure is raised slowly to the desired point (taking approximately 10 minutes) and held at this point for a predetermined time. The peas are agitated during the deaerating and the succeeding periods.

A slow stirring is adequate and it may be done by any suitable means such as a conventional paddle type agitator. After about 10 minutes at 8 to 10 psi, the peas are immediately removed and allowed to cool in the air. They do not appear to soften appreciably and remain separate with very little tendency to stick together. When green peas such as Alaska peas are treated in this manner and cooled they appear unchanged except for a very slight color change. When cool the treated peas may be readily ground into flour without further drying.

The ground Alaska peas produce a flour that can be made into a pea soup by addition of water, salt and seasoning and bringing to a boil. This treated flour can be stored in unsealed polyethylene bags in the dark for periods of up to one year without the development of off flavors. However, the green flour bleaches to white in a day when exposed to sunlight.

Yellow dried raw peas can be treated in the same fashion. However, since the yellow color is not much affected by cooking the treating conditions may be less critical. Dried peas of the First and Best variety (a yellow variety) have been prepared using 10 to 12 minutes time at pressures from 9 to 10 lbs. for optimum debittering without softening.

Dried Split Pea or Lentil Product

The primary use of split peas and lentils today is in the production of soup. In an effort to broaden the acceptability of split peas and lentils, *F. Mader; U.S. Patent 3,738,848; June 12, 1973* has developed a food product to make available the nutritional value of split peas or lentils in a wide variety of diet conditions. This will be particularly advantageous where the human requirements of protein are not met by present diet supplies.

Because of the current trend toward convenience foods and simple cooking techniques, particularly in the more affluent nations, the per capita consumption of split peas and lentils in countries such as the United States has been decreasing over a period of several years. This process attempts to open new markets for these basic farm commodities. It widens the applicability of split peas and lentils far beyond their usual consumption in soup.

Dry peas and lentils do not involve hard shells or skins that must be removed in their util-ization. They are very hard foods in their dry state and are not used as individual food particles. When used in soup, the split pea or lentil is at least partially disintegrated to provide a solution that is rather mushy. In contrast, the food product produced by the process discussed herein is crisp and easily chewable. This process maintains the individual integrity of the split pea or lentil. The end product is most appetizing and answers the need of the human diet for protein values. The product has a very long shelf life.

This process can be carried out in a batch sequence at any desired scale or in a continuous process. The choice involved is dependent upon available machinery and the demand for the product produced. The dry split peas or lentils are first preconditioned by soaking in water without cooking. The soaking of the split peas or lentils is continued for a period adequate to soften and evenly partially expand the cellular structure by swelling and elasti-cally preconditioning the fibers of the individual pea or lentil. The soaking is continued until the split peas or lentils are completely penetrated to saturation; but not oversoaked, causing loss of more than 1% of the carbohydrates.

At saturation the split peas contain about 41% water by weight. Soaking also softens and changes the saponins (glucosides) into a water-soluble state. The soaking preconditions the fibrous and cellular structure so that the split pea or lentil will be able to undergo fur-ther expansion and to subsequently release the moisture under given conditions and to be able to be subsequently hardened or set by heat once full or total expansion has been achieved to absorb a minimum amount of cooking oil.

The soaking water used may be unheated tap water allowed to naturally approach normal room temperature (approximately 70°F.). The soaking period can be shortened somewhat by raising the water temperature, the limit of water temperature being that at which the pea or lentil will disintegrate or lose its individual integrity which is generally considered to be about 180°F.

When using split peas, it has been found that soaking in tap water for a period of 12 to 14 hours without any heating other than normal warming to room temperature accom-plishes the desired result of producing a softened, saturated and partially expanded split pea. When using lentils, it has been found that similar soaking for a period of 7 to 8 hours is adequate.

The preconditioned split peas or lentils are then drained and rinsed in fresh cold water. Excess water is removed by blotting or by passage through a curtain of air. It is not necessary or advisable to completely dry the surfaces. The surface moisture remaining on the individual split pea or lentil should be such that they retain a damp appearance. The split peas or lentils are prepared for the important cooking step. While still saturated, the split peas or lentils are submerged in a cooking oil at temperatures between 375° to 425°F. for a period of less than 3 minutes to rapidly expel the moisture and fully expand the cellular structure and rapidly thermally set and cook the cellular structure in the fully ex-panded condition to form a crisp product having less than 3% moisture and 8% cooking oil by weight. At a temperature of 400°F., the split peas cook in approximately 2 minutes and the lentils cook in approximately 1¼ minutes.

When the saturated or preconditioned split peas or lentils are submerged in the cooking oil at a temperature between 375° and 425°F., the heat from the oil causes the moisture within the cellular structure of the individual split pea or lentil to rapidly form steam. In response thereto the flexible preconditioned cellular structure rapidly expands as the steam pressure increases. The steam pressure increases until the elasticity of the fibers is overcome to re-lease the steam from the cellular structure. At the fully expanded condition the steam is rapidly released without the structure exploding. As the moisture in the form of steam, is rapidly expelled with great force, the steam unites with or links onto the saponins to remove the saponins from the split peas or lentils freeing the split peas or lentils of the bitter or characteistic "raw taste" leaving the split peas or lentils with a pleasant, mild and unique taste.

As soon as the moisture is expelled from the fully expanded split pea or lentil, the heat thermally hardens or sets and cooks the fully expanded cellular structure rendering the expanded split pea or lentil rather brittle giving it a crisp or crunchy characteristic when eaten. The rapid expulsion of the steam from the expanded cellular structure leaves small minute empty cellular pockets throughout the interior of the split pea or lentil.

Preferably, the moisture is reduced to between 1 and 2% by weight in the cooked product. The cooked product preferably has approximately 7% oil content by weight. The product, although somewhat brittle, has sufficient structural body to prevent undue breakage or granulation during normal handling and packaging. The product cannot be considered as being hard or difficult to chew as compared with the starting material. For cooking purposes, a vegetable oil is preferable. One choice that is very acceptable is soybean oil of a conventional quality containing proper additives to preserve freshness in the cooked product.

Any oil plus suitable additives may be used if it is capable of withstanding the high cooking temperatures involved in this process. The choice of a particular oil will vary the flavor imparted to the cooked product. The food product produced by this process is capable of retaining its freshness for several weeks at room temperature in an exposed condition. For prolonged shelf life, vacuum packaging or available preservation techniques can be used.

After cooking has been completed, the peas or lentils are drained to remove excess surface oil. They can be dried either by blotting or passage through an air stream. The product can then be seasoned with salt, pepper, garlic or celery salt, lemon or any other suitable seasoning product. A dry seasoning can be used, which will adhere to the surface of the individual pea or lentil due to the slight oil retained thereon.

The final product is an appetizing nutrient food that can be readily consumed by humans. It is highly satisfactory as a basic nutrient that is readily adaptable to many national diets. It also is highly satisfactory as a snack item to be used as a substitute or replacement for salted nuts, potato chips or the like.

Dehydration of High Maturity Peas and Beans with Added Polyhydric Alcohol

J.P. Savage; U.S. Patent 3,337,349; August 22, 1967; assigned to Lever Brothers Co. has developed a process for the drying of pulses, such as peas and broad beans, of high maturity by perforating the skin of the pulses and impregnating them with a solution of a hydrophilic material.

The maturity of a pea is indicated by its Tenderometer Value (TV) which is a measure of the resistance of its tissue to a crushing force. Herein, a TV greater than 110 is considered a high maturity pea. High maturity broad beans are those having a sugar content of less than 10% by weight of the total bean solids. The hydrophilic material employed is preferably a polyhydric alcohol, for example sucrose (cane sugar) or glycerol. Thus, when sucrose is employed in large amount it is sometimes found to impart too sweet a taste to the pulses, and one way to overcome this disadvantage is to impregnate the pulse with a mixture containing sucrose and common salt (sodium chloride). The preferred temperature of impregnation is from 90°C. to the boiling point of the solution under atmospheric pressure.

The hydrophilic material will usually form from 15 to 60%, and preferably from 20 to 40% by weight of the impregnating solution. In general, the treatment time should be at least such as to allow the solution to pass the pulse skin and diffuse well into the cotyledon; but it should not be so prolonged that the pulses are made unduly tender. Diffusion of the impregnating solution into the cotyledons is facilitated if the pulse skin is perforated prior to impregnation.

The drying step can be perforated by any conventional means, but most suitably by drying in a stream of warm or hot air. After impregnation the drying temperature employed can be high (for example, 110° to 160°C.) in comparison with temperatures normally employed for pulses, because pulses treated according to the process show a reduced tendency for chemical changes such as browning to occur at high temperatures. It is often advantageous to use high drying temperatures because dehydration then proceeds rapidly. Moreover, the use of a high drying temperature leads to the development of an open structure in the pulse, and this markedly facilitates rehydration. The process is illustrated by the following example. All percentages are by weight.

Example 1: This example shows the improvement that can be obtained by applying the process to the drying of pricked peas of relatively high maturity. A batch of peas of the variety "Dark Skinned Perfection," having TV 125, was washed and separated into two fractions by means of a sieve having holes of 10.6 mm. diameter. The fraction retained by the sieve had a TV of 139. This fraction was divided into two lots C and D which were treated separately as described in (a) and (b) below.

(a) The peas forming lot C were mechanically pricked with prickers (pins) of 1.25 mm. diameter. After blanching and rinsing the peas were soaked for 30 minutes in a solution, maintained at about 90°C., containing 30% of sucrose, 2% common salt, 0.04% Na_2CO_3 and 0.2% Na_2SO_3.

After this impregnation, the peas were drained for 5 minutes and then air dried in a through up-draft dryer to a moisture content of about 8%. Drying was carried out in two stages, dehydration to about 30% moisture with an air speed of 3 meters per second (mps) and a temperature of 65°C., and from 30 to 8% moisture with an air speed of 1 mps and a temperature of 50°C.

The dried peas were almost as large as they had been when fresh; they showed little wrinkling of the skin, and had an attractive color. The peas were rehydrated by cooking for 15 minutes in boiling water; after draining they were plump and fresh looking and they were tender throughout without being excessively soft. The total content of sugars (comprising not only those sugars added as above, but also such sugars as the pulses contained before the impregnation step) of these peas, on a dry solids basis, was 41%; of this, 29% was in the cotyledons and 12% in the skin.

(b) For purposes of comparison, the peas forming lot D were treated generally as described in (a), except that no soaking step was employed. The peas obtained after drying were only about two-thirds the size of the fresh pulses, and had the characteristic wrinkled appearance of ordinary dried pricked peas.

The peas were cooked for 15 minutes in boiling water. After draining, they had a less attractive appearance than the sugared sample, and a very large proportion of them were found to have a hard core of tissue in the cotyledons, showing that rehydration was not complete. The total sugars content of the unsugared peas, on a dry solids basis, was 18%, the cotyledons containing 13% and the skin 5%.

Example 2: This example illustrates the application of the process to peas that were partly dried before the sugar impregnation treatment. A sample of peas having TV of 139 was obtained and pricked as described above. These peas were blanched in a hot (98°C.) aqueous solution containing 0.5% Na_2CO_3 and 0.7% Na_2SO_3 (a higher level of Na_2SO_3 being used than in the previous example, where further sulfite was added in the sugar treatment before drying). The blanched peas were dried to 50% of the blanched weight in a through up-draft dryer at 50°C., with an air speed of 3 mps. The peas were then soaked in a hot solution (90°C.) containing 30% sugar, 2% common salt, 0.04% Na_3CO_3 and 0.2% Na_2SO_3 for 30 minutes and, after draining, dried in a through up-draft dryer.

The dried sugared peas obtained were indistinguishable from those prepared in Example 1 and after cooking for 15 minutes in boiling water gave plump, fresh looking peas which were tender throughout. The total sugars content of these sugared peas on a dry solids basis, was 44%, distributed as follows: cotyledons, 26%; skin, 18%.

SOUP MIXES

Ionizing Radiation to Decrease Cooking Time

In the case of dehydrated soup mixes containing dehydrated vegetables, the directions generally require that the contents be added to boiling water and that the boiling be continued for 10 minutes after which the soup is ready for consumption. The relatively long cooking time required detracts from the convenience of using dehydrated vegetables. This is particularly a problem in the case of dehydrated soup mixes, where the feature of quick preparation is attractive to the consumer.

C.W. Schroeder; U.S. Patent 3,025,171; March 13, 1962; assigned to Thomas J. Lipton, Inc. has found that dehydrated vegetables which have been subjected to ionizing radiation such as gamma rays or electron beams may be rehydrated to an edible and desirable texture within a relatively short period of time, for example, 3 minutes. It has also been found that dehydrated soup mixes may be prepared with these irradiated vegetables. These improved mixes may be prepared for consumption within as little as one minute in boiling water, thus making the soup more convenient to prepare.

The vegetables which may be treated according to this process include dehydrated lima beans, okra, corn, potatoes, green beans, celery, green and red bell peppers, peas, carrots, beets, onions, lentils, leeks, and cabbage. Vegetables derived from roots, leaves, stems and some fruit are benefited to the greatest extent when processed according to this process. High-starch vegetables, such as legumes and potatoes, respond to a lesser extent, but are nevertheless benefited in most instances.

The process of dehydration effects a substantial removal of water from the vegetable and in the usual commercial practices the moisture content of a given vegetable is reduced to the point where no significant change in quality occurs during the storage of the dehydrated vegetable at room temperature for a considerable length of time. Thus, the term "dehydration," as used here, does not mean a complete removal of water, and dehydrated vegetables generally contain from 1 to 20% water in their dehydrated state.

Either the radioactivity resulting from the decay of radioactive materials, such as the radioactive isotopes or the electron beam from an electron accelerator, may be employed as a source of ionizing radiation. Where alpha, beta or neutron radiation occurs simultaneously with gamma radiation, the source may be shielded to absorb these less desirable types of radiation and to insure a uniform absorption of ionizing radiation throughout the mass of material being treated. It is desirable, when treating dehydrated food products that the photon or electron energy level be no more than 10 mev.

A higher irradiation level is generally obtained in a shorter period of time by using the electron beam of an electron accelerator such as the Van de Graaff type than with gamma sources. One such electron accelerator of medium size operated at 2,000,000 volts, and a maximum of about 250 microamperes beam current. For convenience, this type of source is preferred, although it should be recognized that any suitable source of ionizing radiation may be employed.

In irradiating dehydrated vegetables, it is preferred that they be arranged so that they are substantially uniformly treated with the ionizing radiation. For example, where an electron accelerator is employed, the samples may be placed in suitable containers and arranged in a layer having a thickness which is no more than the maximum range of the electrons. With a Van de Graaff accelerator operating at 2,000,000 volts, for example, a layer of about 0.8 gram per square centimeter is preferred for economical utilization of the beam energy while maintaining fairly uniform treatment.

The containers of vegetables are loaded onto a conveyor belt moving at 40 inches per minute and passed through the electron beam. The electron beam is preferably scanned back and forth across the conveyor belt at, for example, 200 cycles per second with a 7-inch sweep. These conditions will give a sufficiently uniform 2,000,000 "rep" dose when using the full 250 microampere beam current. A "rep" is a Roentgen-equivalent-physical, which is a unit of absorbed energy equal to 83.3 ergs per gram of irradiated product. One million rep equals 3.8 kilowatt-seconds per pound.

In preparing vegetables for treatment, it is generally preferred that they may be of a uniform particle size and that their moisture content be reduced to a level where the vegetable may be stored at room temperature without undergoing deterioration. The maximum level will vary with each vegetable, and the values are generally well known in the art. Dehydrated vegetables frequently contain approximately 4% moisture.

The dosage of ionizing radiation required for a given decrease in rehydration time will vary with different vegetables. In general, the higher the dosage given, the greater will be the reduction in cooking time. A dosage range of about 1 to 11 million rep has been found satisfactory in most instances. However, certain vegetables require a more restricted range. Dehydrated onion flakes, for example, are satisfactorily irradiated within a dosage range of 0.25 to 2.5 million rep. The following examples illustrate the process.

Example 1: 15 g. of dehydrated green beans, which had been cut into a uniform size prior to dehydration, were placed in a paper-lined aluminum box 6" x 12" x 4" deep. The dehydrated green beans had a moisture content of 2.41%. The beans were then passed through the electron beam of a Van de Graaff electron accelerator operating at 2,000,000 volts. The electron beam was scanned back and forth at 200 cycles per second. The scanned width was 7 inches. A radiation dosage of 4.8×10^6 rep was provided in the beans. After periods of 2, 5 or 8 minutes of cooking in boiling water, the treated beans were more tender than a control of untreated beans cooked for 10 minutes. In addition, the treated samples had substantially the same flavor as that of the control.

Example 2: 15 g. samples of dehydrated cabbage having a moisture content of 4.2% were irradiated according to the procedure outlined in Example 1. The cabbage was subjected to two dosage levels, namely, 0.6×10^6 and 1.2×10^6 rep. For each dosage level, a cooking time of 5 minutes provided a cooked cabbage which was considerably improved in tenderness over the control which was cooked for 10 minutes. Cooking either of the irradiated samples for 3 minutes provided cabbage of about the same texture as the control cooked for 10 minutes, while cooking for 7 minutes provided a product which was too soft. The irradiated samples had substantially the same flavor as that of the control.

Dehydrated Soups Containing Xanthomonas Colloid

R.L. Edlin; U.S. Patent 3,694,236; September 26, 1972; assigned to Kelco Company has developed a process for making dehydrated food products which can be rehydrated or hydrated rapidly and completely, and which upon hydration will possess improved consistency, texture, and mouthfeel.

The process consists of incorporating a relatively small amount of a Xanthomonas hydrophilic colloid into food products which are to be rehydrated. The Xanthomonas colloidal material is preferably mixed in an aqueous solution of food material prior to the dehydration thereof. It may also be admixed with a dehydrated food prior to the hydration or rehydration thereof. Suitable amounts of a Xanthomonas colloid for use in accordance with the process have been found to be in the range of about 0.01 to 1.5% of the weight of the aqueous food suspension to be hydrated, or the weight of a dehydrated food material including the weight of the aqueous medium required for rehydration. A preferred range of such a Xanthomonas colloid for use is from 0.1 to 0.9%.

The colloid used in the process is produced by the bacterium *Xanthomonas campestris.* This colloidal material is a polymer containing mannose, glucose, potassium gluconate and acetyl radicals. In such a colloid, the potassium portion can be replaced by several other cations without substantial change in the property of the material for the instant purposes. This colloid, which is a high molecular weight, exocellular material, may be prepared from the bacterium *Xanthomonas campestris*, by whole culture fermentation of a medium containing 2 to 5% commercial glucose; organic nitrogen source, dipotassium hydrogen phosphate and appropriate trace elements.

Example: In the preparation of a dehydrated vegetable soup mix, selected vegetables are chopped and then dehydrated with the least heat necessary to drive off most of their natural moisture. This keeps them uncooked and retains their raw flavors, colors and vitamin content. These are then mixed as follows:

	Pounds
Dehydrated chopped vegetables	5-10
Vegetable proteins (preferably dried Dunlap, Mandarin or Manchus soy beans)	15-25
Monosodium glutamate	4-8
NaCl	10-15
Sodium nitrate	$^1/_{40}$-$^1/_{20}$
And as desired:	
Commercial milk powder	20-30
Ground sucrose	3-7
White wheat flour (or cornstarch equiv. to this starch)	20-25
Condiments: pepper (black and white), marjoram, thyme, etc.	*
Xanthomonas campestris colloid	**0.02-0.06

*To taste.
**Sufficient to constitute about 0.05% of the weight of the other ingredients.

The resulting product will be observed to have excellent consistency and mouthfeel and viscosity, that is when reconstituted with the conventional amounts of water. It will also exhibit a flavor and color, when dehydrated, that is quite close to the undehydrated soup. The Xanthomonas hydrophilic colloid (i.e., the *Xanthomonas campestris* colloid) used here can be added either prior to dehydration or thereafter, for instance, upon reconstitution. However, adding it prior to dehydration will be especially advantageous since it will require no extra processing steps and will result in a superior product, having all of the necessary ingredients except the water added upon reconstitution.

Green Pea Soup

H. Andrews and S.K. Bedrosian; U.S. Patent 3,752,677; August 14, 1973; assigned to Thomas J. Lipton, Inc. describe a process for making a dry green pea soup mix of improved flavor and color stability that is readily rehydratable on addition of boiling water within one minute to provide a pea soup.

The pea soup mix, which is rehydratable within less than one minute upon the addition of boiling water and which is, at the same time, stable to color and flavor changes during storage over periods as long as a year, is provided by first preparing the bulk of the green pea solids to be used in the mix as green pea flakes, and by thereafter mixing these flakes with a small portion of green pea solids, corn syrup solids and other flavoring ingredients to form a blend. The blend of ingredients is then moistened, agglomerated and dried to form the pea soup mix.

The first step in forming the soup mix is the preparation of the green pea flakes. Washed green split peas, which are dehydrated peas containing only the cotyledons with the seed coat removed, are cooked with approximately three times their weight of boiling water, preferably together with salt, potassium carbonate and locust bean gum. After cooking for about 20 minutes, approximately 4% by weight of the split peas of hydrogenated cottonseed oil or other suitable hydrogenated vegetable oil is added, together with a flavoring mixture and a cold water slurry of raw potato starch. The starch is added at a level of about 4% in an amount of water equal to about ¼ of the weight of the pea solids present. After cooking for an additional 10 minutes, the slurry is pumped to a drum dryer where it is dried, preferably at 70 lbs. psi steam pressure, with a roll speed of 3 rpm. It is preferred that the slurry be dried to provide flakes having an average moisture content of about 5.5%.

The green pea flakes prepared above are then combined with corn syrup solids, pea powder, salt, flavoring, preferably including smoked yeast, sugar, monosodium glutamate, vegetable protein and a small amount of pregelatinized starch. The corn syrup solids, such as maltodextrin, are preferably employed in the range of 10 to 15% by weight of the final mix, a level of 12% being preferred. This material assists in the agglomeration process and is used to improve the stability of the mix in hot water.

The pea powder is preferably added at the level of 3 to 8%, preferably about 5% by weight of the mix. The pea powder consists of green pea solids in the form of a free-flowing powder having a maximum moisture content of about 10.5% and a maximum ash content of 3%. Approximately 99% of the particles will pass through a U.S. 30-mesh screen. The flavorings used are those characteristic of pea soups in general, and include salt, spices, artificial ham flavorings, monosodium glutamate, vegetable protein, a small amount of sugar and smoked torula yeast. A pregelatinized tapioca starch is also used in the mix at a level ranging from 1 to 2%, preferably 1.6%, to help provide the desired consistency in the final product.

The green pea flakes are combined with the other ingredients of the mix with suitable blending techniques, care being taken that the flakes remain substantially intact. After the blending operation, the mix is agglomerated. Either steam or hot water may be used to increase the moisture content of the mix sufficient to cause some degree of clumping. After this has taken place, the mix is dried to its former moisture content, i.e., 3 to 4%, and the clumps are broken up. Preferably the mix is passed through a 12-mesh screen to provide a coarse sandy powder. In one preferred agglomeration procedure, the mixture is airveyed through a stainless steel tube, where it is continuously mixed with wet steam to raise the moisture content to 11 to 12%. The corn syrup solids become sticky at this moisture level and cause particles to lump together.

The lumps are then dried by passing them through an enclosed vibrating conveyor with a fine mesh base allowing a stream of heated air (180°-190°F.) to pass through the material drying it to a moisture content of 3 to 4%. The lumps are then sifted through a 12-mesh screen. The process is further described by the following example.

Example: Green pea flake was prepared from green split peas. 1,000 lbs. of washed U.S. No. 1 green split peas were added to a cooking kettle containing 2,750 lbs. of boiling water, 28 lbs. of salt, 2 lbs. of locust bean gum, and 2 lbs. of potassium carbonate. This mixture was boiled for 20 minutes. 39 lbs. of hydrogenated cottonseed oil was then added, and after 10 additional minutes of boiling, a warm water extract of a spice mixture comprising 340 g. of cloves, 241 g. of thyme and 100 g. of bay leaves were added together with a cold water slurry comprising 37 lbs. of raw potato starch in 250 lbs. of water. After the ingredients were thoroughly mixed, the slurry was pumped to drum dryers and dried at 70 psi steam pressure with a roll speed of 3 rpm. The material was dried to an average moisture content of 5.5%.

The dry material leaving the drying drum was collected and passed through a flaker screen having ⅜" openings. The composition of the green pea flakes obtained as a result of this process was as follows:

	Percent	
Green pea solids	84.76	
Hydrogenated cottonseed oil	3.75	
Potato starch	2.96	
Salt	2.70	(continued)

	Percent
Locust bean gum	0.17
Potassium carbonate	0.16
Clove	Trace
Thyme	Trace
Bay leaves	Trace
Moisture	5.50

A dry blend of the green pea flake is then made with the other ingredients of the mix in accordance with the following formula:

	Percent by Weight
Pea flakes	72.0
Corn syrup solids	12.0
Pea powder*	5.5
Spice, salt and other flavorings	4.5
Smoked torula yeast	3.0
Pregelatinized tapioca starch	1.5
Sugar	0.5
Monosodium glutamate	0.5
Vegetable protein hydrol, 4 BE	0.5

*Green pea solids in the form of a free-flowing powder having a maximum moisture content of about 10.5% and a maximum ash content of 3.0%. Approximately 99% of the particles are small enough to pass through a U.S. No. 30-mesh screen.

35 grams of the above-described mix were mixed with 6 fluid ounces of boiling water and were stirred for 30 seconds. A thick, flavorful, fully-cooked soup was formed which was free of lumps, hard pieces and uncooked material. In addition, the mix retained its bright green color and excellent flavor over an extended period of storage.

SWEET POTATO

Rapid Dehydration

H.J. Deobald and T.A. McLemore; U.S. Patent 3,046,145; July 24, 1962; assigned to the U.S. Secretary of Agriculture describe a process for the preparation of a dehydrated sweet potato product that retains the color, nutritional, and flavor characteristics of sweet potatoes. The dried product may be consumed as is, as a confection, or the product may be reconstituted instantly by the addition of hot water to form a cooked sweet potato puree useful for pie fillings, puddings, baby foods, candy filling, and the like.

Sweet potatoes are unique in certain respects. The carbohydrate content and the carotenoid content of sweet potatoes is very high in comparison to other vegetable products. Because of its unusual characteristics, modifications in dehydration and subsequent storage operations as conventionally practiced for vegetable products are required.

During processing, a significant proportion of the starch is hydrolyzed to sugar. This fact has been established by chemical analysis. Total sugar increases more than twofold in the sweet potato product as a result of the processing herein described. It is believed that if the conversion of starch to sugar was complete in the process, the material could not be dried successfully. However, as a result of the gelatinization of small quantities of starch that remain unhydrolyzed, it is possible to attain a coherent product that can be successfully dried by conventional methods (drying rolls or drums). If, on the other hand, all of the starch was gelatinized and complete cell rupture took place, without subsequent hydrolysis of the bulk of the starch to sugar, it is believed that the dehydrated sweet potato product would case harden and make rehydration of the dehydrated product difficult, if not impossible. An example will best illustrate the process of dehydration and storage.

Example: The sweet potatoes were washed, preheated in water at 160°F. for 30 minutes, peeled by immersion in approximately 20% by weight aqueous caustic solution, heated to 220°F., and subsequent passage through a rotary peeler and washer, and finally trimmed to remove fibrous ends, scar tissue, and other unwanted material. The peeled and trimmed sweet potatoes were then precooked for 25 minutes at 212°F. Water was added to the peeled and cooked sweet potatoes in an amount sufficient to adjust the solids content of the mixture to approximately 18 to 25%. The added water contained sodium bisulfite and sodium sulfite in an amount equivalent to 0.02 gram of sodium bisulfite and 0.06 gram of sodium sulfite per pound of potatoes to prevent discoloration.

The mixture of peeled and cooked sweet potatoes and sodium bisulfite-sodium sulfite containing water was then made into a puree by comminuting in a commercial type beater-mixer. The pureed sweet potato material was fed to a double drum dryer, the rollers of which were internally heated with steam at approximately 75 pounds per square inch gauge pressure. (Temperature of the dryer roll surface under these conditions is approximately 300°F.) Spacing between the dryer rolls was held between the limits of 0.007 and 0.010 inch. Roll speeds were adjusted so that residence time of the sweet potato material on the drying rolls was approximately 17 seconds.

These operating conditions produced a dried sweet potato product containing from 2 to 4% moisture and a product, the physical appearance of which was a more or less intact sheet. The dried sweet potato sheet was removed from the drum dryer by a doctor blade, and packaged immediately in air- and moisture-tight containers in an atmosphere of nitrogen.

The sugar content of the dried material immediately after removal from the rolls averages close to 40% by weight. Softened by heat, this predominantly carbohydrate material is quite thermoplastic and can, if desired, be easily formed. Alternatively, the dehydrated material from the drying rolls may be allowed to cool somewhat, following which cooling the thin sheets of dehydrated product may be easily reduced to flake or powder form. The dehydrated sweet potato product is immediately usable in the dried flake or powder form, but for deferred use, the dehydrated product must be packaged in containers that are impervious to both air and moisture. The oxygen content within the sealed container must preferably be about 2%, and in no case should the oxygen content of the sealed storage container exceed 3%.

Packaging of the dehydrated material in gas and moisture impervious containers is essential for preserving the product. Such packaging is necessary because of the susceptibility of the product to oxidation. The dehydrated product, if exposed to the atmosphere after cooling, takes up moisture and becomes gummy.

Addition of Amylase

A critical problem associated with the production of sweet potato flakes or granules from uncured and/or starch roots is the high starch content in relation to the amount of sugar and other soluble solids present. When sweet potato puree containing a high starch-to-sugar ratio is dried on a drum dryer, the puree fails to stick properly to the drums and the resulting product is a very thin, porous, lacy sheet with a low bulk density. By proper control of the starch-to-sugar or starch-to-soluble solids ratio in sweet potato puree through enzymatic starch conversion, a finished dehydrated product of superior quality can be produced with greater efficiency and with improved flavor and handling characteristics.

A process developed by *M.W. Hoover; U.S. Patent 3,169,876; February 16, 1965; assigned to North Carolina State College* provides a high quality dehydrated sweet potato flake from uncured and/or starchy roots that contains a higher than normal sugar content and possesses an improved flavor, texture and color by supplementing the naturally occurring saccharifying enzymes in the sweet potato with amylase from other sources along with the addition of a color preservative consisting of sodium acid pyrophosphate and tetrasodium pyrophosphate.

Briefly stated, the process comprises first the conventional processing technique of peeling the sweet potato roots, trimming, cooking in water or steam at atmospheric pressure or higher, and pulping through a screen with openings ranging from 0.033 to 0.090 inch to break up the large particles and to remove excessive fiber and other undesirable matter. Sodium acid pyrophosphate or a mixture of sodium acid pyrophosphate and tetrasodium pyrophosphate (generally 3 parts sodium acid pyrophosphate to 1 part tetrasodium pyrophosphate) is added to the cooked sweet potatoes or puree at the rate of 0.05 to 0.8% on a dry weight basis.

Sodium acid pyrophosphate is slightly more effective than tetrasodium pyrophosphate for controlling discoloration in sweet potato products; however, when concentrations of sodium acid pyrophosphate higher than about 0.2% on a dry weight basis are used, the pH of the product is lowered to the point where the finished product has an acid flavor. Although tetrasodium pyrophosphate is not quite as effective in controlling discoloration as the acid form, it serves a dual purpose of helping control the pH of the finished product and at the same time contributes to the preservation of color. The phosphate color preservatives can be added at any time during the manufacturing of sweet potato flakes. However, results indicate that the preservatives should be added and mixed with the product after the roots are cooked.

After the cooked roots are pulped, a portion of the puree is cooled to a temperature ranging from 110° to 170°F., but generally 130°F. Saccharifying enzymes (alpha and beta amylase) are added to the puree at a concentration of 0.01 to 0.2% based on the weight of the puree. The conversion time required after the enzymes are added generally ranges from 10 to 120 minutes, depending upon the starch content, enzyme concentration and temperature of the puree which preferably is maintained at a temperature of from 125° to 140°F.

Since it is difficult to closely control the enzyme action, best results are obtained when only a portion of the puree to be dried is treated with the enzymes and the reaction is allowed to go to at least 90% completion in the treated portion. The treated and non-treated puree is then recombined into the desired proportions prior to drying on the drum dryer. Almost or complete saccharification of the starch in the treated puree is allowed to occur prior to adding it back to the untreated puree. The amount of add-back or treated puree added back to the untreated material depends upon the starch-to-sugar ratio of the nontreated puree. The amount of enzyme treated puree added back to the untreated material generally ranges between 10 and 60% of the total amount of puree going to the drum dryer.

It is desirable, but not absolutely necessary, under all conditions to raise the temperature of the enzyme treated puree to above 190°F. prior to mixing with the untreated material in order to inactivate the enzymes. In the flaking operation, the temperature of the puree reached on the drums is sufficient to inactivate the enzymes.

The mixture of treated puree and untreated puree is then preferably dried on a drum dryer operated with a steam pressure between 65 and 90 psi. The mixture is preferably dried until a flake having a water content of from 2 to 4% by weight, based on the total weight of the dehydrated product, has been produced.

Example: A batch of sweet potatoes of the Nugget variety was peeled for one minute in a steam peeler with a steam pressure of 100 psi. The peeled roots were then washed, trimmed, cut into slices ¾" thick and cooked in steam for 25 minutes at atmospheric pressure. A mixture containing 3 parts sodium acid pyrophosphate and 1 part tetrasodium pyrophosphate was added to the cooked potatoes at the rate of 0.3% based on the dry solids of the sweet potatoes. The cooked product was mashed or pulped through a screen containing 0.06" openings.

The puree was then pumped into two stainless steel tanks. One of the tanks served as a conversion tank to which the puree and saccharifying enzymes were added, the other stainless steel tank was used for the nontreated portion. The sweet potato puree to be treated was pumped through a heat exchanger where it was cooled to approximately 130°F. prior to entering the treating tank. The enzymes were added to the 130°F. puree in the conversion tank at the rate of 0.05% based on the weight of the puree. After 30 minutes the enzyme treated puree was pumped through a second heat exchanger where the temperature was raised to 200°F. in order to inactivate the enzymes prior to recombining the treated and nontreated puree in the desired proportions.

The two portions, treated and nontreated puree, were recombined at a ratio of 1:1 by weight. The puree was then dehydrated to 2% moisture on a drum dryer with 75 psi steam pressure in the drums. The sheet of dried product was broken and ground into flakes or granules. The dehydrated sweet potato flakes or granules were packaged and sealed in cans under nitrogen in order to stabilize or preserve the flavor of the product during extended shelf life.

Elimination of Peeling Before Cooking

It is well known that in the process of preserving foods, or in the preparation of so-called convenience foods from natural (raw) food products which have a peel, this peel must be removed. This is usually done before the cooking operation. Formerly, hand peeling was practiced, but because of its high cost, commercial procedures employ one or more of the following methods: (a) steaming; (b) the use of 20 to 25% aqueous alkaline solutions, such as lye, at temperatures of about 215° to 220°F.; or (c) some method of abrasion. When sweet potatoes are the natural (raw) food being processed, peeling prior to cooking is usually done, because the cooked sweet potato would be too soft to permit steam, lye, or abrasive peeling after cooking.

A process is described by *S.P. Koltun, E.J. McCourtney and J.J. Spadaro; U.S. Patent 3,394,012; July 23, 1968; assigned to the U.S. Secretary of Agriculture* which, by eliminating the peeling of sweet potatoes before their dehydration, prevents discoloration, either by enzymatic activity or by oxidation, reduces equipment and utility costs, reduces labor requirements, and eliminates lye and other chemical costs. Most important, the loss of nutrients is decreased and the yield of edible material is increased. An example will illustrate the process.

Example: The raw sweet potatoes are washed and trimmed very lightly, i.e., the ends are removed and any bad or soft spots are cut out. The washed and trimmed sweet potatoes are then strip-cut into strips of about 0.125" to 0.32". These strips are then added to a kettle containing boiling water, the amount being necessary to adjust the solids content of the sweet potatoes to about 20 weight percent. Cooking time for this mixture is about 10 minutes, after which the cooked sweet potatoes are mixed and fed slowly into the pulper. During passage through the pulper, the peel and fiber are separated from the edible portion of the solids. The peel and fiber are then separated from the pulp by passing the cooked product through a metallic screen. A screen having a mesh size of about 0.045" is preferred. Such a screen is optimum for discharging the bits of peel and fiber from the pulper. The resultant discharged puree is free from peeling or fiber.

It is a critical feature of the process that the pulper be a finisher-type pulper which has a secondary discharge opening for the peel-fiber portion of the cooked product. The sweet potato puree (pulp) is then pumped onto a double drum dryer for dehydration. The drums are heated with steam at a gauge pressure of approximately 75 pounds per square inch (psig) and the clearance between the two drums is adjusted to approximately 0.010" when the drums are hot; the retention, or residence time, of the puree on the drum is about 17 seconds, during which time an orange-colored, crepe paper-like sheet of dried sweet potato product is formed. The moisture content of this sheet is about 2 to 4 weight percent, and the product is commercially acceptable.

TOMATOES

Evaporation of Serum

R.A. Miller; U.S. Patent 3,172,770; March 9, 1965; assigned to Campbell Soup Company describes a product for the preparation, from fresh tomatoes, of a concentrate which, when used after storage, will produce tomato soup, and like products, in which the taste and color will be substantially identical to similar products made from fresh tomatoes.

Ripe tomatoes are first prepared in the conventional manner and passed through cyclones or centrifugal dehydrating equipment to separate the juice from seeds, skins, and other waste material. The tomato juice is then passed through a centrifuged separator to separate the juice into two fractions, the first fraction, representing from approximately five to fifteen percent of the original tomato juice, contains substantially all of the insolubles. The second fraction, representing approximately eight-five to ninety-five percent of the original tomato juice, contains the solubles which, conventionally, are referred to as tomato serum.

After the tomato juice has been separated into the fractions, the first fraction containing the insolubles is cooled, packed in containers of suitable size, and quick frozen. It is important that this fraction, after it is frozen, be maintained and stored in its frozen condition until used.

The second fraction, that is, the fraction containing the solubles, is concentrated to approximately one-tenth of its original volume. Such concentration can be accomplished by well-known low temperature evaporation processes or by freeze concentration processes in which the water phase is frozen into small ice crystals which are then removed by centrifugation. The second fraction, when properly concentrated, will be approximately 50° Brix. The concentrated tomato serum contains the fresh tomato flavor. Where properly preserved, the fresh tomato flavor in the concentrate can be retained for extended periods of time.

To preserve the concentrated tomato serum and to retain the fresh tomato flavor in the concentrate, the concentrate may be stored at a temperature below 32°F., or may be preserved by adding salt to the concentrate in an amount sufficient to produce a salt concentration of approximately 80% saturation of the free moisture. Where it is intended to recombine the concentrate, or second fraction, with the first fraction after a limited period of storage, it may be preferred to leave the concentrated second fraction unsalted and to store the unsalted concentrate at a temperature below 32°F. However, where storage is intended for prolonged periods, it is preferred to salt and fill the concentrate into suitable containers and to store the salted concentrate at room temperatures.

Preferably the tomatoes are processed and the two fractions prepared in the tomato-producing region during the tomato harvest. Only a relatively small amount of equipment is required for this processing and large quantities of concentrate can be prepared. After the first fraction has been frozen and the second fraction has been concentrated and salted, the two fractions may be stored for extended periods of time before use. It is, however, important that the frozen fraction be stored and maintained in its frozen condition until used.

The frozen and concentrated fractions may be withdrawn from storage and used as required, or the two fractions may be shipped to a processing cannery at some distance removed from the growing region. Thus, the tomatoes can be separated, frozen and concentrated into relatively compact units and shipped to a processing cannery at a considerable distance from the growing region and more convenient to the market where the finished product is to be marketed. As required, the concentrated fractions can be recombined and employed, similar to the juices of fresh tomatoes, in preparing and canning the finished product. The two fractions, when recombined, can be employed to prepare tomato soup, tomato juice, vegetable juices, tomato catsup and similar tomato products. When so used, the finished product is comparable to products made with fresh tomatoes in both flavor and color.

Freeze Concentration of Serum

A process for concentration of tomatoes developed by *W.H. Craig; U.S. Patent 3,404,012; October 1, 1968; assigned to Pennsalt Chemicals Corporation* includes freeze concentrating the serum centrifugally separated from the solids and derived from a heated mass of chopped tomatoes, thus forming ice crystals and serum concentrate, the latter being centrifugally separated from the ice crystals and recombined with the separated solids.

According to the process, washed whole tomatoes are presented to a chopper in which the tomatoes are reduced to a size no larger than ½ inch. As is well known in the art, cells of the vegetable are broken during the chopping process, liberating certain enzymes. In order to avoid the enzyme degradation of the valuable pectin in the tomato pulp, from the chopper the tomatoes are delivered as rapidly as possible to a "hot break" tank or cooker in which the pulp is heated to a temperature ranging between about 170° and 205°F., and held at that temperature for a period of about 20 minutes to destroy the enzymes.

As in well known, enzymes if allowed to flourish attack the pectin to deprive the finished product of its body. A "cold break" may be substituted, as is well known in the art. Such a quick heat break while the material is thin does not produce the burnt taste that is noticeable in evaporation procedures in which surface contact of a relatively viscous mass has been relatively long.

From the cooker the chopped tomato pulp, including skins, seeds, stems and a wide range of impurities, is delivered to a screening device which preferably is a centrifuge of the type comprising a perforate frustoconical rotor with a feed inlet at the small end, and containing a screw conveyor driven at a slightly different speed than the rotor so that the flights on the conveyor urge the gradually deliquefying undesirable solids toward the discharge end. Such a centrifuge is available under the trademark Conejector centrifuge. Solids which pass out the larger end of the centrifugal solids dryer comprise the waste pomace which may be used after additional processing for fertilizer or animal food. The liquid and desirable small solids which pass through the perforations in the dryer rotor are delivered to a feed tank where they may be gently agitated to produce as near as possible a homogeneous mass of screened whole pulp.

From the feed tank the screened whole pulp is delivered to a centrifuge. The centrifuge is preferably of the clarifier bowl type comprising a solid bowl containing a central disc stack and of the type available under the trademark, Nozljector centrifuge. The pulp is fed to the bowl from an axial feed tube. As is conventional, the heavy solids discharge passes through peripheral nozzles spaced about the outer wall of the bowl. From the nozzles the solids discharge or concentrated tomato pulp is collected in one line and the light discharge comprising a clear serum having for practical purposes no insoluble solids content discharges over a ring dam centrally above the disc stack is collected in another line.

It will be understood that the serum contains sugars, color bodies, etc., but is approximately 90% water. From the centrifuge the serum may be pumped or otherwise conveyed to the first of a plurality of crystallizers. Each of the crystallizers may be conveniently comprised of a jacketed kettle having brine circulating through the kettle jacket to absorb heat from the serum through the inner wall of the jacket and to thereby lower the temperature in the first crystallizer to a temperature below about 20°F. As is conventional, the crystallizer features a rotating scraper which gently agitates the liquid and scrapes crystal structures from the wall of the crystallizer as these crystals gradually form there.

After a period of time depending on the temperature of the brine in the crystallizer jacket as well as on a number of other features, the dump valve at the drain of the crystallizer may be thrown open automatically or by manual means to discharge the mixture of concentrated serum and ice crystals into a collecting trough.

From the trough the mixture of concentrated serum and crystals is conducted by an insulated tube to a centrifuge. In order not to damage or heat the crystals, the centrifuge is preferably of the pusher type comprising an outer perforate rotor and an inner frustoconical perforate rotor which rotates with the outer rotor and reciprocates in relation to the outer rotor. This centrifuge gently drains and tangentially accelerates the crystals, pushing them towards the discharge open end with a minimum of abrasion. The ice crystals may be held in a heat exchange container through which the brine directed towards the jackets of the crystallizers is led.

The liquid concentrate passing through the perforations of the centrifuge is collected and may be pumped to a second freeze concentrating stage comprising another crystallizer, a trough and again the centrifuge. Virtually the same process is repeated using the second crystallizer, it being understood that in the meantime a new load of serum has been introduced into the first crystallizer and is undergoing the chilling process. The concentrated serum, after becoming chilled to the point of formations of crystals in the second crystallizer, is drained into a trough for subsequent feeding to the centrifuge. After removal of the crystals the more concentrated serum may be sent through a third stage of freeze concentrating, this stage involving a third crystallizer.

Obviously additional stages of freeze concentration, each one involving the use of an additional crystallizer, may be used. It should be understood that the centrifuging operation takes but a fraction of time for the crystallizing process, and thus the one centrifuge may accomplish the crystal separation for the products of a plurality of crystallizers. Moreover, for a limited operation, the same crystallizer may be used for more than one stage in the concentrating process; that is, the same crystallizer may be used for the first and third stages, for instance, by a rearrangement of the piping as will be obvious to one skilled in the art.

The sequence of operations for the various valves at the drain openings of the crystallizers and the valves downstream from the pump in the liquid discharge centrifuge will be obvious to one skilled in the art. Suffice it to say, the crystallizers are dumped serially and the concentrated liquid separated from the crystals in each dumping is pumped to the next adjacent crystallizer for subsequent additional concentration. After the final stage of concentration the concentrated liquid or serum is conducted in a stream to recombine with the concentrated pulp.

The tomato concentrate produced by the process is remarkable for its body and solids content. In addition, the tomato concentrate has the taste of a natural tomato product, with no traces of burnt flavor. The product readily lends itself to use in the forming of any of the conventional tomato products, such as puree, paste, ketchup, chili sauce, tomato juice, etc.

Spray Drying Tomatoes

When attempts are made to spray dry a tomato puree from which the seeds and skin have been removed, the product does not spray dry satisfactorily. *P.P. Noznick and R.H. Bundus; U.S. Patent 3,353,969; November 21, 1967; assigned to Beatrice Foods Co.* have found that spray dried tomatoes having good flavor and texture can be obtained by spray drying a tomato puree containing the seeds. Desirably the puree also contains the skin as well, since a more readily spray dryable product is obtained. It is critical that the seeds be ground so that 100% pass through a 40-mesh screen (Tyler sieve series). Generally there is no need to grind the seeds below 325 microns.

Example: Fresh San Marzano pear tomatoes from California were ground through a series head Urschel Mill (180 blades top, 200 blades bottom). The puree obtained was heated at 220°F. for 15 seconds to cook it, cooled to room temperature, and ground through a 212 blade Urschel Mill. The water-soluble puree solids were 3.5% and the total solids were 4.34% by weight. The puree had a pH of 4.57, was dark red and had a viscosity of 3,000 cps. The puree was then spray dried at an inlet temperature of 310°F. and an outlet temperature of

182°F. The puree dried well to give a product with no tackiness. The material left in the cyclone portion of the spray dryer caked slightly but the cake crumbled to a powder on cooling. In this example the weight of the skins was equal to the weight of the seeds. While it is preferable to cook the puree, this is not essential. Also it is preferable to employ two separate grinding steps, although only one such grinding is required providing the seeds are ground to pass through a Tyler 40-mesh screen.

WATER CHESTNUTS

In drying water chestnuts, a texture difficulty is encountered with the outermost portion or skin covering the meat of the chestnut. This skin has a tendency to harden, thus forming a horny shell-like structure extending over the surface of the meat of the nut. As a consequence, the formation of this horny layer impedes the rehydration of the product at a later date.

A.F. Stagmeier; U.S. Patent 3,023,110; February 27, 1962; assigned to General Foods Corporation has found, however, that when raw chestnuts are dehydrated under controlled drying conditions such that the moisture content of the interior nut meat is maintained substantially uniform throughout the course of water removal, the horny interfering condition of the skin layer does not appear and a readily rehydrated dried water chestnut is obtained. This controlled dehydration of the raw nut or pieces thereof is preferably carried out at normal atmospheric humidity conditions and at temperatures ranging from 105° to 125°F. The preferred relative humidity maintained in the treating vessel is 70 to 80, although relative humidities as low as 50 will present a satisfactory product. The process will be illustrated by the following specific example.

Example: 5 lbs. of raw, unpeeled water chestnuts were placed in a 20% sodium hydroxide solution having a temperature of 225°F. These chestnuts were immersed for 3 minutes and then removed. The caustic treated material had a soft, slimy exterior texture. The chestnuts were then introduced into an abrasive peeler which comprises a chamber having a rotating disc with a corrugated rubber surface which, when in contact with the soft chestnuts, rubbed the softened surfaces of them to effect a removal of the peel, rot and other imperfections of the chestnuts, leaving a substantially skin-free white product. This product was then rinsed in cold water and dipped in a mild hydrochloric acid bath at 50°F. for 20 seconds. After removal from the acid bath, the chestnuts were again rinsed in water to remove residual quantities of acid from their surfaces which, it could be noted, took on the appearance of an extremely white color.

The thus treated chestnuts were diced by standard cutting equipment to produce ⅛ inch cubes. The subdivided cubes of nut meat were then dried on a tray dryer at a temperature of 125°F. and a relative humidity of 70, which was maintained by forcing circulating hot humid air at a temperature of 110°F. through the containing vessel, the method of circulation comprising the upward flow of the moist air through the bed. Drying under these conditions was controlled to between 105° and 125°F. for a cycle of 10 minutes drying time, which results in a moisture content of the final dehydrated product of 8 to 10%.

The product obtained was not horny or case hardened but rather was substantially porous at its surface. The pieces exhibited a slight appearance of a ring or marking which is known as a heat ring or line of demarcation of precooked product. This ring disappears immediately upon rehydrating in cold tap water, and the rehydrated product has a pure white appearance the same as the original nut from which it was derived, and exhibits approximately 60 to 70% moisture.

POTATOES

PREPARATION PROCEDURES FOR MASHED POTATOES

Soaking Raw Potatoes in Water

Various methods have been used for preparing mashed potatoes in dehydrated, precooked form. The processes involve cooking potatoes, mashing them and applying further steps to eventually produce dehydrated products in particulate form, usually as flakes or granules.

C.E. Hendel; U.S. Patent 3,009,816; November 21, 1961; assigned to the U.S. Secretary of Agriculture has developed a process which produces dehydrated potatoes which exhibit enhanced moisture absorption. That is, when the dehydrated products are reconstituted for use, they will absorb more water than will products made by conventional methods. This means that a unit weight of dehydrated potatoes produced in accordance with the process will yield a greater amount of reconstituted mashed potatoes than will conventional products. Thus, dehydrated products are produced which will yield as much as 25 to 35% more reconstituted mashed potatoes as will the same weight of conventional products.

The advantages described above are attained by soaking the raw potatoes in water prior to cooking them. This soaking is extended over a period long enough to cause the potatoes to take up so much water that their weight increases by 7 to 25%. For maximum increase in moisture absorption of the final products, the potatoes may be given a supplementary soak in water following cooking. This supplemental soak restores added water which may be expelled from the potato tissue during the cooking step.

In applying the process, potatoes are first subjected to the usual preliminary steps of washing, peeling, trimming and cutting into suitable pieces such as slabs, slices or dice. The potatoes are then soaked in water until their weight increases from 7 to 25%, usually 7 to 10%. Depending on such factors as piece size, variety of potato, etc., the soaking to attain this effect may require anywhere from one to four hours. The soaking step is a useful point at which to impregnate the potato tissue with sulfite to preserve color and flavor during processing and storage of the product. Hence, the water may contain a small proportion, say 0.02 to 0.1% of sodium sulfite or bisulfite. This addition of sulfite may be used instead of, or in conjunction with, sulfiting at other stages in the process. Following the soaking stage the potatoes are cooked. The cooked potatoes are subsequently treated by any of the various processes which lead to the production of dehydrated potatoes in particulate form. Regardless of the type of product eventually produced, the advantages outlined above will be realized.

Skin Removal Accomplished with Precooking Step

In the production of a dehydrated potato product practical difficulties arise in consequence of the number of necessary steps which have to be applied to the potatoes, including washing, peeling, precooking, cooling, final cooking, mashing and drying. In the well-known method of skin removal by immersion in alkali solution the potatoes are immersed in a heated caustic soda solution of 20% strength and the immersion time is three minutes at 180°C.

R.A.S. Templeton; U.S. Patent 3,314,805; April 18, 1967 has found that it is possible to reduce the number of steps and at the same time to produce an end product having qualitative advantages and economic advantages with regard to the operating cost of the process by combining the precooking treatment with an application of alkali, thereby facilitating the removal of the skins, after the precooking treatment.

There is provided here a process for the production of a dehydrated mashed potato product in which the potatoes to be dehydrated are precooked for 10 to 60 minutes in an unpeeled condition in a dilute alkali solution having a strength of less than 5%, the potatoes thereafter being peeled before final cooking. The strength of the alkali solution will be determined by the time selected for the precooking treatment; the longer the precooking treatment, the less strength is required for the alkali solution. The strength of the alkali is adjusted to give adequate skin softening in the time necessary to complete the precooking treatment. Where it is necessary to continue the precooking treatment for a period of 40 minutes at a temperature within the range of 140° to 180°F., it suffices to use a caustic soda solution of 1% strength. However, where adequate precooking can be achieved within 20 minutes, it is necessary to employ a caustic soda solution of 3% strength to achieve adequate skin softening within the time available.

The skin softening treatment is thus carried out under very different conditions than is customary in the known method of skin removal by the use of alkali referred to above. It is found that the treatment of potatoes for a much extended period, as compared with prior practice, in relatively dilute alkali under the conditions indicated does not lead to the discoloration and the production of off-flavor to any noticeable extent, contrary to ordinary expectation, and furthermore, it is found that the extended treatment in dilute alkali leads to greater penetration of the alkali solution into cracks and splits in the potatoes, with the result that these are more thoroughly cleansed.

By this method the precooking and the gelatinization of the starch is satisfactorily achieved while simultaneously the potato skin is prepared for easy removal immediately following the completion of this heat treatment. The skins may be removed by the use of strong water sprays acting on the surfaces and skins of the potatoes, preferably in conjunction with a tumbling action producing sufficient abrasion at the surfaces to break the softened skins for removal by the water sprays.

By using this process for the softening of the skins at the same time as a precooking treatment is effected it is possible to reduce the loss of potato weight by peeling from as much as 30% weight of the potatoes in the case of abrasive peeling to as little as 7½% of the initial weight, so that the cost gain is self-evident. The qualitative gain is caused by the particularly effective and complete removal of the skin and eyes due to the prolonged treatment in the abnormally dilute lye and to the fact that solids are not leached out during the precooking treatment. Preferably the potatoes, after precooking in the dilute alkali solution, are immersed in or sprayed with a dilute solution of a weak acid, such as a 1% aqueous solution of citric acid to neutralize alkali on the skins of the potatoes.

GRANULES

In producing dehydrated potato granules, it is always the aim to obtain a product which on mixing with water, rapidly forms a mashed potato dish of mealy texture comparable to freshly prepared mashed potatoes. This desideratum, however, eludes attainment in commercial operations. A foremost problem is that the reconstituted mash tends to be sticky and pasty instead of mealy. This undesirable effect is caused by a rupture of cells during processing whereby starch is released from the cells. It is the presence of this extracellular starch which gives the reconstituted product its pasty texture. Cell rupture is principally caused in the step or steps of the process wherein the potato material is subdivided. Thus, to produce a product which reconstitutes rapidly and which forms a mash on reconstitution, it is necessary that the dry product be in finely subdivided form.

Accordingly, somewhere along the line it is necessary to reduce the potato tissue to particulate form. This is very difficult to do effectively. For example, if the potatoes are dried in the form of slices, slabs, or dice then pulverized, the product is totally useless. Pulverization of the dried potato tissue causes such extensive cell damage that the product on addition of water forms an unpalatable paste. On the other hand, it is futile to sub-divide the potato tissue early in the process when it is highly moist because the particles will stick together and dry as agglomerated lumps which will not reconstitute properly; they will form a lumpy unappetizing dish. Moreover, the product cannot be dehydrated uniformly; the lumps will tend to form horny crusts about the still-moist inner portions.

Various procedures have been advocated to attain the desired subdivision with a minimum of cell rupture. One procedure, known as the add-back process, is presently employed on an extensive scale in this country. The process involves these steps. Potatoes are peeled, sliced, cooked in steam, then mashed. The mashed potatoes are mixed with sufficient dried potato granules from a previous batch to give a moisture content of 35% for the composite material. This material is conditioned by holding it at approximately 60° to 80°F. for an hour or more. The conditioned composite material is then dehydrated in a pneumatic drier to produce the dried granules. In this procedure the add-back operation (addition of previously dried granules to the mashed potatoes) is a critical step as it con-verts the sticky mashed potatoes into a free-flowing moist powder which can be dehydrated readily without agglomeration of the individual particles. Although the add-back process is widely used it presents many significant disadvantages, as explained below.

One disadvantage of add-back is that the dehydration equipment must handle about 6 to 10 times as much material as is actually packaged. To further explain, in order to reduce the moisture content of the mashed potatoes (originally 75 to 80%) to a level of 35%, approximately equal weights of dried granules and mashed potatoes must be mixed. Then, when this composite product is dried, 84 to 90% of it must be returned for recycling. It is thus obvious that only one-tenth to one-sixth of the solid material being handled is product; the remainder stays in the system. Naturally, this means that the equipment must be several times as big as would be needed to handle a certain level of output and operating costs are correspondingly increased.

Another point is that this continued recycling means that the product has been through the dehydrator 6 to 10 times. Obviously, any subjection of potato tissue to dehydrating operations will expose it to conditions of mechanical, heat, and oxidative damage and to expose it many times will multiply the amount of quality damage, particularly cell rupture caused by mechanical stresses such as abrasion.

A third problem caused by this continued recycling is that the system is very inflexible and if a bad lot of material is produced it will take considerable production before the effect of this bad run is essentially eliminated. This is caused by the fact that the major proportion of the bad lot must be recycled with the result that each successive lot will contain a proportion of the bad material. Oftentimes it will take many hours of produc-tion to eliminate the effect of one bad lot. Naturally much thought has been given in the industry to schemes for eliminating the add-back procedure. However, despite much in-vestigation and experimentation no practical process has been heretofore devised.

It has been advocated in the prior art that the problem of cell rupture during the subdivision step can be minimized by a conditioning step, involving a holding of the potato mash, preferably after partial dehydration, in a refrigerated state. Such procedure is shown for example, by Barker, British Patent 542,125 (1942). In this process, potatoes are peeled, washed, cooked, mashed, then partially dehydrated. The partially dehydrated mash is then chilled and held at such temperature for a period long enough to equilibrate the moisture content and toughen the cell walls. The conditioned mash is then pressed through a sieve while contacting it with heated air to form a moist powder which can be dehydrated without agglomeration of the individual particles.

The process outlined above offers the advantage that the add-back step is not used. However, the process involves certain disadvantages, explained as follows. For one thing, relatively long periods of conditioning the partly dried mash are advocated. This has the disadvantage that the potato material tends to become gray, develops off-flavors, and suffers a loss in vitamin content. Another point is that the step of forcing the conditioned mash through a sieve causes substantial cell rupture. Consequently the final product tends to form a pasty material when reconstituted. To avoid pastiness the product must be reconstituted with water not over 176°F., which produces a luke-warm potato dish.

Granulated in Frozen State

In a process developed by *G.K. Notter and C.E. Hendel; U.S. Patent 2,959,487; Nov. 8, 1960; assigned to the U.S. Secretary of Agriculture* partially dehydrated, cooked potato mash is frozen and granulated while in the frozen state. In the process, raw potatoes are peeled, washed, cut into slabs or slices, then cooked in a conventional manner. Preferably, the cooking is carried out by contacting the potatoes with steam at 212°F. The cooked potatoes, while hot, are mashed. This operation may be carried out by pressing the cooked potatoes between warm rolls, by pressing them through a screen, or by other conventional potato-mashing techniques.

The mashed potatoes are then partially dehydrated. This may be conveniently done, for example, on a double-drum dryer. The mashed potatoes are fed into the nip between rotating drums and the partially dried potato material removed by scrapers. The drums are generally heated to a temperature in the range from 150° to 300°F. The temperature of the drums, the speed of rotation, and the thickness of the film of potato material are so correlated that the partially dried potato mash has a moisture content from 50 to 70%, preferably 60%.

Next, the partially dried mash is frozen. This freezing step may be accomplished in various ways. For example the mash may be spread on trays and subjected to a current of refrigerated air. In the alternative, the trays carrying the mash may be placed in a chamber which may be sealed and subjected to vacuum. In this case the freezing will be caused by rapid evaporation of moisture. In any event, the temperature of the partially dried mash should be brought to a level at which the mash freezes. This will generally be 27°F., or below, depending on the moisture content of the mash. Preferably, the mass is frozen while in relatively thin layers, that is, a thickness of one inch or less.

Application of the granulation step follows without any substantial delay following freezing of the mash. In the granulation step, the frozen mash is fed into the nip between rolls which are at room temperature and spaced a small distance from one another. Generally, the clearance between the rolls is from 0.01 to 0.02 inch. In this way, the frozen mash is pressed into a very thin layer, essentially one cell in thickness, which is removed from the rolls by the action of scrapers. As the material is dislodged from the rolls it breaks up into small flakes or platelets.

The flakes of potato material from the granulating rolls, may then be subjected to mixing. In this step, the flakes are gently agitated to cause a breaking up of the particles. The use of violent agitation or other excessive mechanical action is to be avoided as it will tend to rupture cell walls which in turn will cause the final product to be pasty when reconstituted. Although this mixing operation is preferably used, it is not an essential step and the material from the granulating rolls may be fed directly to the dehydration step. Thus the film potato material produced by the granulating rolls is so thin that it has essentially no self-sustaining properties and breaks up into small fragments by its own weight. These fragments on subjection to any type of handling as necessary in the subsequent dehydration step break up into granules consisting of individual cells or small aggregates of individual cells. The granulated potato material, directly from the rolls or the mixer is then dehydrated to form the final product, i.e., dried potato granules. This final dehydration may be carried out in any manner as is conventional in the art.

Example: Raw potatoes (Idaho Russets) were peeled, washed and cut into slabs about ¾ inch thick. The potatoes were then cooked by exposing them to steam (212°F.) for 25 minutes. The cooked potatoes, while hot, were then mashed by pressing them through a ¼ inch mesh screen. The mash was then partially dehydrated on a double-drum dryer. The drums were at a temperature of 240°F., separated by a clearance of 0.01 inch and were rotated at one rpm. The partially dehydrated mash had a moisture content of 63%.

The partially dried mash was then placed on stainless steel trays in a layer ¾ inch thick. The loaded trays were placed in a vacuum chamber where they were exposed to a vacuum of 3.5 mm. Hg for 20 minutes whereby the mash was brought to a temperature of 26°F. The vacuum system was then disconnected and the frozen mash removed. Due to evaporation of moisture during vacuum treatment, the moisture content of the frozen mash was 56%. The mash while still frozen was fed into granulating rolls. The rolls were at room temperature and clearance between the rolls was 0.01 inch. The frozen mass was thus granulated into small flakes or platelets. This material was mixed for five minutes in a planetary-type food mixer operating the mixing blade at slow speed, 60 rpm.

The granulated material was then put on trays which were inserted in a tunnel dryer wherein the potato material was subjected to a current of air at 120°F. After two hours the product was dehydrated, moisture content less than 10%. A sample of the product (108 grams) was added to 2 cups of boiling water and mixed with a fork. In less than one minute the product was reconstituted forming mashed potatoes of desirable mealy texture, free from lumps and pastiness. Color of the mashed potatoes was creamy-white and the taste was excellent, indistinguishable from freshly prepared mashed potatoes.

Granulation Under Special Conditions

In a process described by *C.E. Hendel, G.K. Notter, M.E. Lazar and W.F. Talburt; U.S. Patent 3,009,817; November 21, 1961; assigned to the Secretary of Agriculture* the potato tissue is effectively granulated so that on drying it yields a product in the form of fine particles whereby the product can be rapidly reconstituted and directly forms a mash free from lumps or other unrehydrated particles. Moreover, these desirable results are attained without significant cell rupture so that the reconstituted product has a desired mealy texture, completely free from pastiness. As a consequence of these advantages the process enables the production of high quality potato granules without the use of add-back.

A principal feature of the process is that subdivision is accomplished by subjecting the potato material to repeated mild compression and mild shear forces while drying the material through a critical moisture region. The intensity of the mechanical forces is so controlled as to achieve effective separation of one cell from another rather than rupture of individual cells. As a net result the potato material is formed into fine particles so that the end product will reconstitute rapidly and directly forming mashed potatoes free from lumpy or gritty particles.

Further, because cell rupture is kept at a minimum, the reconstituted product is free from pastiness. In preparing granules without add-back, this improved method of subdivision is preferably applied after the potatoes have been subjected to a series of steps usually including cooking, mashing, and partial drying. The subdivision step may be performed as a separate step or as part of other steps in the sequence of operations. Thus, for example, the potatoes may be cooked, mashed, partially-dried, conditioned by known methods, subdivided while drying through the critical moisture region in accordance with the process, and given a final drying. More preferably, the conditioning step is carried out while also applying a part of the subdivision partial drying treatment. In the stage where the potato material is subdivided, the conditioned material is granulated and dried through the critical moisture range. The granulation and drying are carried out either simultaneously or in closely successive operations repeated a number of times. In this granulating step the aim is to subdivide the mash into particles containing not more than ten individual cells, preferably unicellular particles, and it must be done by separating one cell from one another rather than by rupturing individual cells. Were the latter to be done the product would yield a pasty, unpalatable mass on reconstitution. The granulation can be successfully accomplished by applying to the mass repeated mild compression and mild shear forces.

Reference is made to Figures 6.1a and 6.1b which illustrate one modification of apparatus for effecting the granulation and drying through the critical moisture region. The apparatus, generally designated as **20**, comprises a trough or U-shaped chamber **22** provided with a removable lid **23** and a longitudinal shaft **24**. Suitable equipment, not illustrated, is provided to rotate shaft **24** in the direction shown at a low speed, 1 to 5 rpm.

Attached to the shaft are a series of arms **25**, each bearing a paddle **26**. Dimensions are so chosen that the tips of the paddles have a clearance on the order of ¼ to ½ inch from the cylindrical base of the trough. Also positioned on shaft **24** are arms **27** which carry a blade **28** which extends essentially the length of trough **22**. The blade is made of flexible material such as silicone rubber (Silastic), neoprene, Teflon or other elastomer and is so positioned that its edge actually wipes against the cylindrical base of the trough. This base may be provided with small protuberances, as by welding wires longitudinally along it, to increase the shearing effect to the desired level. An inlet conduit **29** is provided for introduction of gaseous media, for example, air for drying.

In using the illustrated device to granulate the potato material, conditioned potato mash is introduced into the trough. The shaft is then caused to rotate and drying air is introduced through the inlet conduit. The resulting action of the paddles and the blade effect the granulation of the potato material. Thus, paddles **26** cause a repeated mixing of the material and a disintegration of the larger aggregates of cells. Blade **28** being actually in contact with the cylindrical base of trough **22** effects a further size reduction of the particles.

FIGURE 6.1: APPARATUS FOR GRANULATION OF COOKED POTATOES

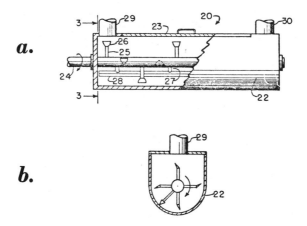

Source: C.E. Hendel, G.K. Notter, M.E. Lazar and W.F. Talburt; U.S. Patent 3,009,817;
              November 21, 1961

The reduction in particle size effected by the device is essentially limited to separation of individual cells one from another as contrasted with rupture of individual cells. Important in this regard is the fact that the paddles and blade exert what may best be termed as mild compression and mild shear forces. Thus the mechanical forces exerted by these elements are of sufficient intensity to separate agglomerated cells but insufficient to rupture individual cells. The action is continued until the potato material forms a well-granulated mass of unicellular particles and small aggregates of unicellular particles that have only a very slight tendency to agglomerate together. Any agglomerates formed are very readily separated from one another. This is at the lower end of the critical moisture region where the potato material contains 32 to 40% water, the precise moisture value depending somewhat on the previous history of the material.

During the granulation, the potato material is contacted with a current of air, for example, at a temperature from 75° to 200°F., to cause the desired reduction in moisture content during granulation. Ordinarily, the product remains at room temperature (75°F.) during the granulation. The warm air introduced does not appreciably raise the temperature of the potato material because of the cooling effect as water is evaporated therefrom. The granulated potato is further dehydrated to produce dried potato granules.

Example: Idaho Russet Burbank potatoes were washed, peeled, trimmed, and cut in three-fourths inch thick slices. The slices were dipped five minutes in a 1.25% aqueous solution of sodium bisulfite. The slices were then cooked 60 minutes in a mixture of air and steam having a temperature of 190°F. The cooked potatoes were mashed by pressing through a one-half inch mesh screen, then blended in a planetary-type mixer for one minute with 10% of their weight of water containing 0.6 gram of sodium bisulfite per 10 lbs. of potatoes. The potato mash was then partially dried on a single drum dryer, drum temperature 250°F., speed of drum 2.5 rpm. The partially-dried mash had a moisture content of 56.5%.

The partially-dried mash was then conditioned. To this end, it was placed in a trough granulator as depicted in Figure 6.1. The shaft was rotated (2 rpm) continuously during addition of the mash (30 minutes), then for the next hour the shaft was rotated five minutes out of each fifteen minute period. During this operation the temperature of the mash decreased from 125° to 65°F. The mash was then friable and ready for the beginning of granulation. It was near the upper limit of the critical moisture region referred to above. To granulate the conditioned mash, it was left in the trough granulator and the shaft was operated continuously (2 rpm) for one hour while air at room temperature was blown through the device. Moisture content of the material was reduced to 50.5%. The material was near the middle of its critical moisture region. It was now granulated well enough that the rate of drying could be increased without resulting in an excessive proportion of coarse dried product.

The potato material, still in the trough granulator was subjected to a current of air at 200°F. while the speed of the shaft was increased to 5 rpm. In 30 minutes, 94% of the dried granules were received in the collector. This product containing 20% moisture was then finished-dried in a fluidized bed dryer to produce granules of 6% moisture content. The product had a blue value index of 17, indicative of very slight cell damage. Bulk density of the product was 0.92 gram per cubic centimeter; moisture absorption was 5.7 cc of reconstituted mash per gram of product. A portion of the product on reconstitution with boiling water formed mashed potatoes of a desirable mealy texture free from both pastiness and graininess.

Dehydrating, Rehydrating, Ricing, Dehydrating

One of the commercial processes for the production of dehydrated potato products is the granule process wherein raw potatoes of appropriate low sugar content are washed, peeled, sliced, steam cooked, mashed, and dried in the form of cooked potato granules. In the process a majority of the granules are recycled back into the stream of cooked mashed potatoes, this mixture being eventually passed to a dryer and the dried product being screened to remove imperfections. This process is characterized by a number of manufacturing advantages, e.g., the ability to remove imperfections without waste of peeled potato; the granule product also has fairly acceptable storage stability. However, as is recognized in the art, the product of the process suffers from the loss of some potato flavor by reason of the plurality of steps and the recycling and reheating of the potato material. It would be most desirable to provide a product which has the advantages of the so-called granule process and at the same time offers a desirable potato flavor as well as ease of recipe preparation.

A process for making a dehydrated potato product as described by *A.F. Stagmeier; U.S. Patent 3,021,224; February 13, 1962; assigned to General Foods Corporation* essentially involves the combination of the cooked potato granule with moisture to produce a mash or mixture containing approximately 70 to 85% moisture. This mixture is subsequently shaped in any suitable apparatus (such as a conventional potato ricer) into the form of elongated filaments wherein the granules are loosely agglomerated. These filaments are subsequently rapidly dried by means of circulating hot air or other suitable means to evaporate the moisture in such a manner that the filaments set in a substantially uncollapsed state to yield a porous rehydratable product characterized by its plurality of cavities and interstices and the substantially intact unruptured condition of its starch granules.

The granule used in the process can be best described as comprising a cooked preswelled and dried potato powder which has not been gelatinized to the point where it has burst. The mixture resulting from the blend of potato granules with the moisture is generally of a nature which permits it to be extruded in such a form that the filaments issuing from an extrusion opening will have the granules arranged so that moisture residing therewithin will readily escape upon evaporation. To achieve this desired condition a minimum of agitation is employed to avoid rupture of the starch granules prior to and during filament formation whereby pastiness is avoided in the final product; the hydrated swollen granules are loosely aggregated to one another in the manner of cooked mashed potatoes.

At this point of the process, various additives may be employed to endow the eventual product with protein supplements or flavors. Included among these additives are such materials as soy flour, peanut flour; proteinaceous extracts in a more purified state from nutmeats like soybean and peanut; vegetable materials in a dehydrated powderous or liquid condition such as tomato paste, onion juices and the like; skim milk solids; deoiled fish; and emulsified plastic fats and powdered fats produced by drying emulsified fats with proteinaceous encapsulating solids.

In addition to or in lieu of these additives other flavoring materials heretofore discarded as potato waste may be added to the mash to fortify the potato mash with desirable proteins and most important a desirable potato flavor; preferred of the sources of this potato flavor are the water-soluble solids of the discarded peel fraction recovered either as a result of steam treatment or a hand or mechanical peeling of potatoes; flavoring materials can be recovered as a supernatant from an aqueous slurry of such materials in which form they are relatively free of rot and other undesirable material.

The rehydrated potato mash is preferably tempered prior to shaping into filaments by holding for a few hours to effect a uniform distribution of moisture throughout and is thereafter introduced to a shredding chamber having orifices or openings of a size ranging from $\frac{1}{16}$ to $\frac{3}{8}$ of an inch, the size of such openings being determined by the end use or form of the dehydrated product. Preferably the product is extruded by means of a conventional ricing apparatus, which apparatus consists of a perforated drum with a suitable internal rolling mechanism to force the potato mash through the orifices in the drum.

Rapid filament dehydration is critical to the process; otherwise undue starch granule rupture will result thereby producing a product which is pasty upon rehydration and, furthermore, is not as rehydratable as desired. A continuous dryer having a series of stages of controllable temperatures wherethrough a wire mesh conveyor belt travels is preferably employed. With such a continuous dryer a circulating body or bodies of hot air are practiced having air temperatures ranging anywhere from 125° to 280°F., the air being circulated through a bed of shreds an inch or less in thickness at a superficial air velocity of 500 to 700 fpm in the early stages of drying at temperatures of 240° to 280°F.; in the later stages of drying when the product moisture content is below 20%, a lower air temperature of 125° to 135°F. is employed until a superficially dry shred of 3 to 10% moisture is achieved. The product so prepared is characterized by a highly acceptable potato flavor and by its highly porous, readily rehydratable and stable character.

Controlled Cooking

In the preparation of dehydrated potato products, it is conventional to cook the raw potatoes at an early stage in the process. The cooking procedure employed in preparation of the various products is essentially the same as ordinary home-cooking methods. That is, the raw potatoes in the form of slabs or thick slices (¾" thick) are boiled in water until tender. Sometimes steam, instead of boiling water, is applied to the potato pieces. In either case, the cooking is continued for 20 to 30 minutes. The cooked potatoes while hot are then mashed and eventually dried in particulate form.

C.E. Hendel, G.K. Notter and R.M. Reeve; U.S. Patent 3,031,314; April 24, 1962; assigned to the Secretary of Agriculture have found that the method of cooking has a profound effect on the properties of the dehydrated product. The interdependence of these variables is best explained by reference to Figure 6.2 in the annexed drawing. In this figure, the minimum and maximum cooking times corresponding to different cooking temperatures are plotted with the temperature on a linear scale and the time on a logarithmic scale. Curve **AB** represents the minimum cooking conditions while curve **CD** represents the maximum cooking conditions. Thus, for example, if a cooking temperature of 180°F. is chosen, the potatoes are cooked from 75 to 420 minutes. Other conditions which may be used are, by way of example: at 190°F., 30 to 180 minutes; at 200°F., 12 to 45 minutes; at 212°F., 5 to 10 minutes.

By operating within the area **ABDC,** many significant advantages are gained. Most important is that the potatoes are properly cooked yet the final dried products are not sticky on reconstitution. Moreover, operations subsequent to the cooking step are simplified and take less time. For example, in the manufacture of granules, conditioning under identical conditions will require one-half or less time than potatoes cooked by conventional methods. Also, the potato material can be readily subdivided without cell damage, thus to produce products free from stickiness on reconstitution.

Researches have shown that where the cooking is carried out under conditions falling below curve **AB**, inferior results are obtained. Under such conditions, there are many uncooked particles and when the potato material is treated in subsequent steps, so much mechanical force must be applied to subdivide these particles that extensive cell rupture occurs.

As a result, the final dried product on reconstitution forms a mass of undesirable pasty texture. Also, if the cooking is carried out under conditions falling above curve **CD** inferior results are obtained, these involving either (a) discoloration and development of off-flavors, or (b) development of stickiness, these respective defects depending on the temperature used. Thus, at the higher temperature ranges, 200° to 212°F., cooking above curve **CD** leads to stickiness so that long conditioning times are required and subdivision without cell rupture is difficult to accomplish. At lower temperature ranges, 175° to 195°F., cooking above curve **CD** leads to discoloration (browning) and development of unnatural, undesirable flavors.

Generally it is preferred to carry out the cooking at a temperature of 190°F. for the reason that at this temperature there is greater leeway than at higher temperatures between minimum and maximum cooking times. Thereby the process can be more accurately controlled and variations due to diffcrences in composition of different batches of potatoes, localized or temporary temperature changes in the cooking medium, and the like, are canceled or at least minimized. Also, at the lower temperature range, 175° to 195°F., the thickness of the potato slices which are to be cooked is immaterial, and can be as much as one or two inches, or moderate-sized potatoes can even be cooked whole, without appreciable nonuni-

FIGURE 6.2: CONTROLLED COOKING CONDITIONS

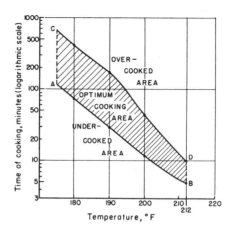

Source: C.E. Hendel, G.K. Notter and R.M. Reeve; U.S. Patent 3,031,314; April 24, 1962

formity of cooking through the material. At higher cooking temperature it becomes more important that slice thickness or piece size be small enough to avoid nonuniformity of cooking whereby the center of the piece or slice is undercooked while the surface is over-cooked, with resultant loss of the beneficial effects of the new cooking procedure. Thus, at the highest cooking temperatures, 205° to 212°F., it is preferred to use slices up to and including ½ inch, or other pieces with similar ratio of surface area to volume, for example, dice up to and including one inch on a side.

The cooking, carried out under the conditions of time and temperature as explained above, is generally effected by immersing the potato slices in a bath of water at the selected temperatures. Another plan is to subject the potato slices to a current of steam or other hot gases. For cooking at temperatures below 212°F., mixtures of steam and air proportioned to provide the selected temperature, are useful.

It is to be noted that in the method of the process, the raw potatoes are given a single cook at prescribed conditions of time and temperature. This procedure is in contrast with methods which have been previously advocated wherein the potatoes are given a precook at relatively low temperatures followed by a cook at boiling temperatures. The potatoes, cooked in accordance with the process may be subsequently treated by any of various processes which lead to the production of dried potatoes in particulate form. Regardless of the type of product eventually produced, the advantages outlined above will be realized. As examples, the cooked potatoes are mashed and then treated by known procedures to produce dried flakes or granules.

Cooling and Aging at Low Temperatures

R. Kodras; U.S. Patent 3,039,883; June 19, 1962; assigned to International Minerals & Chemical Corporation describes a method for producing granules of dehydrated mashed potatoes in which potatoes are cooked until capable of being readily mashed, are then cooled and aged at a relatively low temperature, mashed in water to a thin slurry, separated from the liquid phase, and dried. During the low-temperature aging operation after cooking but before mashing, the potatoes appear to undergo alteration in some unascertained manner, as a result of which they are readily mashed in water without objectionable rupture of cells, and are converted thereby into a slurry which is readily filtered. The resulting solids dry readily, forming particles or granules of high bulk density which are readily rehydrated to form a mashed potato product of superior quality. The cooking operation is carried out at a temperature below 60°F. and above the freezing point, preferably between 32° and 40°F., and the potatoes are held at a temperature within this range for a period of 1 to 24 hours or longer, preferably around 10 to 20 hours. An example will illustrate the process.

Example: Small Idaho Russet potatoes averaging 150 grams in weight were steamed at atmospheric pressure for 35 minutes, then transferred to a refrigerator at 35°F. and held for 14 hours. At the end of this time they were removed from the refrigerator and allowed to stand at room temperature for 7 hours. They were then mashed to a thin slurry in distilled water at room temperature, colandered, filtered, broken up into small granules, and dried in a vertical column with a stream of warm (100°F.) air over a period of 3 hours. The product was a white, crystalline material containing many small granules and having a bulk density of 0.4 g./cc. It rehydrated readily into a potato product of good taste, texture and color.

Addition of Natural Gum

In the production of dehydrated mashed potatoes, an outstanding problem has been the prevention of stickiness in the reconstituted product. Considerable research has been devoted to this problem and various improvements have been made which make possible the production of products which are free from stickiness when reconstituted for consumption. The improved procedures have taken various forms including the incorporation of additives, such as glycerol monostearate or similar emulsifiers, or variation in the procedural conditions as initially cooking the potatoes at limited temperatures, precooking the potatoes at low temperature before the regular cooking, various conditioning treatments of the cooked mash, etc. Although these measures have to a large measure remedied the problem of stickiness, they have given rise to a new problem in that the reconstituted product is grainy, that is, it does not have the smooth texture of mashed potatoes made from freshly-cooked potatoes.

C.E. Hendel, R.M. Reeve and G.K. Notter; U.S. Patent 3,054,683; September 18, 1962; assigned to the U.S. Secretary of Agriculture have described a process which remedies the graininess of rehydrated mashed potato by mixing with the product, after dehydration or during processing, a minor amount of a natural gum. Various gums can be employed as, for example, carragheen, algin, pectin, low-methoxyl pectins, guar, karaya, arabic, tragacanth, agar, locust bean, acacia, or other natural edible polysaccharide gums.

The amount of gum used may be varied from 0.05 to 1% (based on dry weight of the potato material), depending on such factors as the properties of the gum selected, the character of the potato material, and the properties desired in the final product. In general, it is preferred to use 0.5% of the gum. It is to be emphasized that investigations have demonstrated that the natural gums have a unique ability to eliminate graininess in dehydrated potato products whereas synthetic gums such as carboxymethylcellulose do not have this ability. As noted hereinabove, the gum may be added after the dehydrated potatoes are prepared or may be added at any stage in their processing. The gum is preferably incorporated with the potatoes during processing, for example, in the stage when the potatoes, after cooking, are mashed. In this way an added benefit is achieved in that the potato material can be granulated more readily and without rupture of cell walls.

Example: California Russet Burbank potatoes were washed, peeled, trimmed, cut into ¾ inch thick slices, dipped one minute in a 0.1% solution of sodium bisulfite, then cooked 60 minutes in a steam-air mixture at 190°F. The cooked potatoes were divided into two lots, each being riced through a ½ inch mesh screen, then mixed in a planetary mixer (60 rpm) for one minute with an emulsion (composition given below) equal in amount to 10% of the weight of the potatoes at the beginning of the mixing. In run 1, the emulsion was an aqueous emulsion containing 1% carragheen, 1% glycerol monostearate, and 0.09%

$NaHSO_3$. (This is equivalent to approximately 0.5% carragheen and 0.5% glycerol mono-stearate in the final dry potato product.) In run 2, the emulsion was as above except the carragheen was omitted. The two lots of potatoes containing the additives were each processed further as follows. The mash containing the additive was dried on a single drum dryer at 250°F., approximately 2 rpm to a moisture content of 50%.

The partly-dried mash was then placed on a belt equipped with a blower for passing air over its surface. The mash was thus cooled to room temperature and conditioned at such temperature for 3 hours. During this conditioning period the following manipulations were applied. After one-half hour, the mash was subjected to the action of rotating blades to reduce the size of clumps and aid in equilibrium of moisture content. After 2 hours the mash was passed between 12 inch diameter mashing rolls at room temperature, 0.01 inch clearance, 3 rpm. This formed the potatoes into a friable, discontinuous sheet very easily separated into individual particles that did not reagglomerate.

After the third hour of conditioning the potatoes were again passed through the mashing rolls to form a very friable product. This product was then passed in a trough granulator dryer, wherein the material was subjected to a current of air at 130° while rotating paddles at 6 rpm effected a repeated subdivision of the mash under mild compression and mild shear forces. In 20 minutes the product had been dried in the form of granules. The granules were then finish-dried in a fluidized bed dryer to produce granules of 6% moisture content. Samples of the two products were reconstituted and appraised by a taste panel for stickiness and graininess. The panel judged the potatoes of run 1 to be both less sticky and less grainy that the potatoes of run 2.

Improved Add-Back Process

R.W. Kueneman and J.E. Conrad; U.S. Patent 3,085,019; April 9, 1963; assigned to J.R. Simplot Company describe a process for preparing a dehydrated mashed potato product by treating the potatoes at some stage in the process such as after or during the add-back step, with a hot moist atmosphere, with critical limits as to relative humidity of the atmosphere surrounding the mix, temperature, moisture content of the potatoes, and time. The relative humidity may range from 15 to 100%, being 100% at or above the temperature of atmosphere steam. The time may vary from 1 to 60 minutes. Temperatures range from 120° to 250°F. The moisture content of cooked partially or completely dried potato particles such as the seed, mix, or granules, upon which the process is preferably practiced is between 3 and 50%.

Conventional add-back methods of preparing potato granules include various stages, such as mixing the freshly cooked mashed potatoes with the add-back material (sometimes called seed), mashing and adding the seed simultaneously, equilibrating the moisture content of the seed and freshly cooked vegetable, additional mixing, and cooling the mix and holding or tempering the same at low temperatures to obtain starch retrogradation, obtain additional equilibration, and avoid microbiological damage. Other steps include the addition of other food ingredients such as milk, dry milk, and reconstituted dry milk, flavoring agents such as salt and pepper, emulsifiers, and preservatives. In addition various sifting or particle size separation steps and drying operations are performed, the drying methods used being very gentle to avoid impact and abrasion damage to the fragile potato cells.

This process may be performed at any stage of the process after the add-back of seed or upon the seed itself. Thus, the process (which will be referred to as modification in view of the fact that the rehydration characteristics of the potato cells and of the free starch are altered as indicated herein) may be performed upon the wet mix having a moisture content of between 25 and 50%. Of course, seed may be added to reduce the moisture content of the mix to a greater extent, but this is not economically feasible, although technically practicable. The process may be performed upon the dry seed, having a moisture content of 9 to 20%. A reduction to below 9% is not economical, and at above 20%, separation of seed from the end product is difficult, although the seed may have a moisture content outside of this range and still be used in this process. Modification may take place in a mixer-masher simultaneously with mashing and the addition of seed. During drying, the modification may be performed by adhering to the conditions as specified herein, and the same is true as to the finished dry product. Dry granules may be used and have 3 to 12% water.

Mashed Potatoes Dried on a Screen

In another process developed by *R. Kodras; U.S. Patent 3,110,574; November 12, 1963; assigned to International Minerals & Chemical Corporation* freshly cooked potato is spread into the interstices of a screen, and is dried therein at moderate temperature. The dried product is readily removed from the screen by an air blast. The completed product is crumb-like in form, light in color, high in bulk density, and readily reconstituted with water and/or milk to form a superior mashed potato product.

Cooked potatoes for use in the process can be prepared in any convenient way. In a preferred technique, the potatoes are washed, peeled, sliced to a thickness of around one-half inch, rinsed free of nonadherent starch, and cooked in live steam or boiling water for around 35 minutes. The resulting cooked slices can be used as such, or can be mashed in a portion of the cooking liquor in any convenient manner to produce a pulp or mash containing between 5 and 25% by weight of dry solids, preferably between 10 and 15%.

In another embodiment, cooked potato or a mash thereof is fed onto the top of a continuously moving wire screen having a thickness between $1/32$ and $1/4$ inch, preferably $1/16$ inch, with void spaces ranging in area between $1/32$ and $1/4$ inch square, preferably around $1/16$ inch square. The potatoes are fed upon the screen as the latter moves across a flat, smooth surface, and are spread and forced into the interstices of the screen by a properly placed scraper blade. The screen then moves into a drying chamber, where a current of dry air rapidly removes the moisture from the potato. The temperature of the air stream is preferably around 150° to 250°F., but lower temperatures may be employed (e.g., as low as around 75°F.), and higher temperatures (e.g., up to 300°F.) may also be employed, so long as burning and case hardening are avoided. The drying time will of course depend upon the relative humidity of the drying air, the drying temperature, and the water content of the potato composition.

In general, a drying time between 20 and 40 minutes at around 175°F. and a relative humidity around 50% will be sufficient. It is advantageous to dry first at a relatively low temperature (e.g., from 75° to 125°F.) to a solids content above 50%, then complete the drying at 150° to 250°F. The preliminary drying may be conducted at high relative humidity (around 60 to 80%) if desired to minimize case hardening. The completed product should have a moisture content below 12%, dry basis, preferably between 5 and 10%.

The dried potato film on the screen remains loosely adherent thereto, and is readily removed by blowing a stream of air through the screen, by brushing it by mechanically vibrating. The dried potato is obtained thereby in the form of crumb-like granules without damage to the cell structure thereof. Moreover, the product is of high bulk density (around 0.35 gram per cubic centimeter or higher), and is readily rehydrated simply by stirring into water.

Example: Idaho Russet potatoes were washed, peeled, trimmed and sliced to around one-half inch thickness, again washed, and cooked with live steam at atmospheric pressure for 35 minutes. For the drying operation, an apparatus was employed comprising a 14 x 18 mesh per inch aluminum screen passing downward at the rate of 2 feet per minute between a pair of drums rotating in opposite directions. The drums had a diameter of $6^{1/8}$ inches, and were heated to a surface temperature of 285°F. by pressurized steam therein.

The cooked potato slices were dropped upon the moving screen just before the screen passed between the drums. They were broken up and squeezed thereby into the interstices of the screen. As the screen emerged from between the drums, the potato granules adhered to the drums and were thereby removed essentially completely from the screen. The granules were dried on the drums as rotation continued, and were scraped from the drums by a doctor blade, positioned at a point 227° in the direction of rotation from the point of contact of the screen with the drums. The resulting product had a bulk density of 0.4 g./cc. The product was readily rehydrated simply by stirring with water, and the resulting mashed potato was of good texture and flavor.

Potato Granules Eliminating Add-Back

In order to obtain discrete potato cells in the dry form, it is necessary to add-back large quantities of dried finished product during the process. This results in repeated exposure of the cells to mixing and drying equipment and multiplies the possibility of damage to the potato cells before they find their way out of the process. Since 80 to 90% of the end product is recycled, equipment must be necessarily oversized in relation to pounds of product packed.

Furthermore, a possible hazard exists, because of the necessity of holding a damp potato product for an hour or more under conditions which, if not carefully controlled, could allow bacterial growth. Finally, in order to obtain the necessary cell separation, the soluble amylose fraction of the potato starch must be allowed to retrograde. This reduces the solubility of the starch and reduces the amount of water which the dried product is capable of absorbing in reconstitution to a mashed potato. This necessary retrogradation likewise reduces the ability of the product to reconstitute in cold water, an attribute desired by producers of mashed potato products which are to be frozen.

A process developed by *R.G. Beck and J.H. Rainwater; U.S. Patent 3,459,562; August 5, 1969; assigned to American Potato Company* produces dehydrated granular and agglomerated mashed potato products by a one-pass process. These potatoes can be reconstituted in water at lower temperatures than previous potato granules and absorb more liquid in reconstitution. The essential first step is to produce for foaming a damage-free and debris-free slurry of potato flesh, which consists essentially of water containing for example, 18% substantially completely separated intact potato cells and the conventional additives.

Potatoes are washed, peeled, cut if necessary, precooked, cooled and cooked in the conventional manner. The hot cooked potatoes are then premashed at a temperature of 130° to 150°F. in the presence of liquid and additives such as BHT antioxidant, sulfite salts, monoglycerides and chelating agents. This mash, which is in the form of a lump slurry with 12 to 18% potato solids, is then passed through an altered finisher operated at less than 100 rpm and using a screen with 0.023 to 0.060" perforations and a paddle clearance of ¼". This step removes any residual peel, eyes, fibers or other debris and discharges an undamaged, debris-free slurry consisting of more than 90% intact unicellular potato cells. The remainder consists of groups of two or three attached, undamaged cells.

Potato cells toughened by precooking and cooling prior to cooking can be foamed at 18% solids without rupture of the cell walls. The foaming operation is performed, for example, in an Oakes Mixer which rapidly mixes the potato slurry containing additive emulsifiers in the presence of gas, thereby incorporating the gas into the slurry to form a stable foam with a density in the range of 0.4 to 0.8 g./ml. The actual density is controllable and is dependent upon such factors as the amount and type of emulsifier used, the solids of the potato slurry, the temperature of the slurry, and the amount of gas introduced into the foaming unit. It is preferred to use Myverol 18-07 monoglyceride at a concentration by weight of 0.5 to 1% of the weight of the potato solids.

The stable foam is then divided into two parts on roughly a 70-30 basis. Seventy percent of the foam is sent directly to the nip of a conventional double drum dryer where it is evenly distributed and quickly dried to a moisture of 10% at which time it is scraped from the rolls and introduced to a mixer. This intermediate has a snowflake-like appearance and, although it is undamaged, the bulk density is so low that it is difficult to wet it with required reconstitution liquid to produce mashed potatoes. The balance of the stable foam (30%) is fed directly into the same mixer. Although the composite moisture content is

60%, the mix is a fluffy, porous, and workable damp powder or granular mass, substantially the same in handling characteristics as the moist powder initially resulting in the conventional add-back process. Its moisture level, however, is far above that which can be tolerated when mixing cooked potatoes with potato granules in the add-back process. Early workers in the granule field found that a mix of 50% moisture was an absolute maximum and that 40% moisture mix was a practical maximum. By the foam-dried foam mix process, mixes as high as 65% have been handled with no operating problems.

The very damp mix is then fed into a primary dryer where the moisture content is reduced to 30%. Any dryer which does not damage the product is satisfactory for this step. It is preferred to stop the primary drying at the 30% moisture level since this is the level at which retrogradation of amylose is reported to be most rapid. The 30% mix is immediately cooled to room temperature to further encourage its transition to a mealy texture. The cooled mix is then introduced into a mixer to equilibrate as an aid in granulation, after which the equilibrated mix can be fed onto a holding belt if further retrogradation or a finely granulated end product is desired.

The retrograded 30% mix can then again be mixed to promote finer granulation and screened to the desired size. Oversized agglomerates can be broken down to the desired size without damage, but this small fraction is best returned to the first mixer following the primary dryer. The fine fraction from the screening is then fed into a second dryer where the moisture content is reduced to 7%. After cooling, the fully dry product is given a final screening. Dried agglomerates which fail to pass the desired screen are now returned to the primary mixer in which the foam is introduced.

The process as outlined has extreme flexibility. If large agglomerates are desired, initial cooling and subsequent holding and mixing steps with the 30% mix can be eliminated. Such a process produces a product with increased water absorption ability and with the attribute of rehydration in cold water. If a fine end product consisting of many single cells is desired, all the cooling, mixing and holding steps are utilized. This allows complete amylose retrogradation and results in a product comparable to conventional potato granules, but with a bulk density in the range of 0.7 g./ml.

Another advantage of the flexibility is the ability to vary the process to suit the type of potatoes available for processing. For example, a weak-celled potato normally unsuitable for a conventional potato flake or potato granule process, can be satisfactorily processed by using extra emulsifier with such longer or cooler holding steps in addition as may be found necessary to form a mealy product. It has been found that this flexibility is advantageous in the case of California White Rose, Red River Norland, and some Maine potatoes.

Agglomerating Potato Granules

Potato granules constitute a form of dehydrated mashed potatoes that has been marketed for more than 20 years. The granules are free-flowing, need no refrigeration, take up a small amount of space, and can be packaged so that they suffer no flavor degradation for at least one year. They can be easily reconstituted into mashed potatoes by mixing with proper amounts of hot water and milk. Any temperature of liquid above 160°F. is satisfactory for the reconstitution of the granules.

The texture of the reconstituted granules is improved by whipping; and in fact a certain amount of whipping is required in order to produce a white, light-bodied mashed potato. In institutional kitchens, where machine whipping is available and space is at a premium, potato granules are an ideal source of mashed potato. Whipping is however, considered inconvenient by some in a household kitchen where the whipping must be done by hand. Moreover, although it has been found that the need for whipping can be reduced by increasing the porosity of the potato produced in its dehydrated form, prior methods of increasing this porosity have not been entirely satisfactory.

In the process developed by *R.W. Hutchings and C.H. Stringham; U.S. Patent 3,565,636;*

February 23, 1971; assigned to R.T. French Company an agglomerated dehydrated potato product is produced by adding water in a steady stream, or in a dropwise fashion, into a traveling bed of potato granules, and by screening off and drying the resultant moist agglomerates. The dry agglomerates have uniform disc shapes if the water is added dropwise. They are in the shape of rods if the water is added as a steady stream. They are friable; and each has a porous structure containing pockets of air, which are the source of the incorporated air which brings about, upon reconstitution, a light, fluffy, white mash with a minimum of stirring.

Figure 6.3a illustrates diagrammatically the equipment and the successive steps employed in practicing the process of the method according to one embodiment thereof; Figure 6.3b is an enlarged, fragmentary plan view of that part of the equipment which forms moist agglomerates on a bed of potato granules; and Figure 6.3c is a fragmentary cross-sectional view taken along the line **3-3** in Figure 6.3b looking in the direction of the arrows.

Preparation of agglomerated dehydrated potatoes according to the process begins with the step of mixing dehydrated potato granules with water or cooked potatoes for the purpose of adjusting the moisture content of the granules. The granules may be fed to a mixer **10** from a conventional hopper or bowl **12**; and the water may be supplied from a vessel **14**. Other ingredients, such as preservatives, surfactants, flavoring agents, and color, can also be added to the mixer from tank **14** during this mixing step.

The mixture is fed from the mixer onto a conveyor belt **16**, and beneath a plow **18**, which spreads it into a uniform layer or bed **20** on the upper run of the conveyor. The band or ribbon of granules is conveyed by the belt beneath a dropping device comprising a pipe **22**, which is spaced above and extends transverse to the upper run of belt **16**. Pipe **22** is closed at one end, and is connected at its opposite end to a tank **24** containing an agglomerating solution such as water and/or additives of the type noted above. Nozzles or apertures **26** in the bottom of the pipe cause droplets of the agglomerating solution to fall upon the traveling bed **20** of potato granules in such a manner that no drop overlaps another drop. Each drop of liquid soaks into the potato granule ribbon and forms a disc-shaped object **28**, which although moist and soft, is strong enough to retain its shape through the subsequent processing steps.

About five seconds after a droplet of liquid strikes the bed or ribbon of potato granules, the resultant moist disc has achieved its final shape, and has enough strength to undergo screening. The discs can be screened away any time after that, the sooner the better. Therefore, shortly after the discs **28** are formed in the bed **20**, they are dropped off the belt **16** onto a screen **30**. The screen is somewhat downwardly inclined and the discs slide off onto a further conveyor belt **32**, which carries them to a dryer **38**. The fines, or the granules which have not been agglomerated by droplets of liquid, fall onto a return conveyor **33** which transports the fines (granules) to an elevator **34**, which feeds the fines back to mixer **10** for reprocessing.

The wet discs, which are separated from the fines by screen **30**, contain from 40 to 60% solids. These wet discs are conveyed by belt **32** through the dryer **38**, and are dried in any convenient fashion, such as by a current of warm air passing upward through the belt **32** and the discs thereon at a temperature of between 150° to 250°F.

The bulk density of the final dried product can be in the range of 0.3 to 0.5 g./cc. The bulk density can be controlled by the size of the droplets used. When water droplets are released from tubing having an internal diameter of $\frac{1}{16}$ inch about 560 droplets will weigh 1 ounce. The bulk density of the final product made from droplets this size is 0.35 g./cc. When droplets are released from an 18 gauge hypodermic needle, 1,300 droplets weigh 1 ounce, and the product made from droplets this size has a bulk density of 0.37 g./cc. When a 27 gauge hypodermic needle is used to release the droplets, 4,470 droplets are required to weigh 1 ounce, and the product made from droplets of this size has a bulk density of 0.42 g./cc. When small drops are used, such as those formed by a 27 gauge hypodermic needle, the product resembles small spheres and does not have the disc shape

FIGURE 6.3: METHOD OF AGGLOMERATING DEHYDRATED POTATOES

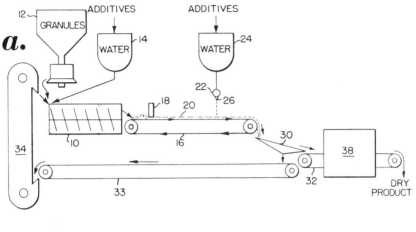

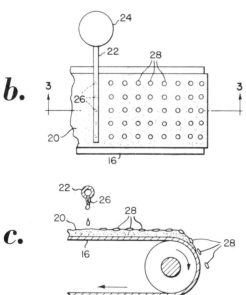

Source: R.W. Hutchings and C.H. Stringham; U.S. Patent 3,565,636; February 23, 1971

that is evident when larger droplets are used. The distance the drop of water falls before
striking the bed of wet potato can be from ½ to 7 inches. At distances greater than 7
inches the droplets achieve such velocity that they splatter into several smaller drops. The
depth of moist potato on the bed into which the water droplets fall should be great enough
that the droplet does not wet the belt. It has been found that a ³⁄₁₆ inch deep bed of moist
potato granules is satisfactory. The particle size distribution of the granules used as starting
material influences the characteristics of the reconstituted agglomerates. The best size

range would be granules which would pass through a 50 mesh screen. Granules larger than 50 mesh yield a grainy-textured product. The water used for forming the droplets can contain various additives such as preservatives, surfactants, flavor, nutritional additives and color. The temperature of the water used for the droplets can vary over a considerable range. The most desirable temperature is in the vicinity of 110°F. Tap water seems to work as well as distilled water.

The moisture content of the potato granules into which the water falls is very important. When the moisture of these granules is below 12% the resultant dry agglomerates reconstitute very slowly and the mash is lumpy. When the moisture content of these granules is above 28% moisture, it is difficult to separate them from the moist agglomerates. The dry agglomerates formed from these high moisture granules are shapeless and irregular. The range of 12 to 28% moisture granule is most desirable for these reasons. The following example illustrates the process. The starting dehydrated potato granules were produced by a conventional process.

Example: Four pounds of dehydrated potato granules having a 10% moisture content were placed in a 10 quart Hobart bowl, and 1 pound of potatoes, peeled and cooked in known manner, was riced into the potato granules. Four-tenths of a gram of a monoglyceride were added, and the mixture was mixed for five minutes at high speed using a wire whip. This potato mixture, which had a moisture content of 23.7%, was spread on a tray ½ inch deep, and droplets of water from tubing of $\frac{1}{16}$ inch i.d. were allowed to fall into the potato mixture from a height of 1 inch. The resultant moist discs were screened away from the damp potato granules with a 12 mesh screen, and were placed in a screen bottom tray and dried for 12 minutes by blowing air at 150°F. upwardly through them. The granules or fines passing through the 12 mesh screen were returned to the tray for the formation of more moist discs. The bulk density of the dried discs was 0.35 g./cc. The dried discs were reconstituted in boiling liquid in seconds to give a mash of light, fluffy texture and good flavor.

Producing Granules from a Mixture of Flakes and Granules

R.G. Beck; U.S. Patent 3,457,088; July 22, 1969; assigned to American Potato Company describes a process for manufacturing potato granules which utilizes a mixture of potato flakes and potato granules to produce a fine granular product with the performance characteristics of potato granules produced by the add-back process. Two types of dehydrated potato products from which instant mashed potatoes can be made by reconstitution in liquid have met with acceptance by housewives, institutions, and the Armed Forces. Potato granules and potato flakes are both produced commercially in large amounts.

The potato flakes have a commercial density of 15 lbs./ft.^{3}. The advantages are as follows: (1) low density is an advantage in housewife applications; and (2) can be reconstituted in cold liquid for specialized applications such as frozen TV dinners. The disadvantages are: (1) will not tolerate boiling liquid reconstitution or whipping; (2) density is too low for Armed Forces and most institutional uses; and (3) density is too low for economical inert gas packaging.

The potato granules have a commercial density of 56 lbs./ft.^{3}. The advantages are as follows: (1) high density is an advantage for Armed Forces and most institutional uses; (2) high density allows efficient inert gas packaging which aids storage stability; and (3) potato granules can be reconstituted in boiling liquid and can stand whipping without creating poor texture or pastiness. The disadvantages are: (1) high density for retail sale and use; and (2) will not reconstitute in cold liquid.

A number of processes have been developed to eliminate the disadvantages of commercial potato flakes and are given in the section on potato flakes. Unfortunately, when the bulk density of the flakes is increased, other desirable characteristics are lost. The object of the process described here is to produce by a continuous and practical method a dried potato product with all the advantages of potato granules, using commercial potato flakes as the principal starting ingredient in the process. A product with all the desirable characteristics of potato granules was made by uniformly moistening standard -8 mesh potato flakes with cold water to 46% moisture. The moistened flakes were then allowed to equilibrate and retrograde for 30 minutes. The wet flakes were then mixed with dry commercial potato granules equal in weight to one-third the solids of the flakes. The moisture of the mix was 40% and the source of the potato solids was 75% flakes and 25% granules. The moist mix was then allowed to stand one hour before mixing again to further granulate the product. The mix was then dried and screened over a 20 mesh screen. Approximately 95% of the dried product passed through the 20 mesh screen. This product had a density of 44 lbs./ft.^{3} and more than 30% passed a standard 80 mesh screen. If a -20 mesh product is the desired final product the small percentage of coarser fraction could be rewet along with the potato flakes in a continuous process.

The end product from this process has excellent flavor and texture and has all the advantages of potato granules, being capable of boiling water reconstitution and capable of whipping without texture damage. This product is much finer than could be obtained by partially reconstituting potato flakes alone and processing them in a comparable manner. By adding all the reconstitution water to the potato flake fraction alone, the advantages of higher moistures were obtained, namely, easier flake breakdown to potato cells and small agglomerates and faster retrogradation. By adding the dry potato granules after partial retrogradation, the moisture content of the mix is reduced thereby creating a friable mix which does not tend to agglomerate and results in a larger percentage of finer particles.

In another patent by *R.G. Beck; U.S. Patent 3,458,325; July 29, 1969; assigned to American Potato Company* a potato product is described which consists of small porous balls having an appearance which is completely different from either potato flakes or potato granules. This product is made from the same ingredients as the potato product in the previous patent, in the same proportions, 3 parts by weight of -8 mesh potato flakes and 1 part by weight of potato granules, but the two types of dehydrated potato are mixed dry, then rewet to a moisture of between 25 and 30%, mixed, allowed to stand for an hour, mixed again, and finally dried.

This product has the desirable characteristics of both potato granules and potato flakes, with a density of $\frac{2}{3}$ that of granules and about twice that of commercial potato flakes. It can be economically gas packed. The product reconstitutes fully in cold water, can be reconstituted in boiling water with proper care, and is capable of absorbing considerably more water than potato granules in both cold and hot water reconstitution. The larger porous particles result in slower reconstitution than in the case of potato granules thereby eliminating a major reconstitution problem often encountered by the housewife. That is to say, if the reconstitution liquid is added to a mass of granules or a mass of granules is added to a pool of liquid at rest, the very fast rehydration of the granules at the limited liquid-granule interface interferes with the penetration of the granule mass, and balls with dry or unreconstituted centers can result.

The mixing of potato flakes and potato granules gives results which could not be obtained by the use of either product alone. The flake constituent is necessary because the amylose fraction is in the soluble or unretrograded form which is capable of absorbing cold water and larger amounts of hot water. The potato granule fraction is necessary since this is the only way to get the desired small particle size without having ruptured cells which would cause pastiness. Potato granules contain retrograded amylose and do not contribute any stickiness to the mix. This promotes the necessary granulation required to produce products with the improved rehydration characteristics. The flakes and granules used in this process are the standard products of commerce containing the usual additives such as emulsifiers, sulfites, and antioxidants.

Multicellular Product Dehydrated by Continuous Spray Drying

The disadvantages of potato flakes and add-back potato granules have been reviewed at the beginning of the previous patent. It is obvious that a product having all the assets of potato granules and potato flakes and none of their disadvantages would be extremely desirable.

J.H. Rainwater and R.G. Beck; U.S. Patent 3,764,716; October 9, 1973; assigned to American Potato Company have developed a process relating to the preparation of a damage-free and debris-free slurry consisting substantially entirely of unicellular potato particles suspended together with the desired additives in sufficient water to be sprayable without damage, to a continuous spray dry process for drying such a slurry, and to a resulting product which meets these exacting criteria.

The process will be more clearly understood by reference to Figure 6.4, which illustrates a flow sheet of the preferred process. The potatoes are washed, peeled, sliced, and prepared for cooking in a conventional manner. In order to produce all the desired objectives in a product having the characteristic flavor and color of mashed potatoes produced from boiled potatoes, the potatoes must be peeled as at **10**, since peeling results in improved color, both dry and after rehydration, as well as the desired bland flavor.

Immediately following the removal of the peel, the potatoes are sliced, rinsed to remove free starch, and immersed in water. If precooking is used, the rinsed potatoes may preferably be introduced directly into the precooking water. The water precooking **11**, cooling **12**, and final steam cooking **13** may follow accepted prior art. It is preferred to precook and cool since this yields, after cooking, a potato cell more resistant to possible damage in the following steps.

Immediately following the cooking step, the cooked potatoes are continuously mixed or mashed with the addition of water, conveniently containing the desired additives in solution, to produce a free-flowing slurry. This slurry should contain 17% or less in potato solids, and is preferably in the 10 to 14% potato solids range. The purpose of this mixing step is to make a substantially homogeneous slurry the solids of which consist of a small fraction of debris in the form of peel, eyes, rot, fibers and undercooked spots of flesh and the balance of wholesome, cooked potato flesh, approximately 70 to 90% in the form of intact single potato cells, the balance being small lumps of unmashed cooked potato in the order of one-fourth inch diameter. This can be accomplished in mixers of various designs. A mixing step is necessary in order to prepare the slurry for subsequent processing steps. The following must be observed:

    (1) Water must always be present when mashing takes place. It serves as a lubricant and prevents damage to the fragile cooked cells.

    (2) Additives include sulfite salts, mono- and diglycerides, sodium acid pyrophosphate, and antioxidants such as BHA and BHT. Tests indicate that this is the most advantageous point to incorporate these additives.

    (3) In this mixing step, temperature is controlled in the range of 130° to 150°F. in order to best assist in cell separation and beneficial effects of the additives.

    (4) Where a complete prepared mashed potato product is desired, making it unnecessary for the user to add milk, this process permits an improved product to be produced by adding either fluid milk in lieu of all or part of the water or dried milk solids at this mixing stage.

The final step in the production of this sprayable slurry is the finishing step which substantially completely separates the potato flesh into individual, discrete, undamaged cells and simultaneously removes all debris. It was found that, surprisingly, this can be done continuously in a conventional finisher or pulper **26** of the type which is widely used in the canning industry for removing seeds and skin from fruits and vegetables, if the operation of the device is properly controlled. Rotary paddles **27** operating with proper clearance break up substantially all the wholesome potato flesh into unicellular particles or particles consisting of clumps of only a few cells which flow in the thus finished slurry through perforations **28** of predetermined size into a receiving hopper **29**. Debris is retained on the screen **31** and is discharged continuously from a separate outlet **32**. It is a surprising fact that this debris contains all the rot that may be in the cooked potato even when test runs are made with potatoes having commercially wholly unacceptable proportions of rot even in excess of 50%. The speed of rotation of the paddles **27** and size of perforations **28** used in the finisher must be selected so that the small remaining potato lumps of wholesome flesh are separated into individual cells without damaging the delicate potato cell walls.

Even with the preceding mixing step which is included primarily to reduce cell damage, it is found that the operation of this finishing step is quite critical. Speed, screen hole size, and paddle clearance must be optimized for any given slurry solids content. As a specific example, for a slurry of 12% potato solids in the equipment, it has been found that 60 rpm, 0.045 inch perforations, and one-fourth inch paddle clearance gives a slurry consisting of more than 90% intact unicellular cells. The remainder consists of groups of two or three attached, undamaged cells. With slurry as low as 10% solids, perforations as small as 0.023 inch are satisfactory; at 17%, 0.060 inch are needed. The pulper or finisher **26** is intended to be operated at rotation speeds in the range from 680 to 780 rpm. Such specified speeds would be wholly unacceptable in producing the slurry of this process,

FIGURE 6.4: PROCESS FOR PRODUCTION OF MULTICELLULAR DEHYDRATED POTATO

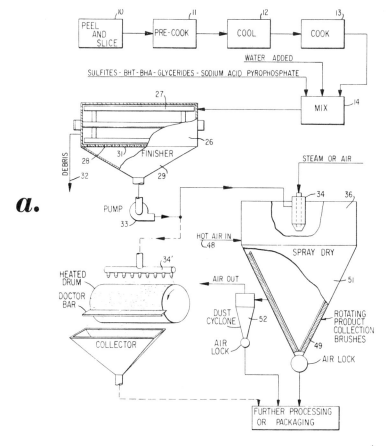

(continued)

FIGURE 6.4: (continued)

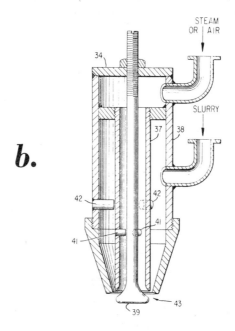

b.

Source: J.H. Rainwater and R.G. Beck; U.S. Patent 3,764,716; October 9, 1973

for cell damage in undesirable amounts at speeds as low as 100 rpm has been found. A slurry of the undamaged debris-free quality produced by the process already described is essential to the subsequent production of an acceptable dehydrated mashed potato product by spray drying. The spraying can be either into a stream of hot air or onto a heated surface.

The slurry discharged by the finisher **26** can be pumped without damage by a conventional lobe-type pump **33** to a dispersing device **34**, which is shown mounted in the top of a drying chamber **36**. Unacceptable cell damage using a centrifugal pump at this stage has been experienced. Some dispersing devices used in conventional spray drying were found to cause excessive potato cell rupture. Nozzle or disk design and atomizing fluid pressures and velocities are critical because potato cells can be damaged by excessive pressure, abrasion, or fluid acceleration.

Figure 6.4b shows the preferred nozzle design. The slurry flows in the annulus between two concentric pipes **37** and **38**. The dispersing gas, most conveniently air or steam, flows through the inner pipe **37**. The longitudinally adjustable mushroom-headed stem **39** is located concentrically within inner pipe **37**. It is held centered and steadied against vibration by a plurality of locating pins **41**. Similarly, the inner pipe is spaced and held in relation to outer pipe by a plurality of locating pins **42**. At the exit **43** of the nozzle, the two fluids mix so that the potato slurry is dispersed in a conical-shaped pattern. The size and shape of this pattern can be controlled by changing the position of the head of the stem **39**. Thus, the nozzle can be adjusted so the dispersed slurry does not contact the walls of dryer **36**. The spray is introduced into the dryer supplied with air at **48** heated in the range of 250° to 400°F. in sufficient quantity to dry at the desired rate. It is desirable that the air flow be uniformly distributed in the dryer and that the dryer be of such height as to give approximately 15 seconds or more contact time between the air and dispersed particles. The bottom of the dryer, which may be either sloped or flat, is continuously swept by nylon brushes **49**. To produce a predominantly unicellular product similar in size to add-back granules, conditions of slurry feed rate, air temperature, and air flow are adjusted so that all particles are dried below the adhering stage before reaching the floor **51** of the dryer.

To produce a predominantly multicellular product, dryer conditions are adjusted so that some damp particles fall on the dryer floor where they serve as nuclei for multicellular units formed by the rolling action of the brush sweep. By moisture interchange, these multicellular units equilibrate at a moisture of 6 to 14%. If this final moisture is higher than desired for finished product storage, dryer conditions can be altered or the product can be further dried by conventional means such as a fluid bed dryer (not shown). In all cases, the finest potato particles (5 to 30% of the potato solids, depending upon dryer design and operating conditions) are entrained in the air leaving the dryer and must be recovered from the air by cyclone separators **52**, settling chambers, or bag collectors. This fraction is unicellular and can be combined with the balance of the product swept from the dryer floor.

To produce a product consisting of practically 100% multicellular units, an additional step is used. The fine fraction from the collection system **52** is combined with the unicellular material which has been separated by screening from the fractions deposited on the dryer floor. Water is uniformly added to bring the moisture content to the range of 20 to 40% while the product is subjected to a rolling and tumbling action. It is preferred to do this continuously by spraying a fine water mist into the product as it passes through a rotating horizontal drum (not shown) equipped with lift flights and baffles.

At this elevated moisture, the particles adhere to form multicellular units. If milk solids are desired in the multicellular product, and have not already been completely added in the earlier mixing step, milk solids as desired can be included in the rewetting liquid and when so included assist in promoting particle adhesion. The size of the units can be controlled by varying the moisture level; as moisture increases, unit size increases. The multicellular units thus formed are then subjected to a gentle final drying operation such as in a vibratory or fluid bed dryer.

The above rewetting and drying operation also improves the mealiness of the finished mashed potato product. A further improvement yet in mealiness can be obtained by instituting a holding step between the rewetting and final drying steps. Longer times, lower temperatures, and moistures in the range of 30% all promote this improvement in mealiness. This rewet multicellular product can then be combined with the multicellular fraction from the dryer floor **51**.

As shown by the dotted line flow arrows from pump **31** in Figure 6.4a, the slurry discharged from the finisher **26** can alternately be fed to a plurality of discharge devices **34'**, which can be of the type shown in Figure 6.4b, and by them sprayed onto the surface of a heated rotating drum **60**. Slurry sprayed on the surface of the drum is dried thereon to the desired moisture level and transferred therefrom to the collector **61** by the doctor blade **62**.

Product from the collector goes as shown by the further dotted line flow arrows to further processing or packaging as desired. With a drum surface speed of 65 feet per minute and drum surface temperatures of 295°F. a fully dried unicellular product in the collector can be obtained. By appropriate relative adjustment of spraying rate, drum speed, and temperature this may be varied from a fully dried, unicellular product ready for packaging to a product only dried down sufficiently not to adhere and in the form of damp lumps which can be advantageously processed by conditioning and further drying to a low-density multicellular granular product. In comparison with the drum drying of the flake process, no applicator rolls are required as the coating thickness is not critical since it can be readily varied as the moisture of the product leaving the drum is varied to avoid any significant cell damage or indeed even flavor variation from cell to cell. Further, in comparison with spray drying in an air stream, problems of cell damage during product collection are avoided. Thus, it is seen that by selecting proper conditions, procedures, and particle sizes, there is a flexible process whereby it is possible to produce products ranging from all unicellular units to all multicellular units or any desired combination of the two. The selection of the exact process would be dictated by the proposed end use. For example, the Armed Forces would surely prefer the unicellular form because of savings in container costs and shipping space. On the other hand, the housewife might prefer the more bulky multi-cellular form which can be simply reconstituted without mixing or special equipment.

Products produced by drying the slurry of this process have the distinctive character though granular, whether multicellular or unicellular or mixed, of absorbing in the range of 5.26 to 6 times their weight of liquid upon reconstitution, thus equalling or exceeding the performance of flakes in this respect, and of doing so without pastiness or loss of true mashed potato texture regardless of liquid temperature, thus exceeding the performance of flakes.

FLAKES

The basis potato flake process was developed at the Agricultural Research Service in Philadelphia by *J. Cording, Jr. and M. Willard, Jr.; U.S. Patents 2,759,832; 2,780,552; and 2,787,553; assigned to the U.S. Secretary of Agriculture.* A preferred process included the steps of peeling, trimming, washing, and slicing the potatoes, immersing them in water at 155° to 180°F. for 30 minutes, cooking in live steam at atmospheric pressure for 16 minutes, mashing the potatoes with a planetary mixer, diluting the potatoes with water to a solids content of 20%, spreading the resulting slurry in the form of a thin film (0.005 to 0.015 inch thick) on the rolls of a double drum dryer, the rolls of which were heated internally by steam, and dehydrating the potatoes until they contained only 4% moisture. Subsequent patents substitute a single drum dryer of the type used for flour production for the double drum dryer.

Additive for Improving Texture

Cooked, dehydrated potatoes have been made by forming in effect a mashed potato and incorporating into the wet mash various conventional components such as an antioxidant, preservative, and parting agent to improve separation of the material from rolls or the like during processing. The mash is then dried as in a thin layer on steam-heated rollers. The dried sheets so formed are broken into flakes and the product packaged for shipment. The main disadvantages of reconstituted potatoes produced in this manner are related to the texture of the product, gumminess and stickiness being common complaints.

J.F. Hale, R.A. Klein and E.M. Bradway; U.S. Patent 2,980,543; April 18, 1961; assigned to The Borden Company have attempted to eliminate these disadvantages by applying to the potato, after cooking and drying, a texture improving and antisticking agent, hereinafter referred to as the improver.

In general, the process is as follows. Cooked and disintegrated potato containing approximately its natural proportion of water and, if desired, the conventional additives are applied as a thin layer to a drying roller or drying rollers and there dried. The resulting dried thin sheet is then cut or broken into particles, thin flakes being illustrative. In a subsequent step, the improver is introduced. In the commercial embodiment, the improver is usually glycerine monopalmitate (GMP), conveniently mixed with a hydrogenated vegetable fat, and it is sprayed upon the dried flakes. An example will illustrate the process.

Example: Maine Katahdin potatoes of water content 17.5 to 20% are peeled as by pressure steam at 115 psi, de-eyed, washed, sliced, and cooked in water or in steam or both, all in conventional manner. Steam cooking, if used, is ordinarily with atmospheric pressure steam for 17 minutes. The potato slices so cooked are then disintegrated by passage through

0.4 inch holes such as those of an ordinary type of meat grinder from which the cutting knives are removed. The resulting potato mash is then mixed with 0.075% of Tenox 6 antioxidant and 0.1% of GMP parting agent on the dry weight of potato and with sodium sulfite (3 parts) and bisulfite (1 part) in amount to provide 0.02% of SO_2 in the final dried product. The whole is then applied as a layer 0.005 to 0.01 inch thick over the exterior of a steam-heated drying roller. As the roller is rotated to the position at which the sheet has been dried, the dried sheet is removed by scrapers. At this stage it contains 3.5 to 4% of water although proportions of water as high as 7% or so are considered permissible but not desirable. The dried sheet is then passed through breakers or cutters which form it into small flakes about a half inch square, resembling somewhat ordinary soap flakes.

The dried potato flakes are treated with the texture improving and antisticking agent, as by being sprayed therewith at the rate of 0.55% of the glycerine monopalmitate (Myverol 16-00) dissolved in an equal weight of hydrogenated cottonseed oil, this rate of addition corresponding to a total weight of GMP and fat of 1.1 to 1.3% on the dry weight of the potato flakes.

The product, when reconstituted by being stirred at the rate of 1 part of the treated flakes to 5.5 parts of a mixture of 1.5 of boiling water to 0.5 of milk, was ready to serve after less than 1 minute's stirring into the heated liquid at 160° to 170°F. The reconstituted material was mixed with a small proportion of butter and salt for seasoning. The product was not sticky in the vessel in which the reconstitution was effected or during eating. It had the desired mealy consistency and closely resembled the consistency, taste, and acceptability of home prepared mashed potatoes.

Cooling After Precooking

J.F. Sullivan, J. Cording, Jr. and M.J. Willard, Jr.; U.S. Patent 3,012,897; December 12, 1961; assigned to the U.S. Secretary of Agriculture found that the inclusion of a cooling step between the precooking operation and the final cooking of the potato causes a change in certain characteristics of the potato. The result is an improvement in texture of the cooked potato and in the potato products made therefrom. The improvement in texture is so significant that potatoes of very low solids content, which are ordinarily not mealy in texture, when processed into dehydrated products according to the process have a desirable mealiness upon reconstitution to mashed potato. Another advantage is that the potatoes, after dehydration, can be rehydrated with liquids at high temperatures without materially breaking down the cell structure. A further advantage is that the dehydrated flakes can be made of smaller size, thus giving a higher density product, than was heretofore possible and still give a reconstituted mashed potato of the desired texture.

In the amended process the potatoes are peeled, washed and trimmed by any conventional method, the raw potato pieces precooked at 140° to 180°F. for 10 to 60 minutes, then the precooked potatoes are cooled by contact with cold water (for example, by spraying with water, immersion in water, or a combination of spray and immersion), after which the cooked potato is cooked, as with live steam or boiling water, until soft enough to mash (rice), and then dehydrated. Dehydrated potato flakes are produced, using a single drum flour type dryer.

The cooling of the precooked potatoes can be achieved by using a water pressure wash, by holding in a tank containing water, or by a combination of these procedures. In general, the lower the temperature of the water and the longer the period of cooling, the greater the increase in mealiness of the cooked potato product. The extent (degree) of cooling and the duration of cooling can thus be varied according to the type and history of the particular batch of potatoes being processed and the mealiness desired in the product. Even with low-solids potatoes a cooling treatment of 20 minutes immersion in water at 65°F. or lower or a 4 minute wash with tap water, 65° to 70°F., plus holding for 16 to 20 minutes in water at the same temperature has sufficed to improve significantly the texture of the cooked potato and the flakes and granules made therefrom.

Upgrading Low Solids Potatoes

It is known that the texture of some waxy type potatoes or nonmealy varieties can be changed by a gelatinization of the starch by rather low temperature cooking followed by cooking at temperatures of 212°F. This texture change makes possible the production of a mealy product from nonmealy potatoes and makes possible the production of a suitable product for use as mashed potatoes by means of flake dryers. This change of temperature, however, has not been effective for all types and solids contents of potatoes. It is known that the higher solids content potatoes are more amenable to production of a satisfactory dehydrated product.

The natural solids content of potatoes varies between 16 and 24%. Solids content varies among varieties of potatoes, geographical areas where the potatoes are grown, growing conditions such as weather, soil, fertilizer used, and the like. Heretofore, the production of dehydrated potato flakes of satisfactory texture and flavor has been largely limited to the use of relatively high solids potatoes having 20% solids or more. Paradoxically, those potatoes which produce the best dehydrated product are also in greatest demand for shipping. Those potatoes which are least adapted to dehydration are in least demand for shipping and therefore more readily available for processing in the areas in which they are grown.

A process has been developed by *A.M. Cooley; U.S. Patent 3,016,303; January 9, 1962; assigned to the University of North Dakota* to provide a method of drying potatoes wherein low solids content potatoes are upgraded to produce a dehydrated product of satisfactory flavor and texture.

The improvement in the process is due to a treatment during the cooking cycle of the potatoes, before they are drum dried. The potatoes, which have been washed, peeled, trimmed, sliced, precooked and cooled in accordance with the process described in the previous patents, are fed into a cooker, which is desirably of the screw conveyor type. It is connected to a steam supply line and also to a vacuum line for the intermittent application of steam pressure and vacuum. The cooker is desirably tilted slightly with its entrance end lower than its exit to permit withdrawal of steam condensate.

In the cooker the potatoes are subjected to steam cycles of from 4 to 6 minutes' duration and then are subjected to vacuum for periods of 2 to 4 minutes' duration, after which the steam and vacuum cycles are repeated in sequence several times; i.e., four to six times. Under preferred operating conditions the potatoes are subjected to 5 minute steam cycles and 2 to 3 minute vacuum cycles and the potatoes are permitted to remain in the cooker to undergo a total of five steam and vacuum cycles. Cooking time varies somewhat depending upon the slice thickness but the potatoes should be subjected to a minimum of 20 minutes' cooking time.

Introduction of steam to the closed cooker results in a temperature between 212° and 220°F. The length of the vacuum cycle depends somewhat on the efficiency of the vacuum pump. A vacuum between 26 and 27 inches of mercury is relatively easily attainable and has been found to produce satisfactory results. During the vacuum cycle the temperature is reduced to 125°F. The alternate steam and vacuum treatment results in an overall reduction of water content in the potatoes with a consequent increase in solids content. The greatest water loss is in the first vacuum cycle. The steam condenses on the potato in subsequent cycles in reheating the potato and apparently a considerable amount of water remains on the surface of the potato slices and reenters the potato. The net water loss is still sufficient, however, to upgrade potatoes of low solids content so as to permit the production of high quality flakes at good production rates. The alternate steam and vacuum cycles may be automatically timed by means of commerically available timers.

After the cooking treatment, the potatoes are mashed, dehydrated, flaked and packaged. The process is illustrated by an example.

Example: A batch of Kennebec variety of potatoes was processed. This is a medium solids content potato grown in the Red River Valley of Minnesota and North Dakota. The potatoes processed have a solids content of 20.2% and contained 79.8% water. The potatoes were washed and peeled by submersion for 3 minutes in a 20% solution of sodium hydroxide maintained at 180°F. The peeled potatoes were washed, trimmed and cut into half inch slices. The slices were precooked in water for 20 minutes at 160°F. and then cooled in water for 20 minutes at 50°F. The precooked and cooled potato slices were then subjected to five alternating cycles of steam at 220°F. for 5 minutes each followed by 26 inches of vacuum for 2 minutes. Each steam cycle was followed by a vacuum cycle.

The potato weight upon introduction to the cooker was 57.4 lbs. and at the end of the first vacuum cycle this batch weighed 51.9 lbs. for a weight of 5.5 lbs. of water. The water loss is greatest during the first vacuum cycle. In each of the following four cycles the water loss varied between 1.2 and 1.4 lbs. After the fifth cycle the final weight of the batch was 46.8 for a total water loss of 10.6 lbs. This represented a weight loss of 18.4% with the result that the solids content of the potatoes was upgraded to 23.6%. The product was mashed and dried to produce a flake of good flavor and texture. The result of the alternating steam pressure and vacuum in the cooking step upgraded the potato from one of only medium solids content to one of relatively high solids content. The upgrading of the raw material resulted in increased dryer capacity.

Increasing Bulk Density of Flakes

Potato flakes normally have a package density of 15 lbs./ft.^{3}. This can be increased at most to 30 lbs./ft.^{3} by grinding the substantially dry flakes. Finer grinding ruptures too many potato cells and destroys body so that on reconstitution an unpalatable pasty or pudding-like mashed potato results.

R.K. Eskew; U.S. Patent 3,021,223; February 13, 1962; assigned to the U.S. Secretary of Agriculture describes a process in which potato flakes having a moisture content in the range of 5 to 20% and a bulk density from 5 to 30 lbs./ft.^{3} are hydrated to a moisture content from 25 to 50%, the hydrated flakes are mechanically manipulated to achieve the desired form and are then dried to a moisture content of from 5 to 8% to produce a product having a bulk density from 35 to 58 lbs./ft.^{3}.

It has been found that when potato flakes are hydrated to a moisture content of between 25 to 50% they can be subjected to mechanical action which separates them into smaller pieces by cleavage primarily between individual cells rather than by fracture of cell walls. If the mechanical action is applied at a moisture range between 36 and 45% the cells may separate substantially completely into individual cells or small clusters thereof so that the identity of the flake is lost. Such a product when dried may have a bulk density between 50 and 57 lbs./ft.^{3}.

On the other hand if the moisture range during manipulation is between 25 and 35% the flakes may not lose their identity. Some separation into individual cells or small cell clusters may occur but in the main the dry product will consist of flakelets mostly smaller in size then before hydration. Some may be aggregated, laminated, or slightly curled which with the interstitial flake fragments and individual cells will contribute an increase in bulk density above that possessed by the flakes before treating by the process. Operation in this lower moisture range results in a product possessing the desirable flavor and texture properties of the initial flakes, at the same time retaining the appearance of very small flakes or flakelets yet possessing a relatively high bulk density. The following examples will illustrate the process.

Example 1: Russet Burbank potatoes containing 21.4% solids were lye peeled and trimmed in a conventional manner. They were then sliced to ½" slabs and precooked in water at 160°F. for 20 minutes, followed by cooling in water at 35°F. for 30 minutes to retrograde soluble amylose. The slabs were then steamed at atmospheric pressure for 30 minutes. These were divided into two parts, one to be made into flakes and the other to be employed as shown later. The flakes were prepared as follows. The cooked slabs were riced and there was incorporated an emulsion of glycerol monopalmitate and butylated hydroxyanisole and butylated hydroxytoluene in water to give the equivalent of 0.3% glycerol monopalmitate and 100 ppm each of butylated hydroxyanisole and butylated hydroxytoluene on the solids in the mash. A solution of Na_2SO_3 and $NaHSO_3$ was incorporated into the riced material to give 400 ppm equivalent SO_2 based on solids in the mash. This was dried on a single drum dryer operated at 4 rpm and employing steam in the drum at 55 psi. The sheet of product was cut in a sharp knife hammermill through holes of $\frac{1}{8}$" diameter. The flakes contained 7.5% moisture.

In order to obtain material of higher density, the flakes prepared as above were mixed with the previously prepared slabs of cooked potatoes using a planetary mixer operated at 100 rpm. The proportions were calculated to give 40% moisture. There was added to this enough solid glycerol monopalmitate to bring the total in the mix to 1.0% on the basis of potato solids present. Mixing was for 3 minutes followed by 30 minutes holding with 10 second stirrings at 10 minute intervals. The container was covered during this period. The cover was then removed and agitation was continuous for 30 minutes.

During the entire 60 minutes holding and mixing period the temperature dropped from 150° to 70°F. and the moisture from 40 to 36%. The mixture was sieved through a 20 mesh vibrating sieve. The oversize aggregates which separated were forced through a rotary screen having holes 0.023" diameter to subdivide them. This screened material was combined with that passing through the sieve and the entire product was dried to 12% moisture in an airlift dryer at 250°F. and air velocity of 2,000 ft./min. The 12% moisture product was further dried in a fluidized bed dryer to 6% moisture. The product from the foregoing example had a density of 56 lbs./ft.^{3}. All of it passed a 30 mesh screen and 80% passed 50 mesh. The entire product could be reconstituted with hot water and milk to a product of good flavor and texture.

Example 2: Same as Example 1 except:

    (a) California White Rose potatoes of 19% solids were used.
    (b) Steaming at atmospheric pressure was for 40 minutes instead of 30 minutes.
    (c) 0.7% glycerol monostearate was added in preparing the flakes instead of 0.3%.
    (d) The slabs of cooked potatoes were cooled to 80°F.
    (e) The proportions of flakes and cooked slabs were such as to give a mix moisture of 28% instead of 40%.
    (f) No extra glycerol monoglyceride was added to the moist mixture.
    (g) The total cycle in the moist state consisted of mechanical manipulation for 45 minutes in an uncovered bowl of a planetary mixer.
    (h) The moist product was dried without sieving or screening.

The product from this example consisted primarily of very small flakes. It had a bulk density of 43 lbs./ft.^{3} and 99% of it passed a 12 mesh screen. This could be reconstituted in 2 minutes to a mash of good consistency and flavor using hot water and milk.

Flaked potato products prepared by the drum drying processes previously described have a relatively low bulk density, sometimes as low as 11 lbs./ft.^{3} if the potatoes from which they were made had a low solids content. In contrast, potato granules have a density in the range of 50 lbs./ft.^{3}, which is very advantageous in packaging them.

M. Pader; U.S. Patent 3,067,042; December 4, 1962; assigned to Lever Brothers Company has developed a high bulk density product of exceptionally fine quality which can be obtained from low solids content potatoes by first comminuting the dehydrated potato flakes obtained according to the Department of Agriculture processes previously described, removing the fines, i.e., the very fine particle sized material, and returning these fines to a subsequent batch of mashed potatoes prior to drum drying thereof. When this process is followed, the first batch of dehydrated product has a considerably higher bulk density than drum dried materials prepared in a conventional manner, and at the same time, reconstitutes to mashed potato product having nonstarch and acceptable consistency. The fines are not discarded, but are added to a subsequent batch. Although it would be expected that the fines would cause subsequent batches to be starchy when reconstituted, it was found that the quality of subsequent batches was at least equal to and in some instances superior to the dried product from which the fines had been removed.

Fines as defined herein is meant to describe potato flakes or portions thereof which will pass through a 6 mesh or smaller screen. These fines, if present in a substantial amount, cause the texture of the reconstituted potato flakes to be excessively pasty and starchy. It is recognized that larger size potato flakes can be added back to subsequent batches of mashed potatoes. Such a practice is not usually advantageous, however, as dehydrated potato flakes having a particle size of larger than 6 mesh generally provide a product of acceptable mealiness and texture without further processing. On the other hand, particle sizes as fine as those found in potato flour may be added back to a subsequent batch of mashed potatoes according to this process.

The amount of fines which are added to a subsequent batch of mashed potatoes is preferably between 0.5 and 5%, based on the weight of the wet mashed potatoes being dried. This amount, however, will vary according to several factors. In the first place, the amount of fines added back depends upon the solids content of the tubers undergoing processing. The lower the solids content of these tubers, the larger the amount of fines that should be added. In the second place, the amount of fines added back will depend on the total quantity of fines recovered after the flakes have been subdivided to the desired bulk density. It is understood, of course, that if only a very small amount of fines is recovered, it may be advisable to add either larger pieces of flake or more finely comminuted pieces of flake. In other words, if the dry solids concentration of the mash needs to be increased in order to improve the reconstituted quality of the flakes and the supply of fines desired from the process is insufficient, larger sized particles may, on occasion, be employed.

In addition to the density consideration, the percentage of fines removed from a given batch of dried flakes will usually be the minimum amount required to provide a reconstituted product which has acceptable texture, and is of a nonstarchy and nonpasty consistency. These factors are in turn related to the solids content of the raw potatoes employed in the process. Flakes from higher solids potatoes can usually have a larger percentage of fines than flakes from low solids tubers and still be reconstituted to mash with good texture. It has also been found that flakes prepared with added fines can sometimes have a larger percentage of fines than flakes prepared from the same tubers without added fines.

The process for producing the flakes is the conventional, previously-described one until after the sheet of drum-dried mashed potato is broken into approximately ½" x ½" flakes. Then the flakes obtained above are comminuted to increase their bulk density. A comminuting mill equipped with fixed blades and a punched hole or mesh screen with openings appropriately sized can be used for this purpose. The fines obtained after comminution are removed by screening the dried product over a suitably sized screen, preferably 12 mesh. Those retained on the screen are employed as the final product. The fines passing through the screen are added back to a subsequent batch of mashed potatoes prepared as described above, uniformly mixed, and the batch is subjected to drum drying.

The preferred particle size which characterizes the products which have been found to be acceptable ranges from a maximum which will just pass through a No. 3½ mesh screen to a minimum which will just be retained on a No. 20 mesh screen. However, acceptable products have been obtained where the minimum particle size is that which is just retained on a No. 28 mesh screen.

Although it is known to the art that flakes prepared from high solids potatoes (over 20%) can usually be comminuted to a higher bulk density than flakes from low solids potatoes, it is sometimes advantageous to apply the technique as set forth above to high solids potatoes so as to increase even further the bulk density of flakes made therefrom. In such cases, it has been found desirable to add a suitable amount of water along with the fines to facilitate both mixing in of the fines and subsequent drum drying of the mash.

Emulsifier Sprayed onto Potato Mash

J.W. Cyr; U.S. Patent 3,056,683; October 2, 1962; assigned to H.C. Baxter & Bro. describes an improvement on the basic Department of Agriculture process for producing potato flakes which consists of spraying onto the monocellular sheet of the cooked potato mash just after the mash is applied to the drum, a melted emulsifier in an amount in the order of 1 to 3% of the dried product. The emulsifiers found particularly satisfactory for this purpose are polyoxyethylene sorbitan monostearate, polyoxyethylene sorbitan monooleate, polyoxyethylene stearate and sorbitan monostearate. The emulsifier is sprayed on the back of the sheet after it is applied to the heated drum because it has been found that if the emulsifier is incorporated in the potato mash it will act as a parting agent and cause the sheet of mashed potatoes to fold away from the drum before it is dried and flaked. Instead of spraying the heated emulsifier against the back of the sheet it may be wiped on by a felt covered roll or belt or applied by a wick from a suitable heated reservoir.

The process may be somewhat modified by mixing with the emulsifier a fat or oil such as commercial shortenings prepared from cottonseed and other vegetable oils. When such shortenings are mixed with the emulsifier in the order of 60% shortening and 40% emulsifier a smaller quantity of the emulsifier is required for the same amount of the product than when the emulsifier is used alone.

The above described process may also be advantageously modified by dusting Methocel (methylcellulose) on the dried sheet to which the emulsifier or the mixture of emulsifier and shortening has been added just before the sheet is removed from the drum or on the flakes made from the sheet. The added Methocel should be about the same in quantity as the added emulsifier or the added mixture of shortening and emulsifier. Other emulsion stabilizers such as pectin and Irish sea moss may be used instead of Methocel.

Addition of Monoester of Glycerin and a Fatty Acid

M. Pader; U.S. Patent 3,163,546; December 29, 1964; assigned to Lever Brothers Company has found that a highly acceptable dehydrated potato product can be prepared by cooking raw potatoes, mashing the cooked potatoes, and drying the potatoes, provided that a high level of a monoester of a polyhydric alcohol containing at least three hydroxyl groups and a saturated higher fatty acid is added to and thoroughly mixed with the cooked potatoes before drying. The level at which the monoester is employed should be from 1 to 5% by weight, based on the dry solids content of the cooked potatoes.

In addition, it has been found that the quality of the reconstituted potato product is further improved by the addition of at least 0.3% of an edible protein material to the cooked potato before drying. The addition of the protein permits the use of a drum drying operation where high levels of the monoester of a polyhydric alcohol and a saturated higher fatty acid are employed. Examples of commercially prepared products suitable for use according to this process include Myverol 18-00, which has a monoglyceride content of 95% and which is a substantially pure molecularly distilled fraction of fully hydrogenated lard, and Myverol 16-00, which is essentially glyceryl monopalmitate.

The edible protein employed may be sodium, calcium or potassium caseinate, whole milk solids, skim milk solids, isolated soybean protein, and the like. The edible protein improves the drum drying characteristics of the mashed potato containing the higher levels of mono-ester within the range indicated above. Drum drying of cooked, mashed potatoes containing relatively high levels of glyceryl monostearate is difficult because the wet mash does not adhere in a satisfactory manner to the drying drum. This deficiency is corrected by adding an edible protein to the potatoes before drum drying. The adherence of the sheet of dried potatoes to the drying drum is vastly improved, thereby increasing the heat transfer between the drum and sheet. This use of an edible protein also serves to increase the rate or production of the drum dried product since the sheet is dried more rapidly.

For practical reasons, when the edible protein is added solely to improve the drum drying characteristics of the mashed potatoes, the level of protein is usually about 0.3 to 5%. Where the edible protein comprises milk solids, however, it may be added in amounts of from 0.3 to 16% and more, since the milk imparts other advantages to the mash as will be pointed out. The following example illustrates the process.

Example: Russet Burbank potatoes containing 21.7% total solids were peeled in an abrasion peeler and trimmed. The potatoes were then sliced into pieces ¾ inch thick, washed, and placed in boiling water. The water was maintained at a very gentle boil for 35 minutes. At the end of this time, the cooked potatoes were drained of excess water and gently mashed in a Hobart mixer by mixing at low speed. After mashing was nearly complete, the following materials were added.

Additive	% Basis Mash Solids
Myverol 18-00	2.0
Nonfat milk solids	1.0
Tenox IV	0.05
SO_2	0.05

The Myverol 18-00 is a molecularly distilled fraction of fully hydrogenated lard, and contains approximately 95% of monoglyceride. The Tenox IV is a food grade antioxidant containing 20% butylated hydroxyanisole and 20% butylated hydroxytoluene as active ingredients. The SO_2 in the above table is added as a mixture of 7.5 parts Na_2SO_3 and 2.5 parts $NaHSO_3$ in aqueous solution.

Mixing was continued until the mash took on a whitish, whipped appearance. The mash was then dried as a film on a single drum dryer, 18 inches long and 12 inches in diameter at 40 psig steam pressure. A continuous, dense, uniform dehydrated sheet averaging 0.007 inch thick was obtained containing 7% moisture. The sheet was broken up crudely by hand and then comminuted in a comminuting mill equipped with sharpened blades and a one-quarter inch mesh screen. The comminuted product had a bulk density of 18 lbs./ft.^{3}.

The dehydrated potato flakes prepared above were reconstituted as follows. 1½ cups water containing ½ teaspoon salt was brought to a boil and removed from the flame. ½ cup of milk was added immediately to the water, followed by 0.2 lb. of flakes. The mixture was allowed to stand for one minute and then mashed with a fork. The reconstituted mashed potato product had a desirably mealy and firm texture and an excellent flavor. It was very much like fresh whipped potatoes in appearance and was not starchy either in taste or appearance.

Keeping pH Between 6 and 8 During Cooking

M.S. Cole; U.S. Patent 3,219,464; November 23, 1965; assigned to The Pillsbury Company has provided a process for producing dehydrated mashed potatoes having increased tolerance to mechanical mixing and high temperatures during rehydration, and one which does not require a cooling step to produce a product having a quality at least as high as the quality of the product produced by a process having a cooling step.

These objects are attained by conducting at least a portion of the cooking step in a process for producing dehydrated potatoes in a buffer system of sufficient concentration to maintain the pH of the system between 6 and 8 during cooking. It has been discovered that such control of pH provides a dehydrated product of very superior texture properties while eliminating the need for such prior techniques as recycling of product or prolonged cooling of potatoes during the process. Although any buffer system which is capable of maintaining the pH within the above-prescribed range, without producing inedible residues in the dehydrated product is suitable for use in the process, the alkali metal phosphates are preferred. Of this group, the potassium and sodium phosphates, especially the latter, are preferred. A sodium or potassium phosphate concentration of at least 0.01 M is sufficient to maintain pH control throughout the cooking of most potatoes. An upper limitation on the concentration of buffer is imposed only when an undesirable chemical flavor is produced in the final product. In general, the concentration is most desirably maintained between 0.01 and 0.05 M. Within the pH range of 6 to 8 mentioned above, two narrower ranges are preferred. These narrower ranges lie equidistant from a pH of 7 and are the ranges of 6 to 6.5 and 7.5 to 8.0.

In the preferred process, the precooking step is conducted at temperatures within the range of 140 and 180°F., preferably between 160° to 165°F. for times ranging from 15 to 45 minutes. At this point the potatoes are taken from the buffered precooking water and placed in hot water or in steam, preferably the latter, for a time sufficient to produce a product which can be mashed (approximately 20 to 30 minutes). Although the product resulting from utilization of a phosphate buffer system in the cooking water produces a satisfactory product, it has been discovered that the presence of a small amount of at least one halide salt of an alkali-metal selected from the group consisting of sodium chloride and potassium chloride in the cooking water in addition to the phosphate buffer surprisingly produces an even superior product. To facilitate a more complete understanding of the process, the following illustrative examples are offered.

Example 1: Run A — Fifty pounds of Idaho potatoes were peeled, trimmed and cut into slices. The sliced potatoes were cooked in water having a temperature of 165°F. until tender enough to mash and were immediately mashed without cooling below the cooking temperature. The potatoes were then conventionally dried on a single drum dryer and broken into flakes.

Run B — An equal amount of potatoes were peeled, trimmed and sliced as above described and cooked until tender at 165°F. in a water solution containing sodium chloride at a concentration of 0.060 M and sodium phosphate at a concentration of 0.012 M (0.001 M in disodium phosphate, 0.011 M in monosodium phosphate which provided a pH of 6.4 in the cooking water). These cooked potatoes were immediately mashed and dried on a single drum dryer to produce dried flakes.

Samples of the dehydrated potato flakes produced by each of the processes described above were rehydrated using 90 grams of flakes per two cups (472 ml.) of water having a temperature of 168°F. and tasted. The product resulting from Run B exhibited a natural flavor whereas the product of Run A manifested a pronounced starch flavor. Moreover, the Run A product had a very pasty texture in comparison to the mealy texture of the product from Run B.

Samples of the flakes of each process were also rehydrated and mixed with a Hobart mixer until a pasty texture developed. The product of Run B withstood mixing for 5 minutes before a pasty texture was detected. However, the product resulting from Run A developed pastiness within one minute. The product of Run A was judged wholly unacceptable for sale whereas the product resulting from Run B was satisfactory for sale, although as will be illustrated hereinafter, a more suitable product can be made. It should be noted that acceptable potato flakes were produced in Run B in a single stage cooking step without the need of a cooling step.

Example 2: Run C — Fifty pounds of raw potato slices were precooked for 30 minutes at 165°F. in a water solution containing sodium phosphates at a concentration of 0.012 M (0.001 M in disodium phosphate, 0.011 M in monosodium phosphate which provided a pH of 6.4 in the cooking water), and sodium chloride at a concentration of 0.060 M. The precooked potatoes were then transferred to a steamer and cooked in atmospheric steam (212°F.) until tender. The resulting cooked potatoes were mashed and dehydrated by conventional means using a single drum dryer. Approximately 8 pounds of dehydrated potato flakes resulted. The product of Run C was rehydrated with water at a temperature of 185° to 190°F. As soon as the water had been absorbed by the flakes, the mash was mixed mechanically with a Hobart mixer and examined periodically. The product of Run C subjected to this treatment of high water temperatures and mechanical mixing remained acceptable after 10 minutes of beating, a period far in excess of what would normally be required in home or institutional use. When lower water temperatures were employed (165° to 170°F.) in combination with the mechanical mixing described above, the product of Run C again was acceptable after 10 minutes of mixing. The above demonstrates the increased cell strength which has been brought about by the combination of buffer and salt in the cooking process employed in Run C. As regards flavor, the product of Run C was judged far superior to the products of Runs A and B.

Variation of the Add-Back Process Using Flakes

M.J. Willard, Jr.; U.S. Patent 3,275,458; September 27, 1966; assigned to Rogers Brothers Company has developed a process which produces potato flakes with a higher bulk density than is common by a variation of the add-back method, used for commercial production of potato granules or powder, but which does not require the continued recycling of a portion of the finished product or, in fact, any recycling at all.

Broadly speaking, the process comprises admixing cooked mashed potatoes with dehydrated potato flakes in an amount sufficient to produce a moist, friable mixture having a moisture content of 30 to 50%, preferably 35 to 45%, permitting the mixture to equilibrate its moisture content for at least 20 minutes, remixing, and then drying to the desired final moisture content. Mixing of the moist and dry components to homogeneity is accomplished with the addition of a fatty acid monoglyceride, preferably in admixture with a fatty acid triglyceride.

The cooked mashed potatoes, for use as the moist component or for preparation of the flake, can be prepared in well-known conventional fashion, with care being taken not to overcook or to mash in a manner so vigorous as to disrupt the cells. The potato is preferably sound, mature, and of the high solids content, mealy type, such as the Idaho Russet, although potatoes of lower solids content, such as the Katahdin, Kennebec, Cobbler, or Pontiac, having a solids content of at least 18%, can also be used. The potatoes are peeled, sliced and cooked, as by steaming, and mashed in any suitable and convenient manner, as by ricing or by means of flaking rolls. A suitable antioxidant, such as a sulfite or butylated hydroxytoluene, can be incorporated into the mashed potatoes, if desired, to extend the normal shelf life of the dehydrated product.

Mealiness of the finished reconstituted product, particularly in the case of potatoes of lower solids content, can be improved, if desired, by subjecting the raw potato prior to cooking to a precooking step at a temperature of 140° to 180°F. for 10 to 60 minutes, by immersion in a hot water bath or by treatment with steam or hot air, and then washing the potato pieces to remove surface starch.

The dehydrated flake can be prepared by applying the cooked mashed potato to the surface of an internally heated drum dryer and spread into a thin layer, as for example, by means of spreader rolls in the case of a single drum dryer, or by proper adjustment of the nip in the case of a double drum dryer. The layer of potato dries on the drum without rupture or other damage to the cells. After drying to at least 30% moisture content, preferably 6 to 15%, the film is removed from the revolving drum in any suitable fashion, as by doctor knife, and broken into flakes of suitable size. The thickness of the dried film or flake should be no more than 20 mils and preferably in the range of 4 to 10 mils.

When the dried film is broken down into flakes, some cell rupture occurs at the point of cleavage, releasing free starch. Where such cleavage is minimized by the production of relatively large flakes, e.g., ½ to ¼ inch in the longest dimension, the flakes can be reconstituted without any noticeable pastiness. Further comminution tends to produce an undesirable stickiness. When the flake is admixed with cooked mashed potatoes, the flake can be comminuted to pieces having lengths in their largest dimension as small as 40 to 45 mils (16 to 18 mesh), in a single cycle operation, namely where the finished product is prepared by the single mixing of flake and moist mashed potato and dried.

The smaller flake sizes have the advantage of promoting easier distribution of the flakes in the cooked mashed potatoes and reducing any tendency to lumpiness. It should be noted, however, that such problems can be overcome, to a substantial extent, by adjusting the conditions of the mixing operation, as, for example, by increasing mixing time. There is no critical limit with regard to maximum size, since the larger particles, e.g., ¼ to ½ inch or larger, are comminuted during mixing with the moist mashed potato.

The dispersion of the flakes in the cooked mashed potatoes to form a homogeneous non-lumpy mixture requires a surface active distributing agent in the form of a higher fatty acid monoglyceride, such as glyceryl monopalmitate, glyceryl monolaurate, glyceryl mono-stearate and the like. The monoglyceride should generally constitute at least 0.05 to 0.1% and, preferably, 0.3% on a finished dry weight basis of the mix. There is no critical upper limit.

The effect of the monoglyceride distributing agent is improved by introducing it in admixture with a higher fatty acid triglyceride, preferably saturated. The fatty acid components of the triglyceride can be, for example, lauric, myristic, palmitic, stearic, arachidic or behenic acids and mixtures thereof, as found, for example in various hydrogenated vegetable oils, such as cottonseed, corn, soybean, peanut and coconut oils, or animal fats of low iodine number. Substantial improvement is obtained with amounts of triglyceride as low as 0.4 to 0.5% on a dry weight basis of the mix. The preferred minimum is 1 to 1.5%.

The monoglyceride can be dispersed in the triglyceride as, for example, by melting the two materials together, and employed as such or in aqueous emulsion. Some products are available commercially which are mixtures of mono- and triglycerides. The distributing agent can be incorporated into the potato mixture in any of several ways. A melted mixture of the mono- and triglyceride can be sprayed onto the flakes with concomitant agitation of the dry solids to obtain uniform distribution throughout the dry mass of material. The monoglyceride, in its fatty carrier as such or in aqueous emulsion, can be added to the moist cooked potato prior to, during or after mashing with sufficient mixing to ensure uniform distribution.

The dry flake component is mixed with the cooked mashed potato by any suitable means which ensures thorough blending, without excessive cell rupture, as, for example, by properly designed ribbon or paddle type mixers. The cooked potato can also be simultaneously mashed and mixed with the dry component by addition of the flakes prior to mashing in a suitable mixer. In this case the emulsifier is preferably previously added to the flakes.

After the moist and dry potato components have been blended to a moisture content of 30 to 50%, preferably 35 to 45%, the mixture should be given a conditioning or aging period of at least 20 minutes, preferably 30, during which it remains undisturbed or given at most an occasional gentle mixing. During this rest period, the moisture equilibrates between the wet and dry particles and retrogradation of a portion of the starch occurs to some extent. Longer conditioning periods can, of course, be employed if desired.

After the conditioning period, it is desirable again gently to mix the potato blend, since this increases separation of the potato particles without cell rupture and contributes toward a lump-free, uniform mashed potato product after rehydration. The potato blend is now a loose, moist, friable mixture which can easily be dried in any conventional dryer, preferably one using heated air as the drying medium. Types of dryers which can be employed include the airlift-cyclone type, the porous bed dryer with or without a vibratory means for moving the material through the dryer, a fluidized bed in which the entraining means is hot air, a directly or indirectly fired rotary dryer, or a simple tray or cabinet dryer. Drying is continued until the desired final moisture content is reached, e.g., 4 to 8%.

If the dried potato flakes employed are at least 40 mils (18 mesh) in their largest dimension, a finished product or superior quality is produced by a single component mixing and drying procedure as described. The product is a granular, free-flowing material having a high bulk density. Bulk density is rarely less than 28 lbs./ft.^{3} and is as high as 48 lbs./ft.^{3} or more. Viewed under magnification, the product comprises a mixture of fine potato flakes, monocellular particles, and multicellular aggregates or clumps of cells only or attached to the fine flakes. The preferred maximum size of the flakes in the finished product is ⅛ to ¼ inch. The product reconstitutes rapidly upon the addition of a liquid, such as water or milk, to a mashed potato of superior, uniform, mealy texture retaining substantially unimpaired the flavor and color of fresh mashed potatoes.

Addition of Vitamin C

While it has been desirable to include additives in dehydrated cooked mashed or riced potatoes, it has been found that when the additives, particularly the most important desired additive vitamin C, are incorporated, the quality of the resulting product is adversely affected. In particular, it has been found that the flavor is adversely affected and the resulting product is not consistent either in flavor or texture. In addition, it has been found that there is a loss of the additives upon storage of the product. Again in particular with respect to vitamin C, in the past it has not been possible to maintain an adequate concentration of the additive under storage conditions approximating normal product shelf life.

T.F. Irmiter and G. Rubin; U.S. Patent 3,027,264; March 27, 1962; assigned to Salada-Shirriff-Horsey Ltd., Canada have developed a process, however for incorporating additives in dehydrated cooked mashed potatoes without any deleterious effects on the product, and to maintain the additive concentration throughout the shelf life of the product. It has been found however, that additives, particularly vitamin C, can be added or restored to dehydrated cooked mashed potatoes by incorporating the additive into a sulfite solution and feeding the sulfite solution into the potato mash separately from the antioxidant and emulsifiers, whereby it has been found that the additives cannot interact with the antioxidant and emulsifiers to upset their intended function, and at the same time the sulfites are maintained at the requisite additive level.

In addition to the enhanced nutritional value occasioned by the incorporation of the vitamin C in the product, it can be further enhanced if there is incorporated into the potato mash as an additive prior to drying, a quantity of nonfat dry milk solids which result in an improvement in color in the product and which render the resulting potato flakes more opaque.

In addition to the ability to incorporate vitamin C into the product without adverse effects, it was also found possible to incorporate other additives such as vitamin B$_2$ and niacin without adverse effect on the flavor, color and texture of the product, providing the emulsifiers and antioxidant were delivered into the potato mash to a point separate from the sulfites, and the vitamin C when added was incorporated into the mash admixed with the sulfites. It has also been found that the fat soluble vitamins such as vitamin A, vitamin D, vitamin E and vitamin K can be successfully incorporated providing they are admixed with the emulsifier prior to addition to the mash. Thus the process has not only enabled the naturally occurring nutritional values of freshly dug raw potatoes to be restored but further has enabled additional values to be incorporated into a dehydrated cooked mashed potato. A typical application of the process is illustrated by the following example.

Example: Five pounds of Russet Burbank potatoes were peeled and trimmed and then sliced into slices approximately one-half inch thick and blanched in water for 20 minutes at 160°F. Following precooking, they were cooked in the autoclave or pressure cooker for 6 to 10 minutes with the pressure cooker operated at a steam pressure of 15 psi. Following cooking, the potatoes were mashed, then according to the process there was added to the mash 17.5 cc of the following additives in the following proportions:

Emulsifier, glycerol monopalmitate	0.8488 g.
Antioxidant	0.7377 g.

(continued)

Nonfat dry milk solids (instant milk powder)	1.6997 g.
Sodium sulfite	0.5543 g.
Sodium bisulfite	0.1794 g.
Water	7.90 cc
Vitamin C	1.58 g.

The emulsifier, antioxidant and nonfat dry milk solids were mixed into the mash separately from the sodium sulfite and sodium bisulfite into which were incorporated the vitamin C. Following the mixing of the additives into the mash, the mash was again dried in a double drum atmospheric dryer operating at a steam pressure of 60 psi and at a speed of 4 rpm. Following drying the moisture content of the resulting flakes was measured and was found to be 5.5%. The vitamin C content was also measured and was found to be 168.7 milligrams per 100 grams of dried flakes.

To reconstitute the dried potato flakes into the table product, 86 grams of flakes are added to one-half cup milk, one-half cup boiling water and one tablespoon butter, to produce four normal servings of potatoes. Therefore, translating the vitamin C content of the potato flakes into the required vitamin C level for one serving of fresh cooked mashed potatoes, the vitamin C level required in the flakes is 167 milligrams per 100 grams of flakes. In the example given above, the vitamin C content actually measured in the flakes was 168 milligrams per 100 grams of dried flakes, so that it will be seen that the vitamin C level was fully restored.

Enrichment with Vitamin B_1

A serious problem arose when attempts were made to incorporate a vitamin mixture containing vitamin B_1 into potato flakes by the procedure described in the previous patent (U.S. Patent 3,207,264). Sulfites have a destructive effect on vitamin B_1 (thiamine) and the latter ingredient could not be incorporated into potato mash mixed with conventional sulfite stabilizers without destruction of a very large proportion of the vitamin.

M. Pader and A.S. Hall; U.S. Patent 3,343,970; September 26, 1967; assigned to Lever Brothers Company have developed a process for preparing potato flakes in which vitamin B_1 and sulfites can be incorporated without any appreciable loss in the level of the vitamin and without any adverse effect on the quality of the reconstituted potato. This has been accomplished by incorporating vitamins and potato mash at a point during the processing of potatoes as far removed in point of time as possible from the addition of a solution containing sulfite stabilizers. By thoroughly blending the sulfites into the potato mash before adding the vitamins, the destructive effect of the sulfites upon vitamin B_1 is avoided in large measure, without adversely affecting the qualities of the potato flakes.

The potatoes are prepared by the normal process for preparation of flakes. In a batch operation, the solution of sulfites is added to the potato mash in a mixer and thoroughly blended in. The emulsifier, antioxidant and milk solids can then be added separately or together in an aqueous slurry. After thoroughly admixing these ingredients in the mash, the vitamins are incorporated generally as an aqueous solution and intimately blended into the potato mash which is then dried.

In a continuous operation, the order of addition of the ingredients is unchanged and a maximum amount of time is allowed between the incorporation of sulfites and the addition of vitamins. In certain instances, it was necessary to incorporate the vitamins into the mash as the latter was applied to the drum dryer. The vitamins are preferably added in the form of an aqueous solution. Individual vitamins can be incorporated if desired. However, it is preferred to add a mixture of water-soluble vitamins including B_1, B_2, C and niacin. The proportion of vitamins added will vary depending upon the ultimate levels desired in the final product.

Whether a batch or continuous operation, the time delay between addition of sulfite and of vitamin solution is dictated by practical considerations. In a batch operation, excessive holding of hot mashed potatoes prior to dehydration can result in deterioration of texture, flavor and color. The time lapse must not be so long as to cause such a defect. In a continuous operation, the same factors must be considered, in addition to rate of producton, capacity of equipment, floor space available, etc. Thus, the shortest time delay is used that is compatible with obtaining the desired result.

Example: In a pilot operation, Michigan Russet Rural potatoes were abrasion-peeled, cut into ½ inch slices and dipped in a 0.5% solution of citric acid and sodium bisulfite. The slices were precooked in water for 20 minutes at 160°F. and cooled to 70°F. for 20 minutes in flowing water. The slices were cooked in steam (atmospheric pressure) for 20 minutes and mashed in a Hobart mixer for 1 minute. A sulfite solution (30 ml.) of 3 parts Na_2SO_3 and 1 part $NaHSO_3$ was added to 30 lbs. of mash and mixed in for 2 minutes. During the mixing 5.5 grams of nonfat milk solids, 5.5 grams of monoglyceride (Myverol 18-07) and 1.6 ml. stabilizer (Tenox IV) were added. Lastly, 25 ml. of a solution of thiamine (1 mg./ml.), riboflavin (0.4 mg./ml.), niacin (12 mg./ml.) and ascorbic acid (164 mg./ml.) was added and mixed in the mash for 2 minutes. The mash was immediately applied to the drum and dried.

The drum dryer used was a modified double-drum dryer, each drum 18 inches long and 12 inches in diameter, heated by steam. Mash was applied to the nip of the two drums, which were set ¼ inch apart. Only one drum was heated, the other serving as an applicator roll. Mash was rolled onto the heated drum by means of a hand-operated roller to ensure good dried potato sheet formation. The drums were rotated towards each other at 1 rpm. The steam pressure was maintained at 40 psig, and changed slightly as necessary to ensure the formation of a uniform sheet with not more than 6 to 7% moisture. The sheet was removed from the drum by means of a doctor blade. The sheet was then comminuted coarsely. The comminuted dehydrated product (93 to 100 grams) was reconstituted with hot water (1½ cups) and milk (½ cup) at 160°F. The mashed potatoes had a desirable appearance and an excellent taste.

Two-Stage Cooking Process

An improved potato dehydration process developed by *J.A. Barnes and H.C. Nora; U.S. Patent 3,355,304; November 28, 1967; assigned to The Pillsbury Company* includes precooking the potatoes while in a whole condition at a relatively low temperature until the outsides are cooked, subdividing the potatoes and cooking them at a higher temperature, further subdividing and partially mashing the potatoes, thereafter breaking down the remaining lumps of cooked potato and then drying the potatoes. The remaining lumps are preferably broken down by pressing the material onto the surface of a drying drum to form a thin layer thereon and the temperature of the drum is preferably sufficiently high to cause further cooking.

The potatoes are first peeled, inspected and cleaned. They are then transported to a cooking tank and precooked in a whole condition at a temperature of from 170° to 190°F. and preferably from between 175° to 185°F. for a period of from 10 to 40 minutes and preferably from 20 to 30 minutes. While the temperature of the tank and time required for the precooking are interrelated, sufficient time and temperature should be provided to cook or gelatinize the outside while leaving the inside substantially raw. The center of most of the potatoes should reach a temperature of 140°F. A suitable oxidation retarding agent such as sodium acid pyrophosphate in the amount of 0.05 to 0.10% is preferably added to the initial cooking solution to prevent graying of the potatoes during the dehydration process.

The potatoes are then sliced into flat slabs or discs from ⅜ to ¾ inch thick. Following the slicing operation, the potatoes are transported to a cooker such as a steam cooker and are held at a temperature of 195° to 215°F. and preferably between 200 and 210°F. for a period of from 10 to 35 minutes. The cooker is provided with a horizontal auger for transporting the potatoes through it and a vertical auger for removing them into a masher consisting of a cylindrical casing and an auger driven by means of a motor. When the auger is turned, the potatoes are forced through a segmenting device or grill which consists of a plurality of fixed members such as parallel vertical bars which are supported at the top and bottom by a retainer ring. The potatoes are forced through the bars by the action of the auger and mashed. During the mashing operation, a suitable emulsifier such as acetylated monoglyceride or other suitable emulsifier is preferably added to the potatoes in an amount of from 0.3 to 0.5% for the purpose of reducing the extent to which starch granules are ruptured. The product which emerges from the outlet side of the mashing device comprises about half and half unmashed chunks and mashed material.

The final mashing, cooking and drying is carried out with a drying apparatus which includes a hollow dryer roll or drum supported on a shaft for rotation by a horizontal axis. Steam is admitted to the inside of the drum under pressure of from 90 to 100 psi. The surface of the drum will then be maintained at a temperature of from 330° to 340°F. The drum can, of course, vary in size, but a diameter of 5 feet and a length of 16 feet has been found satisfactory for operation on a commercial scale. In a typical operating run, the drum is rotated at 4.0 rpm. A number of horizontally disposed and vertically spaced rolls are positioned parallel to the axis of the drum and spaced a slight distance from it to maintain a layer of partially dried potatoes on the lower surface of the drum at a thickness of 0.008 to 0.012 inch.

The mashing is then completed by means of application rollers which squeeze the partially mashed material onto the surface of the drum while at the same time mashing the chunks remaining after the potatoes have passed through the segmenting apparatus. The maximum temperature of the drum surface should be 330° to 340°F., but because of evaporation, the layer itself will have a temperature of from 155° to 205°F. It was found that the temperature of the layer when removed from the drum was 180°F. when the drum surface was maintained at a temperature of 335°F. The sheet or layer is removed from the drum by a doctor knife and cut into a plurality of strips by means of a powdered cutter.

With the drum rotating at a speed of 4.0 rpm and the material layer covering approximately ⅔ of the surface of the drum, the layer will remain on the drum for a period of 15 seconds. The time on the drum can be accurately controlled by regulating the speed of the rotation thereof. Generally speaking, the layer should remain on the drum for a period from 5 to 20 seconds, but preferably from 8 to 12 seconds. For satisfactory performance, the moisture should be reduced to no more than 5 to 8%.

In evaluating the process, four distinct test procedures demonstrated a reduction in damage to starch granules. The first of these is the marked improvement in texture as determined in standard taste panel tests. The product was found to be less pasty and more grainy than potatoes manufactured in accordance with the prior art. It was found that they had less tendency to dry the mouth. Less starch granular damage was also illustrated by immersing the dehydrated flakes in an iodine solution. In the flakes manufactured in accordance with the prior art, the iodine diffused to the center of the flake in a period of 30 minutes causing the entire flake to turn a blue color. In the flakes manufactured in accordance with this process, the blue coloration was present only at the periphery of the flake. These potatoes were also found to exhibit less flavor loss as determined by taste tests. Finally, the finished product exhibits a somewhat yellow color, rather than pure white color indicating a reduction in the loss of pigment from the starch cells. To facilitate a more complete understanding of the process, the following illustrative example is offered.

Example: Field run potatoes were washed and lye peeled in a conventional manner. The peeled potatoes were inspected and damaged sections were removed. The potatoes were then precooked at a temperature of 170°F. for 10 minutes in a solution of 0.05% sodium acid pyrophosphate. They were then immediately sliced into slabs having a thickness of ⅜ inch. They were then cooked in a steam cooker at a temperature of 200°F. for a period of 15 minutes to give a coarse mash consistency. The potatoes were then transferred to a masher of the type described and mashed to a consistency of 50 parts smooth mash and 50 parts chunks. As the potatoes passed through a masher, a solution of 0.3% acetylated monoglyceride was sprayed into the masher. The partially mashed material was then transferred to a roller dryer having a surface temperature of 330°F. and retained on the drum for 5 seconds, then removed and chopped into flakes.

Formation of Thicker Flakes

The disadvantages of the prior art processes for producing potato flakes center around the type of flake which is produced. These flakes have a maximum thickness of 0.010 inch, thus occupying a large volume when they are packaged. More important is the fact that flakes of this thickness cannot be reconstituted by boiling liquids since then the reconstituted product has a pasty texture caused in part by free starch liberation from potato cells that expand too rapidly in the boiling liquid. Liquids heated to no more than 180°F. must thus be used for reconstitution of the flakes. Thicker flakes which would rehydrate more slowly so that boiling liquids could be used for reconstituting the product without cell breakage would therefore be most desirable.

M.J. Willard; U.S. Patent 3,417,483; December 24, 1968 has developed a process for forming a potato flake from a plurality of food cell layers in the dehydrated state. The flakes have a thickness of at least 0.15 inch which enables the flake to be reconstituted in hot water with substantially less cell rupture than occurred with thinner flakes.

The process includes the steps of passing cooked, mashed potatoes between a moving drying surface and a series of spreader rolls until enough layers of potato cells are applied to the drying surface so the total thickness of potato layer is at least 0.015 inch in the dehydrated state. As the potato layers are applied to the drying surface, they are heated to drive water out. The apparatus for dehydrating the potato flakes includes an endless belt being disposed to move over a pair of rotatable spaced drums. A plurality of spreader rolls are disposed along the belt and a means for passing moist food between the belt and the rolls is used to deposit a layer of the food on the belt. The apparatus also includes means for applying heat to the food on the belt to drive moisture out of the food.

The process will be more fully understood from the following detailed description and the accompanying drawings, in which Figure 6.5a is a schematic front elevation of potato mash being applied between a movable endless belt and several spreader rolls; and Figure 6.5b is a schematic cross section taken on line 2-2 showing the interior of a spreader roll. Raw potatoes (not shown) to be dehydrated are peeled by either an abrasion process, a steam peeling process, or a treatment with lye at 200°F. for 4 minutes. The potatoes are washed, hand trimmed to remove blemishes and other irregularities and thereafter sliced into slabs approximately one-half inch thick. Excess starch is washed from the sliced potatoes before precooking them for 20 minutes in water at 160°F.

The precooking step gels the starch in the potato cells at a low temperature before the final cooking step to improve the texture of the reconstituted product and to permit the use of low solids potatoes. The precooked potatoes are cooled to a temperature of 50° to 60°F. to retrograde, polymerize, or cross-link amylose and reduce the stickiness of the potatoes. The cooled, precooked potatoes are thereafter cooked at 200° to 212°F. for a period which varies with the amount of solids in the potatoes. Approximately 20 minutes is required for high solid potato varieties, whereas 40 minutes is sometimes required for low solid potatoes. At any rate, the potatoes are cooked to the point where they are soft enough to be mashed easily with minimum cell rupture. The cooked potatoes can be mashed by inserting the potatoes in a ricing machine to prevent the bursting of potato cells and minimize liberation of free starch.

At this point, additives such as emulsifiers, diglycerates, sodium acid phosphate, polyphosphate, sulfur dioxide, antioxidants and the like are added to the potato mash as required. Referring to Figures 6.5a and 6.5b, cooked potato mash 10 is dropped from a hopper 12 upon a moving flexible stainless steel endless belt 14 that is spaced below the hopper and moves around a pair of spaced cylindrical rotatable drums 16 and 18. Drum 18 is spaced below and slightly to one side of the drum 16 so the endless belt is at about a 70° angle to horizontal. The potato mash passes between the endless belt and a plurality of elongated

FIGURE 6.5: APPARATUS FOR FORMING THICKER POTATO FLAKES

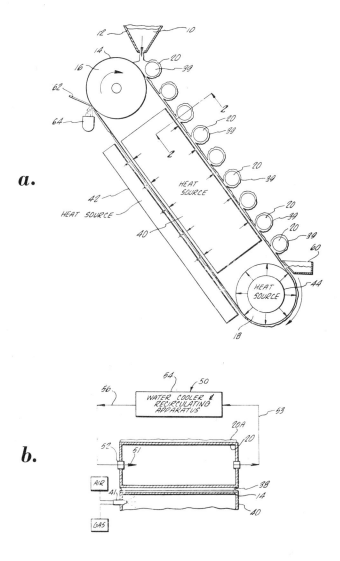

Source: M.J. Willard; U.S. Patent 3,418,142; December 24, 1968

cylindrical spreader rolls **20** that are disposed close to the outside surface of the upper portion of the belt with their respective axes parallel to the axes of both drums. A conventional driving means (not shown) is connected to the spreader rolls and the drums. A relatively thin layer **38** of potato mash is deposited on the endless belt and a relatively thicker layer **39** is picked up by each of the spreader rolls. The mashed potato layer on the uppermost spreader rolls is periodically scraped off by a conventional rotatable knife (not shown) and permitted to fall by gravity against the adjacent rotatable spreader roll where it accumulates around the roll. Alternatively, the potato mash drops of its own accord from the upper to the lower rolls. The position of endless belt and the spreader rolls is such that the potato mash will naturally fall from an upper to a lower roll. Each of the spreader rolls applies a monocellular layer of potatoes upon the endless belt and generally the total number of potato cell layers on the endless belt is equal to the total number of spreader rolls. By extending the distance between the two drums, extending the endless belt, and adding additional spreader rolls, a layer of dehydrated potatoes of almost limitless thickness can be produced.

A first heat source **40** is disposed between the rotatable drums and serves to drive out water from the potato layer on the endless belt. The heat from the source can be generated from any type of conventional device as, for example, steam, electric resistance wires, direct-fired gas burners, radiant heat, and the like could be used either alone or in combination. A gas burner nozzle **41** is mounted in the side of the first heat source **40**. Air and gas are supplied to the nozzle to burn within the heat source. This form of direct gas-fired heating is preferred because it is most economical.

A second heat source **42** is directed toward the endless belt on the side opposite the spreader rolls and also opposite the first heat source **40**. Thus, the potato layer on the endless belt is heated from both sides of the belt which tends to uniformly dry the potato layer. A third heat source **44** is disposed within the drum **18** and serves to drive the water from the potato layer **38** as it passes around the drum such that the potato mash layer is constantly heated as it progresses around the endless belt. The drum is preferably heated by steam.

Referring to Figure 6.5b, each separator roll **20** has a means for cooling the relatively thick potato mash layer **39** on the roll. The cooling means is a water dispensing device **50** disposed within the roll and is, for example, an inlet nozzle **51** journaled through a swivel **52** in one end of the drum. The water discharged within the roll is recirculated out of the roll through an outlet pipe **53** through a water cooler and recirculating apparatus **54**. The temperature of the water is regulated in the apparatus **54** by a conventional thermostat (not shown) and the water is redirected into the spreader roll through an inlet pipe **56** connected to the nozzle. The spreader roll is constantly cooled to a temperature of approximately 180°F. to prevent the potato layer on it from being overheated. Water vapor and steam issues from the potato mash layer, and to prevent water from condensing back onto the potatoes, a warm air blast (not shown) is directed between adjacent separator rolls.

An open trough **60** is spaced below the lowest spreader roll and adjacent the endless belt so that any matter which falls or is cut from the bottom spreader roll, such as any peelings, corky tissue, or foreign matter, may fall directly into the trough. The rolls tend to collect and concentrate such undesirable matter, and the large number of spreader rolls used in accordance with this process improve the automatic removal of the matter from the final product.

After the potato layer has been dehydrated by the heat sources, it is cut from the endless belt by a conventional doctor knife **62** bearing against the endless belt as it contacts the rotatable drum **16**, causing the potato layer to fall as a sheet into a product conveyor **64**. The dehydrated potatoes are broken into flakes and packaged in a moisture-tight container.

An important advantage of this process is that the relatively large number of spreader rolls deposit a fairly thick sheet of dried potatoes or other food product on the endless belt. The thicker flake permits packaging at a higher density so that more pounds can be placed in a package of a given size, thereby reducing the overall cost of the product.

Another advantage of the thicker potato flakes is that they can be rehydrated in boiling water instead of being limited to rehydration with water of 170° to 180°F. The increased thickness of the product results in a longer hydration time, even through hotter water is used, thereby reducing the rate of diffusion of water in most of the potato cells with a corresponding lesser degree of cell breakage. This improves texture of the final rehydrated product without the user having to take special care to avoid rehydration at temperatures below the boiling point of water.

In another patent, *M.J. Willard; U.S. Patent 3,418,142; December 24, 1968* describes a mashed potato dehydration method which includes the steps of spreading a thin layer of the mashed potatoes on a moving drying surface with the help of at least one spreader roll, heating the layer on the drying surface and cooling the spreader roll to lower the temperature of the mashed potatoes disposed thereon. The apparatus includes a movable drying surface adjacent to and spaced apart from the spreader roll which is rotatable about an axis transverse to the direction of movement of the drying surface. A surface of the spreader roll is cooled to maintain the potatoes disposed on it at the temperature which is less than the temperature of the potatoes disposed on the drying surface.

The spreader rolls perform two functions. The first is to deposit succeeding layers of mashed potatoes onto the hot drum, thus building up the thickness of the dried flakes to an economical size. Thinner flakes have a light bulk density, and are expensive to package. The second function of the spreader rolls is the progressive removal of defects from the mashed potatoes. During the time the mashed potatoes stay on the spreader rolls, defective materials, such as particles of skin, rot, corky tissue, etc., are progressively concentrated and transferred with the mash from the first roll to the last roll. The material on the last roll is periodically removed by an operator with a scraper, and is discarded. Thus, the rolls serve as a separation device for removing undesirable portions of the potato from the mash prior to drying. An apparatus for carrying out the process is illustrated and described in detail.

Addition of Monoglyceride plus Lactylate

R. Weiner and G.R. Hegarty; U.S. Patent 3,447,934; June 3, 1969; assigned to The Pillsbury Company have developed a rehydratable and whippable dehydrated potato product which has a greater tolerance to whipping when ultimately used by the consumer without developing a concomitant pasty texture. The product comprises: (a) potato solids, (b) 0.05 to 2.0 weight percent based on the dry potato solid weight of an edible monoglyceride of a fatty acid, and (c) 0.05 to 2.0 weight percent based on the dry potato solid weight of an edible lactylic acid ester salt having the formula:

$$[R-C-O-(O-CH-O)_n-O]_x M$$
$$\underset{CH_3}{|}$$

wherein M is a metal cation, RCO is the acyl moiety of a fatty acid or mixture of fatty acids having from 12 to 22 carbon atoms, n represents the average degree of polymerization of the lactyl group, the value of n being greater than 1 but less than 4 and x is an integer having a value equal to the valence of the cation. Typical metal cations are sodium, potassium, magnesium, calcium and aluminum. Preferably, M is a number selected from the group consisting of alkali metals and alkaline earth metals having an atomic weight greater than 20 but less than 41. Preferably, n is an integer of 2. The lactylic acid esters (i.e., lactylate) employed herein and the method of preparing the same are disclosed in U.S. Patent 2,789,992.

Advantages resulting from employment of the combination of edible lactylic acid ester salts and monoglycerides of fatty acids include greater tolerance to whipping and improved appearance and texture in the ultimate whipped product. Reconstituted and whipped potato products containing the combination of surface active agents are stiff but yet fluffy in character without accompanying pastiness and heaviness. Processing of dehydrated potato products is enhanced by the combination (e.g., less tendency of the product to stick to the drum dryer). Suitable edible monoglycerides of fatty acids include the saturated and unsaturated fatty acid monoesters of glycerol having monoester groups from 12 to 22 carbon atoms. The preferred glyceryl monoesters are glyceryl monostearate and glyceryl monopalmitate.

The preferred amount of glyceryl monoester is from 0.3 to 0.6 weight percent of the weight of the dehydrated potato. Exemplary edible metal salts of lactylic acid esters of fatty acids include calcium stearyl-2-lactylate, calcium monopalmityl-2-lactylate, magnesium stearyl-2-lactylate, potassium stearyl (3) polylactylate, and the like. Superior results are achieved by the addition of calcium stearyl-2-lactylate, especially in combination with glyceryl monostearate. The amount of lactylic acid ester salt ranges from 0.3 to 0.9 weight percent of the dry potato solid weight. The potatoes may be dehydrated by various methods, particularly the Department of Agriculture process described in previous patents.

Example: Field run potatoes (Kennebecs and Pontiacs) were lye peeled in a conventional manner, sliced into ⅜ inch thick slices and rinsed to remove free starch. Employing conventional potato cooking equipment, the potatoes were cooked at a temperature of 170°F. for 30 minutes and then steam cooked for an additional 30 minutes. The potatoes were then transferred to a conventional potato ricer and mashed. An aqueous solution of calcium stearyl-2-lactylate (0.46 g./100 ml. of water) and glyceryl monostearate (0.46 g./100 milliliters of water) was metered onto the mash such that the resultant dehydrated product contained 0.43 weight percent glyceryl monostearate and 0.43 weight percent calcium stearyl-2-lactylate (weight percents based on the dry potato solid weight).

The mashed potatoes were then conducted to a hollow drum dryer via a screw conveyor (20 feet in length with a 12 inch diameter). The conducted potato product had a 20% by weight solids content. Dehydration of the potatoes was accomplished by a typical commercial hollow single drum dryer roller with a diameter of 5 feet and a length of 16 feet which was provided with a doctor blade and 4 feet rollers positioned parallel to the axis of hollow dryer roller at a clearance of 0.50 inch. The hollow dryer roller surface temperature was maintained at 330°F. with a 5 to 8 second contact time. The resultant dried flakes had a thickness of 0.009 inch and a moisture content of 6 weight percent. Uniform distribution of the lactylate ester salt and glyceryl monostearate was accomplished by means of the screw conveyor and mastication of the mashed product on the drum dryer.

Several portions of the dehydrated potatoes thus prepared were reconstituted and whipped. The potatoes were prepared by placing in a saucepan 1¼ cups of water, ½ teaspoon of salt and 2 tablespoons of butter and heating the same to boiling. After boiling, the heating thereof was discontinued and ⅓ cup of milk was added to the saucepan. Seventy grams of the dehydrated potato product flakes were then added and stirred gently. The potato flakes after becoming soft and moist (30 seconds after addition) were then whipped according to the following methods:

Test Procedure 1 — Whipped lightly with a household fork for 30 strokes.
Test Procedure 2 — Whipped 40 seconds with a standard household Hamilton Beach mixer at speed 7 (750 rpm).
Test Procedure 3 — Whipped lightly with a household fork for 30 strokes and then whipped for an additional 100 seconds with a standard household Hamilton Beach mixer at speed 7 (750 rpm).

The reconstituted and whipped products prepared according to Test Procedures 1 to 3 were then evaluated by comparative tests. The mashed potatoes, even after Test Procedure 2 and 3, had good appearance and texture, with much less pastiness than products containing the glyceryl monostearate alone.

Preheating Potatoes and Treating Them with Caustic

A.L. Lewis; U.S. Patent 3,574,643; April 13, 1971; assigned to Overton Machine Company has developed a process for the production of high-quality, uniform potato flakes from all varieties of potatoes with substantial reduction of processing equipment, time and cost. In this process washed raw whole potatoes are heat-tempered and contacted with a caustic solution, with subsequent removal therefrom to allow impregnation to occur sufficient to effect at least a partial digestion of the outer portion of the potato before peeling, trimming, slicing, precooking, cooling, cooking, ricing, drying and packaging. A system and apparatus for continuously filtering the wash water, the caustic solution and the heating fluids so as to recirculate them for reuse within a constant temperature cooling, precooking and cooking apparatus provides economy. The process is carried out as follows. A washer receives raw potatoes from the harvest fields, etc. and subjects them to a thorough washing with clean water to remove all dirt and other adhering foreign material contaminants. The washer may be of the barrel-type with relatively high pressure water jets directed on the potato or may be a high-frequency vibrator type.

The cleansed potatoes are then suitably conveyed to a preheater to temper the potatoes prior to further processing. The preheater tempers the potatoes by uniformly heating them at a temperature of 155° to 185°F. for a period of time ranging from 1 to 5 minutes. The tempering treatment that the raw potatoes undergo not only increases the potato reactivity or response to further processing but also tends to change the cellular structure thereof so that a mealy texture, substantially duplicating the texture of freshly cooked potatoes is obtained. This is thought to be primarily due to the change in the form of starch from amylopectin to amylose, i.e., from soluble to insoluble starch.

The tempered potatoes are then conveyed to an impregnator where they are immersed in a caustic aqueous solution. The preferred caustic is sodium hydroxide, although other readily available bases are also suitable. Caustic solution is maintained at a pH of at least 7.5 and generally higher, so that 4 to 10% by weight of caustic material (preferably sodium hydroxide) is continuously maintained within the caustic solution by an appropriate control unit.

The impregnator is constructed to allow a relatively high volume of caustic aqueous solution to circulate throughout the potatoes and to heat the solution so as to maintain it at a temperature of 185° to 212°F. The heated caustic solution facilitates interaction between the tempered potatoes, particularly the skins and outer portions thereof with the caustic. Generally, sufficient impregnation of the caustic into the outer covering of the potatoes is achieved within 1 to 4 minutes, after which the impregnated potatoes are removed, without washing, and suitably passed into a digester. The digester maintains the potatoes at a relatively constant temperature for 5 to 30 minutes and allows the caustic time to react with the skins, etc. of the potatoes. The digester materially reduces the amount of caustic consumed in the impregnation step and thereby reduces the overall cost of processing potatoes.

Further, the utilization of impregnation and digesting as separate steps materially reduces the waste disposal problems generally incurred from the amount of caustic present in waste water. The at least partially digested potatoes are then suitably passed through a peeler wherein water, under relatively high pressure, impinges upon the potatoes and removes the digested portion thereof. Digested potatoes may be defined as potatoes that have been impregnated by caustic material to cause a degradation of the impregnated portion thereof. Thus, the digested portions of the potatoes generally include the skin, spoiled or similarly damaged portions of the potatoes, etc. The waste water from the peeler may be filtered and recirculated to reduce the total water consumption of the overall process.

The potatoes, after having undergone a thorough washing and peeling process, may optionally be passed through a neutralizer. The neutralizer is primarily used to neutralize any excess caustic solution that may still be adhering to the potatoes subsequent to the peeling operation. Generally, a pH of 5.5 to 6.5 is sufficient to adequately neutralize any excess caustic in the potatoes.

The potatoes are next suitably conveyed from the peeler (and optionally passed through a neutralizer) to a trimmer, where inferior units and portions thereof not removed during the peeling operation, are removed. This double system of removal of inferior portions increases the ultimate quality of the finished product by allowing only wholesome potatoes to proceed to the final operations. The trimmed and culled potatoes are then passed into a metering control unit which feeds the potatoes into a slicer where the potatoes are reduced in size to substantially uniform slices, cubes or particles having a thickness substantially in the range of 0.50 to 0.625 inch. After the potatoes have been substantially uniformly reduced to particle size, they are passed into a precooker.

The precooker utilizes a relatively hot aqueous bath to heat the potatoes to a temperature in the range of 145° to 185°F. for a period of time ranging from 10 to 50 minutes. The precooking materially aids in controlling the mealiness of the potatoes, thus, if the starting stock was a relatively poor quality (low solids) potato, then higher temperatures (within the range specified) and higher and longer immersion times (within the range specified) convert these poor quality potatoes into one substantially similar to the highest quality potato. On the other hand, if high quality potatoes are used as starting stock, then lower temperatures and shorter immersion time yield a superior product. Without the precooking step, it is difficult to control the mealiness of the final product and generally impossible to utilize lower quality starting stock.

After the uniformly sized potato particles have been precooked a sufficient time in accordance with the solids content of the starting stock, they are passed into a cooler where the temperature of the potatoes is reduced to 40° to 80°F. Preferably, the potatoes leave the cooler at a temperature of 68°F., after an immersion time within the cooler in the range of 10 to 30 minutes. The cooler is provided with a continuously circulating aqueous bath to insure uniform exposure of the potato particles to the cooling media. The precooking and cooling steps tend to aid in converting the soluble starch within the potatoes into substantially insoluble starch that give the final product a more desirable taste.

The cooled potato particles are then suitably conveyed to a cooker where the potatoes are cooked at a somewhat higher temperature. The cooker subjects the potatoes to heat in the range of 185° to 212°F. for a period of time ranging from 15 to 40 minutes so as to sufficiently cook the potatoes for the subsequent operations. The cooked potatoes are then passed through a ricer where the potatoes are mashed with a minimum of cellular rupture taking place. Preferably the ricer is an auger enclosed within a tubular body having closely spaced rods at the discharge end thereof forming a grill through which the potatoes are forced. A rod-type grill is preferred over a screen or other sharp-surfaced grill, as less cellular rupture occurs when a smooth-surfaced grill is utilized. Cellular rupture releases additional starch, imparting an undesirable pasty taste to the finished product. Various chemical additives may be added in the cooked potatoes at the intake end of the ricer, so as to become uniformly intermixed therewith at the discharge end, prior to the drying operation.

Generally, the chemical additives utilized to prevent oxidation of the final product are monoglycerides and butylated hydroxyanisole. In addition, sodium acid pyrophosphate is utilized to bleach and remove grayness from the product. As a potato preservative, solutions of sodium sulfite and sodium bisulfite may be utilized. Further, in most locations, it is desirable to add relatively small amounts of citric acid, or the like, to buffer the chemical additives to a pH of 5.5 to 6.5 before addition of the chemicals to the potatoes.

The mashed potatoes are then passed to a dryer where they are spread in a substantially uniform relatively thin layer on a conventional single or double drum dehydrating machine. Generally, such dehydrating machine is internally steam heated and equipped with a multiplicity of applicator rolls containing the peripheral surfaces of the drying drum to insure uniform application of the mashed potatoes.

The rotational speed of the dryer drum and the internal temperature thereof is suitably controlled so as to give a final product having a moisture content of 4 to 10%. The dehydrated potatoes are removed from the dryer drum by means of a suitable doctor blade in the form of substantially monocellular sheets and then broken into various sized flakes depending on the desired end quality. The dehydrated potato flakes are then passed to a packager where they are suitably packaged for storage, transportation and ultimate use. The mashed potatoes are applied to the drying drum in a sheet thickness ranging from 0.005 to 0.008 inch so as to avoid cell rupture and increased amounts of free starch while

at the same time, allowing rapid dehydration to take place without excessive heat requirements. It has generally been found that the greater the amount of free starch present in
the dehydrated potato flakes, the more pasty the reconstituted product tends to be. The
dehydrated potato flakes may be reconstituted into a product having a flavor, texture and
appearance generally equal to a product provided from fresh potatoes by the addition and
intermixing of 4.5 to 5.5 parts by weight of boiling water or a mixture of boiling water
and milk with various flavor agents, as desired. Drawings and descriptions of a neutralizer,
a cooker, a cooler and a preheater, precooker or impregnator are contained in the patent.

POWDER

Screening to Remove Skin, Eyes, and Defects

It will be obvious that retention of the highest possible degree of natural potato flavor in
the rehydrated product is to be greatly desired. A large portion of the flavor characteristics
of a potato are concentrated in the skin and in those zones of the potato core near the
skin. Consequently, it has been attempted to remove only the outer peel portions of the
potato prior to cooking and dehydrating, or even process whole, unpeeled potatoes. Such
processes have heretofore produced off-color products because the browning precursors are
also concentrated in and near the skin of the potato. Further, attempts to screen or otherwise mechanically separate potato peel, eyes, and defects such as rot from potato solids
after cooking and mashing of the potato have brought about undesirable rupture of the
potato cell walls and the liberation of viscous starch. The result is a rehydrated, mashed
potato that has a pasty texture. Failure to separate out the peel and other undesirable
potato portions yields a product which has an uneven texture and little consumer appeal.

In a process described by *F. Hollis, Jr. and B. Borders; U.S. Patent 3,220,857; November 30,
1965; assigned to General Foods Corporation* potato material containing a substantial portion of the peel fraction in the form of peel, eyes and defects is cooked under carefully
controlled conditions so that the core portion as well as potato meat portions circumjacent
the skin of the potato can be recovered after the potato material is mashed by creating a
slurry of the potato mash containing such imperfections and thereafter screening this slurry
under conditions which achieve mechanical removal without converting the usable potato
solids present into undesirable free starch or other forms such as contribute to pastiness
in the product.

Thus, raw potatoes may simply be washed to remove field dirt or superficially peeled,
leaving a substantial balance of the peel fraction of the potato. The potato is subjected to
carefully controlled precooking conditions, whereafter the potato cells are conditioned for
the rigorous treatment to which they are subjected during slurrying and subsequent mechanical removal of imperfections; this conditioning is achieved by quenching the potato following precooking, as by reducing its temperature to below 70°F. Thereafter the quenched
potato is more fully cooked, preferably under steam, and mashed to a slurryable consistency,
slurried with water and screened.

The necessity of a precook-quench series of process steps is even more accentuated when
the slurried potatoes which are subsequently screened to remove the discrete peel imperfections are dehydrated by means of spray drying. In a spray drying process, during the
course of passage of the potato slurry through the atomizer, potato agglomerates are subjected to further frictional forces and to further subdivision into particles of a smaller size
than those which are incurred incident to screening. These forces also tend to rupture
the cell walls and induce pastiness in the reconstituted product.

The precook-quench steps greatly mitigate this tendency toward rupture. In order to
augment the unusual effect of the precooking and quenching operations on a subsequent
slurrying step, it has been found advantageous to incorporate other materials in the process
at various stages thereof in order to further harden the walls of the potato cells in the slurry.
Thus, calcium chloride is preferably incorporated in the slurry in an amount sufficient to

harden substantially the walls of the potato cells therein. Calcium chloride may also be incorporated in the precook liquid or the medium which is used to quench the precooked potatoes.

Further, a monoglyceride emulsifier which is an ester of a polyhydric alcohol and a higher fatty acid is preferably added to the slurried, screened potato product in order to further enhance the texture of the rehydrated product. An inert cellulose composition is advantageously added to the slurried, screened product to aerate the final product after dehydration and give it fluffiness. All such improvements, however, are in addition to the basic combination of steps which comprise precooking and quenching raw potatoes which either have not been peeled or only incompletely peeled, cooking and mashing the potatoes, then slurrying the mashed potatoes, screening undesirable potato portions from same, dewatering the slurry, and finally dehydrating the slurry to a stable moisture content. An example will illustrate the process.

Example: Field run Idaho Russet Burbank potatoes (utility grade) having a solids content between 18 and 26% were selected. These potatoes had been stored prior to processing at a temperature of 55°F. and an ambient relative humidity. The potatoes had the customary amount of rot for a utility grade batch of potatoes. Immediately prior to processing they were conditioned for two weeks by storage at 70°F. in an air-conditioned room having 50% relative humidity; after two weeks of storage the potatoes had a total sugars content of less than 5% and a level of reducing sugars measured as dextrose of 1.9% by weight of the potato solids (dry basis).

The conditioned potatoes were brush scrubbed, 20 pounds at a time, and subjected to treatment for two minutes in an abrasion peeler modified to include a stiff rotating brush and a rubber-covered side wall. The potatoes were then subjected to a direct spray of water. As a result of this scrubbing and washing, the potatoes had removed therefrom that amount of the outer skin which sufficed to remove all surface dirt. The clean potatoes still contained substantially all of the peel, eyes, surface bruises, and rot wherever it occurred. The thoroughly cleaned potatoes were sliced to $^{11}/_{16}$" and then soaked in a 0.2% sodium metabisulfite solution for 30 minutes to inhibit post-slicing browning and for darkening of the potatoes.

The raw, sulfite-treated potato slices were then precooked by immersion in a 2% sodium acid pyrophosphate aqueous solution at a temperature of 160°F. The potatoes were treated for a period of approximately 20 minutes or until the internal temperature of the potato was 150° to 160°F. During this step of the process the potato slices rapidly assumed a temperature of 150°F. and were maintained at a temperature somewhat below 160°F. for the balance of the treatment.

Incident to this precooking operation, sugars, starch, protein, and fat were partially extracted; thus, at the end of this precooking step potato slices were leached to a point where the total sugars present were 1.85% and the reducing sugars content measured as dextrose was 0.83% by weight of the potato solids (dry basis); the solids content of the potatoes had not materially altered during this period. The sodium acid pyrophosphate acts in the manner of a chelating agent, minimizes the creation of post-cooking discoloration and arrests penetration of color precursors from the peel into the interior of the potato during precooking and subsequent steaming.

The precooked slices were discharged onto an inclined belt conveyor on which they were subjected to cold water sprays directed on the belt by means of spray nozzles. The temperature of the water directed from the spray nozzles was 50°F. and the potato slices remained on the conveyor belt for 25 minutes or until the internal potato slice temperature was reduced to less than 70°F. After cooling the potato slices were immediately transferred into a continuous belt stream cooker, wherein saturated steam was introduced at atmospheric pressure for a period of 22 minutes or until the potatoes were cooked to a mashable condition. As a result of steam cooking, the potato cells retained starch which was substantially gelatinized, and a cooked flavor was produced. This flavor was enhanced by reason of the flavor values in the jacket and shell portions of the potatoes as well as the underlying flavor values in the peripheral meat portions of the potatoes being infused throughout the potatoes. The solids content of the potatoes remained substantially constant during mashing.

The cooked potatoes were then conveyed by a belt to a set of opposite rotating stainless steel crushing rolls which had a clearance between them of 0.040 inch, the rolls having a diameter of 12 inches and rotating at 22 rpm. By virtue of the gradual and gentle reduction of the cooked potato slices as they are passed through the pinch between the rollers, the potatoes are gently crushed so that the potato cells are damaged to a minimum extent. The potato mash still contains all of the peel, eye and defects present on the slices together with the subdivided potato solids which have undergone a cleavage of the bonds aggregating them; the mash still contains cell aggregates in addition to individual potato cells. The heterogeneous mash of potato cell is not mashed, however, to a point where a dough-like consistency is produced, or wherein the potato cells are homogeneously subdivided. The mashed potatoes have the appearance of clumps of a baked potato which have been spooned or otherwise removed from the jacket.

The resulting mash was blended continuously with an equal weight of water at a temperature of 50°F. in a slurry mixer to yield a slurry having a solids content of between 9 and 12%. Included in the slurry water was a quantity of antioxidant and that amount of metabisulfite solution measured as 50 ppm sulfur dioxide. The homogeneous mixture thus produced had a temperature of 110°F. and was immediately pumped to and passed over a vibrating 8 mesh screen. The slurry was immediately subjected to screening to prevent flavor loss and conditioning of the potato solids to an extent that results in graininess of the reconstituted mash. Preferably the slurry of potato solids is not elevated to a temperature above 130°F. to mitigate cell rupture. Screening of the slurry removed pieces of peel, eyes and rot, which can either be discarded or reclaimed.

After passing through the screen, the slurry is directed to a pulper, which is a screen having 0.033 inch openings. Most of the peel, the large eyes and defects having been retained as overs on the 8 mesh screen together with a minimal quantity of potato meat which is separately handled and preferably recycled to the homogeneous mixture of potato solids, the thrus fed to the pulper may be described as a relatively clean suspension of potato cells including some solubilized starch and other solubilized chemical constituents including fats and sugars, protein, and amino acids, as well as free starch.

In pulping the thrus, small peel, specks, eyes, and residual rot are retained on the pulper screen and by-passed. The slurry solids are thus impelled through the screen openings which retain the undesirable portions as described above. In carrying out the pulping operations, care is exercised to avoid too vigorous agitation of the slurry solids, such as would damage the potato cells, hence the pulper is operated at an rpm sufficient to impel the slurry solids through the screen openings of the pulper, but not sufficient to materially alter the physical condition of the potato cells themselves, although a minimal amount of cell rupture takes place.

The pulped slurry was then dewatered to approximately 17% solids by passage over a continuous, straight-line filter. Dewatering is necessary to remove significant portions of fat, protein, free and solubilized starches, sugars, and other browning precursors and off-flavor reactants which seriously detract from stability of the dehydrated potato product.

The filter cake from dewatering was blended with Myverol 18-07, methylcellulose (Methocel), an antioxidant (butylated hydroxytoluene, propylene glycol and citric acid in solution of propylene glycol) and a combination of sodium sulfite and sodium metabisulfite, to yield 0.5% of the mono- and diglyceride emulsifier, 1.25% methylcellulose, 50 ppm antioxidant, and 50 ppm sulfur dioxide, all on a dry weight potato solids basis. The thoroughly blended filter cake was diluted to 10% solids by the addition of fresh cold water (60°F.) and the resultant homogeneous slurry at a temperature of 60°F. was transported to a holding tank directly above a spray drying tower, a vertical elongated cylindrical tower

41' high and 12' in diameter. Twelve feet from the bottom of the tower the tower assumes a conical shape, the cone ending in a discharge opening at the bottom of the tower. The spray tower was equipped with a cyclone dust collector for collection of fine potato particles which are combinable with the dehydrated product collected at the bottom of the tower. The potato slurry was pumped to a smooth, three tier, 9" bowl atomizer rotating at 4,600 rpm. The atomizer was located at the center of the 12' tower, 2'3" from the top of the tower and operated to form a pattern whose outline covered the entire cross-sectional area of the tower. The droplets formed descended cocurrently with a downward draft of hot drying air at an inlet temperature of 425°F., which air was introduced at the top of the tower at 3,000 to 6,000 scfm.

The slurry was fed from the top of the tower at one and two-thirds gpm. The droplets were dehydrated as they descended for collection at the trough of the tower bottom. The outlet temperature of the dehydrating air was maintained at 230° to 245°F. dry bulk and measured at the outlet point of exit for the product. The product at 6% moisture was collected from the tower and the cyclone communicating with the point of exit and combined. The product was then passed over a vibrating 16 mesh screen with the thru mesh fraction being retained as the final product, or optionally further dried in a funnel dryer.

The product is a creamy, yellow-colored powder having a density of 0.4 to 0.5 g./cc. The product is generally in the form of individual potato cells and small agglomerates of potato cells, which cells have a mean diameter of 150 microns. When 82 grams of the dry product are dry-blended with 6% nonfat milk solids and reconstituted in 390 cc of boiling water containing ½ teaspoon salt and 2 tablespoons butter, a very dry, mealy mash was produced quite like the texture of baked potatoes. With the subsequent addition of 60 cc cold milk followed by vigorous mixing, a very light, fluffy, nonpasty product with a creamy color was prepared. This product was devoid of a scorched flavor and free of specks of unrehydratable particles and clusters; the product was quite like freshly mashed potatoes.

Two Slurry Steps Before Spray Drying

B. Sienkiewicz and F. Hollis, Jr.; U.S. Patent 3,261,695; July 19, 1966; assigned to General Foods Corporation have found that when a combination of process steps are utilized subsequent to cooking and mashing the potatoes, which steps include slurrying the mashed potatoes, screening the slurried potatoes to remove defects, dewatering the potato slurry to a filter cake and then reslurrying the dewatered potatoes, the subsequently spray dried product will reconstitute to a mashed potato product having a highly desirable texture which is neither grainy or pasty.

By virtue of this combination of three steps: slurrying cooked, mashed potatoes, dewatering the slurry, and reslurrying the dewatered filter cake, spray drying is able to be carried out in a manner which leads to the formation of a highly acceptable product in comparison with spray dried products of the prior art. It is believed that this is because the reslurry is substantially free of soluble materials in addition to the discrete potato particles. Unless the potato slurry is dewatered, the solubles therein tend to form a coating around the potato cells during spray drying. This coating interferes with the diffusion of water from within the cells and results in an improperly dried product. Thus the three steps set forth herein are critical with respect to a subsequent spray drying operation.

The potatoes are slurried directly after they have been cooked and mashed. Sufficient aqueous medium is added to the cooked potatoes to form a fluid slurry of the potatoes in the medium. While the precise degree of dilution of the potato mash with an aqueous medium will vary, it has been found that a general range of 7 to 16% solids content of the slurry after dilution is most preferred. Generally, the amount of water or aqueous medium added should be such as to put the slurry in a fluid state in which it can easily be subjected to mechanical separation of portions thereof, while limiting the quantity of water to a level at which there will not be an excessive amount of solubilization of starches and sugars and at which it will still be economic to dewater the screened slurry at a later step in the process.

A slurry temperature range between 60° and 180°F. has been found operable. The slurry is then subjected to a mechanical separation in which eyes, rot, and parts of the skin, if the potatoes have not been completely peeled, are separated from the slurry. This process has been more completely described in the previous patent.

Following the mechanical separation of undesirable potato portions, the slurry is dewatered to remove large quantities of free and solubilized starch, soluble reducing sugars and some of the proteins which have been solubilized in the slurry medium. Also removed are fats and free amino acids. The more water that is drawn off, the more undesirable factors that are removed from the potato slurry. Most advantageously, dewatering is carried out until the solids content of the dewatered potatoes is substantially equal to the solids content of the raw potatoes being processed. While dewatering may be effected by various types of vacuum filtration, centrifugation and the like, the use of a horizontal vacuum filter is preferred.

In any case, dewatering should be carried out at least until the solids content of the potato filter cake which issues from the dewatering equipment is 14 to 22%, preferably 18%. This filter cake will most advantageously contain only individual, clean potato cells which are substantially free of loose or excess soluble and insoluble starches and sugars and at least partially free of fats, soluble proteins and free amino acids. After dewatering the clean potato cells and small aggregates thereof are reslurried by mixing the filter cake with an aqueous medium until the reslurry contains 7 to 16% potato solids by weight. A desirable product was found to be obtained when the potato concentration in the reslurry was 10%.

The reslurry step has been found to be a suitable place for blending additives with the clean potato cells which are obtained from dewatering. At this time, such additives as an aerating agent, such as methylcellulose commercially available under the trademark, Methocel, may be added. Methocel is a water-soluble inert, colorless, odorless, tasteless, nontoxic cellulose ether which, when mixed with potato granules, imparts fluffiness. About 0.25 to 3.0% Methocel per weight of potato solids is a preferred range.

Other preferred additives are emulsifiers, typically those which are identified as mono-glycerides and which materially offset any pasty, gummy texture that might otherwise be produced in a reconstituted potato product. Generally 0.10 to 2.50% of emulsifier is used, based on the weight of the potato solids. Typical of those monoglycerides which are available commercially and useful as emulsifiers in this process are those series of products identified by the trademark, Myverol.

The reslurried potatoes with additives are then spray dried. The specific spray drying equipment used may vary, for example, the drying gas, ordinarily heated air, may be directed cocurrent or countercurrent to the flow of the liquid being dehydrated. In a cocurrent spray drying operation, it has been found that a preferred inlet temperature of the air used to dry the atomized slurry is 250° to 550°F., most preferably 380° to 420°F. The maximum outlet temperature of the air is 260°F., with an outlet air temperature of 225° to 235°F., say 230°F., being preferred. The outlet temperature of the product should generally not exceed 200°F. to avoid excessive browning and toasting. The product from the spray dryer is screened to remove clumps. The preferred moisture content of the potato granules which are to be packaged and marketed as a final product has been determined to be 5 to 6.5%.

Continuous Cooking to Produce a Powder

H. Griffon; U.S. Patent 3,425,849; February 4, 1969 has developed a process for the de-hydration of potatoes in which potatoes are cut into a stream of uniform pieces while in water, then cooked in boiling water, immediately cooled in water, mixed with liquid into a puree and then dehydrated preferably while below 37°C. Carrying out the cooking in a continuous manner avoids the preparation of a large quantity of potatoes at the same time, thus rendering unnecessary the use of reducing agents such as sulfur dioxide in order to avoid the blackening of the tubers. The continuous preparation also permits the rapid cooling of the tubers between the cooking phase and the pulping phase, this cooling having been recognized by experience as being favorable to obtaining a consistency and a good flavor of the reconstituted potatoes. After the cooling operation, there is preferably added a quantity of water corresponding to about a third to half of the weight of the cooled tubers and it is advantageous to beat the pulp in order to obtain by emulsion a product of creamy appearance which will then be subjected to dehydration. This preparation, combined with the use of dehydration at low temperature, makes it possible to avoid the phenomena of rancidness and hydrolysis and to obtain a potato powder of particularly good quality.

The apparatus for carrying out the process is characterized in that it comprises a boiler of tunnel form, the inlet and the outlet of which are raised relatively to the middle section by an amount sufficient to maintain a good level of water, this tunnel-type boiler being traversed by a flexible endless conveyor serving as carrier for the potatoes or the like, the speed of the conveyor being regulated so that the period of immersion in the water of the boiler corresponds to that necessary for the cooking operation. The tunnel-type boiler is preferably equipped in its central section with an exhaust conduit for the steam which is liberated, so as to prevent this steam entering the room in which the apparatus is located.

The flexible conveyor advantageously comprises a perforated bottom and extensible lateral walls, the upper side being free; it may be made of any appropriate material, for example of a plastic material which is resistant to the maximum temperature provided or of metal. In one simple embodiment, it is formed by a bottom comprising a metal grid or lattice, of which the transverse wires are bent up perpendicularly to the bottom in order to form the sides. When the apparatus is designed for certain applications, such as the preparation of a pulp or puree adapted to be subjected to a dehydration process, the tunnel boiler is followed by a cooling tank through which the conveyor also extends, the conveyor then discharging the cooked and cooled tubers into a pulping press which converts the tubers into a pulp or puree which is then transported to the dehydration section. An example will illustrate the process.

Example: The potatoes, which are washed and scrubbed and preferably peeled mechanically or by steam by the conventional methods, are cut into relatively small pieces in any conventional cutter device and it is operated in water so that the tubers, during this phase, are protected from any oxidation and are washed, thus extracting the starch and freeing them from any skin debris, earth, etc.

As the potatoes are cut up, they are placed on a conveyor. They travel through the water, brought to boiling point, in the tunnel-type boiler at a speed calculated so that each piece of potato is kept in the water for a period of time necessary for the cooking thereof, i.e., from 12 to 15 minutes, according to types. At the outlet end, the conveyor with its load of potatoes enters a cold water tank, and then it tips the potatoes into a continuously running pulping press.

In order to obtain a pulp of a suitable consistency, a certain quantity of water is added, this quantity being approximately between a third and a half of the weight of the cooled tubers. The pulp with a suitable consistency is beaten so as to provide an emulsion by occlusion of air in order to obtain a creamy product. It has been established that this formation of an emulsion is favorable to the subsequent operation of dehydration. This beating can be carried out with any suitable apparatus, for example of mechanical type; this operation can be combined in a simple apparatus with all or part of the addition of water provided above. The creamy pulp thus obtained is then dehydrated in any appropriate apparatus.

After dehydration, a solid product is obtained in the form of flakes or powder. This product is collected and then if desired it is ground and screened (preferably a number 60 screen). There is thus obtained a dry pulverous substance, the content of which is 10 to 15% of the dry material. It has been found that good results are even obtained if the dehydration is only carried out as far as this latter value, and that the products obtained with this water content are preserved satisfactorily under these conditions. This powder, to which water or milk or the mixture of these two liquids is added, is reconstituted after appropriate heating according to the respective proportions either as mashed potato boiled in water, or as traditional potato puree.

DICED POTATOES

Blanching with Hot Salt Solution to Prevent Breakage

Dehydrated potato units in various sizes and shapes are produced and sold in substantial quantities. Popular sizes are ⅜" cubes and dice of ⅜" x ⅜" x ³⁄₁₆" dimensions. These units are used in making stews, chowders, and the like, and in such products it is important that units be in their original unbroken shapes. In freshly dehydrated potatoes very little breakage occurs but after ten days or two weeks in storage the dehydrated units begin to develop cleavage lines and break apart. Such breakage continues throughout the storage period.

J.W. Cyr; U.S. Patent 2,973,276; February 28, 1961; assigned to Snow Flake Canning Co. discovered that this breakage can be substantially reduced by treatment with a weak sodium chloride solution and to a somewhat lesser extent by treatment with solutions of sodium nitrate and ammonium chloride. Mixtures of such solutions may also be used particularly where it is desirable to avoid significant addition of salt to the food.

In carrying out the process, there is preferably substituted for the blanch with hot water or steam to which the freshly cut potato units are subject in the process of dehydration now in common use, a five minute blanch with an aqueous solution of sodium chloride. For dice ⅜" x ⅜" x ³⁄₁₆" a concentration of 2% has been found to give satisfactory results with no detectable difference in flavor. The solution used for blanching should be heated to near boiling. A temperature of 210°F. has been found most satisfactory.

Example: ⅜" cubes given a plain water blanch showed 68% fractured cubes in 39 days; yet material from the same sample given a 2% sodium chloride solution blanch showed but 14% over the same time interval, and that given a 3% sodium chloride solution blanch, 0%. However, these last were checked again at the end of 116 days and showed 2.6% fractures against more than 72% (at this point many of the fractured pieces were broken into two or more pieces) for the plain water blanched sample.

Shorter Rehydration Time

J. Cording, Jr. and R.K. Eskew; U.S. Patent 3,038,813; June 12, 1962; assigned to the U.S. Secretary of Agriculture have discovered a process for the preparation of fruit and vegetable pieces for both domestic and commercial use capable of rapid rehydration to a product ready for immediate consumption. A further advantage of the process is that if the product is to be stabilized by dehydration it enables a great reduction in the overall time required for dehydration.

In an embodiment of the process potatoes are peeled and cut into pieces of the desired size and shape. They are then either steam or water blanched in the conventional way or are precooked with or without an additional cooling step. Whether blanched or precooked and cooled, the pieces may be sprayed with a solution of calcium chloride and sodium bisulfite to minimize sloughing on reconstitution and to improve keeping properties.

The pieces are then partially dehydrated by any conventional method as for example in belt or tunnel dryers. Moisture content at the end of this stage of drying should be 20 to 50%. The partially dehydrated pieces are then placed in a chamber and sealed. The pressure within the sealed chamber is then brought rapidly to 20 to 60 psig by the application of external heat. In general, higher pressures are used with the lower moisture. In the preferred method the chamber is rotated and heat is applied with a gas flame. Alternatively, heat may be applied by other means, as for example with steam in a jacket, in which case rotation might not be necessary. Upon reaching the desired pressure the chamber is opened instantly to the atmosphere, releasing the pressure and discharging the still intact pieces.

The pressure-cooking effect combined with the flashing from liquid to vapor of a portion of superheated water from all parts of the piece forms a porous structure. The pieces may then be subjected to further drying to a moisture content appropriate for storage, for example, about 7%. This can be done by any conventional means. The time required for this normally slow stage of drying is greatly reduced because of the porosity of the piece. Alternatively, the porous pieces, instead of being finally dried, may be frozen for storage and use. The finished product retains the general shape of the raw piece and does not exhibit the case hardening found in the conventionally prepared dehydrated products. On immersion in boiling water dehydrated vegetable pieces rapidly imbibe water, swell to approximately their original size and are cooked ready for serving in 5 minutes.

Example: Sebago variety potatoes of specific gravity 1.074, corresponding to a solids content of 18.6%, were peeled and cut into cubes ⅜" on a side. The cubes were precooked in water at 160°F. for 15 minutes followed by cooling in water at 38°F. for 20 minutes. Then they were dipped in a water solution containing 0.5% by weight of sodium bisulfite and 0.5% by weight of citric acid followed by drying in trays by circulating heated air through the bed. The air was at 160°F. (dry bulb), and 100°F. (wet bulb) and the cubes were dried to 31% moisture. The partially dried cubes were placed in a sealed pressure chamber equipped with a hinged lid capable of instant opening.

While rotating, the chamber was heated externally until the internal pressure reached 60 psig. This required 9.6 minutes, whereupon rotation was stopped and the pressure instantly released by tripping the lid. The cubes were discharged from the chamber. They were then returned to the tray dryer and dried under the same condition as were employed in the first stage drying except that they were dried to a final moisture content of 6%. The cubes, when placed in boiling water, rehydrated in 5 minutes to a uniformly cooked condition ready for eating and regained their cubical shape.

Low Temperature Irradiation

It is known to subject dehydrated potatoes to irradiation such as gamma rays, for the purpose of reducing the rehydration and cooking times. The moisture content of such dehydrated potatoes is conventionally between 1 to 20% by weight of water, but it has been found that such irradiation imparts undesirable properties to the dehydrated potatoes with resulting production of off-color, flavor and/or loss of texture.

D.S. Gardner and C.K. Wadsworth; U.S. Patent 3,463,643; August 26, 1969; assigned to the U.S. Secretary of the Army have found that such undesirable side effects may be substantially reduced or even eliminated by irradiating dehydrated potatoes in an environment having a temperature below -100°C.; the level of irradiation, which may be either gamma or beta rays, should lie between 1 to 10 megarads, the rad being the unit of absorbed dose and is equivalent to 100 ergs per gram. Dehydrated potatoes irradiated under these conditions may be prepared for consumption within as little as 2 minutes boiling in water while retaining to a significant degree their original flavor and color.

A comparison illustrating the effect of irradiation at a temperature below -100°C. was made using sliced ½" x ½" x ⁵⁄₃₂" potatoes which were obtained from one variety, viz. Cherokee (1.075 to 1.080). One group of canned dehydrated sliced potatoes were treated at ambient temperatures with irradiation levels of 4 and 6 megarads and the other group treated by immersing the cans in liquid nitrogen with irradiation levels of 4, 6 and 8 megarads. Both groups were then rehydrated and cooked along with a control. Even with the dosage of 8 megarads the flavor is fair and the resulting product is palatable. Even with a dosage at a level of 10 megarads irradiation given at the requisite low temperature, an improvement in flavor was found over much lower dosages of irradiation given at ambient temperatures.

FRENCH FRIES

Use of Toasted Flakes

French fries produced in the conventional manner from raw potatoes are disadvantageous in several respects. First, the quality of french fries produced varies from species to species of potatoes. Moreover, even within one species the quality varies depending on the age of the potato.

Secondly, the frying time required to produce a palatable french fry varies between species and also is a function of age. Thirdly, cut potato strips, from which french fries are made, necessarily vary in size because of the irregular shape of potatoes; therefore, a batch of cut strips does not fry uniformly.

To solve the inadequacies of french fries produced from regular potatoes, attempts have been made to produce fabricated french fries from an extruded dough. In this manner the composition of the french fry can be carefully controlled. The raw material can be made just prior to frying, thus eliminating the quality variation induced by time. Moreover, the size of the pieces can be carefully controlled to assure uniformity of frying.

Although palatable fabricated french fries have been produced by prior techniques, the quality attained has never equalled the quality of a high quality french fry produced in the normal manner from raw potatoes.

Two problems have been encountered; namely, the flavor of a high quality natural french fry has never been duplicated, and secondly, a dry, mealy, friable texture has never been achieved.

It has been proposed that french fries be formed by extruding a dough made from dehydrated mashed potatoes, such as potato flakes of the type which are normally used to produce instant mashed potatoes.

Although, by means of such a process a palatable french fry, the quality of which may exceed the quality of french fries produced in the normal manner from many types of potatoes, may be made, there is much room for improvement as regards flavor and texture.

Heretofore french fries, including those produced by conventional means from raw potatoes, have manifested a tendency to become limp and soggy when cooled. Moreover, in the frozen french fry field where potatoes are prefried and frozen before distribution to the consumer who merely heats the frozen product in an oven, a product which is crisp after reheating has never been available.

E.L. Fritzberg; U. S. Patent 3,282,704; November 1, 1966; assigned to the Pillsbury Co. has developed a process for preparing a high quality french fry produced from a dehydrated potato-flake dough which is less susceptible than formerly to starch cell breakdown during handling. The process provides a fabricated french fry in which at least a portion of the dehydrated potatoes used are toasted. In french fries made according to the process a dry, mealy friable texture, which is fully equivalent ot the texture of the best french fry made from raw potatoes, may be easily attained: starch breakdown does not occur as readily during handling as in dough made with nontoasted flakes; and the flavor of fabricated french fries is materially enhanced.

The term toasted, as applied to dehydrated mashed potatoes, is hereby expressly defined to mean the slightly browned product resulting from heat treating dehydrated potatoes at elevated temperatures. To produce the brown color, it is necessary that the water content of the dehydrated potatoes, which is normally 6 to 7% by weight, be reduced to less than 1%. It is believed that moisture content of less than 1% indicates the total absence of unbound free water. To minimize the time required for this treatment, the temperature should be in excess of approximately 200°F., preferably within the range of 200 to 300°F.

Within this temperature range, toasting of dehydrated mashed potatoes containing 6 to7% moisture is achieved within 4 hours time. Toasting may be conducted in an oven under atmospheric conditions or under a vacuum, the efficacy of the resulting product as regards improvement of flavor and texture in formed french fries being comparable when either method is used.

Dehydrated potato flakes are currently made by drying mashed potatoes in a thin film on a heated drum dryer, the drying process usually being terminated by scraping the dry film from the drum when moisture content of the potatoes has been reduced to approximately 6%.

Toasted flakes may be conveniently produced by allowing the drum drying operation to proceed beyond the point of 6% moisture. When toasted flakes are produced in this manner, the rate of rotation of the drum dryer should be reduced to the point where the drying potato sheet has been reduced in moisture to less than 1% and a slightly brown color has developed as noted above by the time the sheet is scraped from the drum.

Although potato flakes are preferred for use in extrudable dough for making french fries, granules may also be suitably used. In this connection, it should be noted that while starch breakdown is less of a problem in granules than in flakes, the flavor and odor of untreated granules detract from their use in formed fried products.

The essential ingredients of an extrudable french fry dough are dehydrated mashed potatoes, preferably in the flake form, and a sufficient amount of water to afford a moldable consistency. In general, depending on the properties of the starting material, such doughs comprise approximately 25 to 55 weight percent dehydrated mashed potato and 45 to 75 weight percent water. In terms of parts, this percentage range can be expressed as about 80 to 300 parts of water for each 100 parts of dehydrated mashed potato. It should be realized, of course, that the term water as used herein includes equivalents such as milks which comprise primarily water.

To be easily extrudable the dough must be cohesive. Cohesiveness can be attained by adding to the basic ingredients mentioned above, binder materials such as free starch or a soluble protein, such as egg albumen. Alternatively, and preferably, the requisite cohesiveness in the doughs of the process is attained by physically mixing the dough to such an extent that a portion of the starch cells of the dehydrated mashed potato contained therein are broken down, thereby providing free starch.

Potato flakes are altered by toasting to such an extent that breakdown of starch cells is difficult to produce by mere physical mixing. For this reason, when potato flakes are used in french fries produced in accordance with this process, it is preferred that at least a portion of the flakes be untoasted. The ratio of toasted to untoasted flakes may be varied to meet mixing capabilities; however, in order that maximum flavor and texture improvements noted above may be enjoyed, the percentage of toasted flakes should be as high as possible. This percentage should be above 50% by weight, more preferably above 90%, even though any amount of toasted flakes will enhance flavor and texture.

In addition to the components mentioned above, other ingredients may be added to improve keeping quality and color or to impart other flavors. Flavoring agents such as salt, cheese, onion and garlic may be easily incorporated in the dough. The use of a small amount of dextrose (i.e., up to 0.2 weight percent), improves the browning characteristics of the dough during frying. The use of dextrose will be illustrated in the examples below.

The dough may be used in a variety of ways. It may be formed by any technique (e.g., extrusion) and fried in hot oil immediately. The resulting product may be eaten immediately or it may be frozen and distributed to consumers who merely reheat the product before serving. Potatoes made according to the process reheat to a very crisp condition. Alternatively, the doughs may be formed and frozen without frying and distributed to the consumer in that state. In other variations, the formed food product may be prefried and frozen prior to distribution, or distributed without freezing or frying of any type.

Example 1: Fifty pounds of dehydrated potato flakes containing about 7 weight percent moisture were placed in a steam-jacketed ribbon mixer supplied with 60 psig steam (308°F.). After 1¾ hours in the mixer at this temperature, the flakes began to change from their normal white color to a slightly brown color. After 10 to 15 minutes had elapsed, a light brown color had developed fully and the flakes were removed from the oven and allowed to cool. An analysis revealed that less than 1% moisture remained in the treated flakes.

Example 2: A batch of dough containing 5,220 grams of ordinary potato flakes, 8,955 grams of water, 140 grams of salt and 14 grams dextrose was mixed until cohesive. The dough was extruded and cut into elongated pieces having the dimensions three-eighths by three-eighths by two inches.

A second batch of dough containing 4,410 grams of toasted flakes produced as described in Example 1, 490 grams of ordinary potato flakes, 9,275 grams of water, 140 grams salt and 14 grams dextrose, was extruded and cut into similar pieces. The pieces from each batch were fried separately in vegetable oil at 350°F. for 15 seconds. The resulting pre-fried potatoes were frozen and placed in storage.

After two weeks in storage, samples were taken from each batch and fried in vegetable oil at 350°F. for approximately one minute. Upon visual examination, the fried potatoes made from toasted flakes had a more mealy, friable texture than did those produced from regular flakes. Samples from each batch were also eaten and in this test, the toasted flakes manifested a less pasty texture than did the product which contained no toasted flakes.

Addition of Cellulose Ether Binders

M.J. Willard, Jr. and G.P. Roberts; U.S. Patent 3,399,062; August 27, 1968; assigned to Rogers Brothers Company describe a method for producing heat processed food products formed from a mixture of dehydrated, comminuted, starchy vegetables and a thermal gelling cellulose ether edible binder.

In general the process comprises providing dehydrated, starch-containing vegetables that may be composed of aggregates, such as crushed dice or granules, and the optional addition of flakes, and also the optional addition of raw vegetable pulp, adding water at a temperature between 45° and 130°F. for rehydration of dehydrated vegetables, adding a thermal gelling cellulose ether binder to form a mixture, the binder being either dry and added before the water, or in aqueous solution, and providing at least part of the water of rehydration, mixing the mixture to rehydrate substantially all of the dehydrated vegetable, extruding the mixture between 45° and 130°F. without the necessity of artificially cooling the mixture, and deep fat frying the mixture to form an edible vegetable product.

The preferred binders are polymeric ethers of cellulose that are thermally gellable. Included within this term are: (a) Alkylcellulose ethers, such as methyl and ethyl cellulose ethers and the the methylcellulose ethers produced as Methocel MC type. (b) Alkyl hydroxyalkylcellulose ethers, such as hydroxyethyl methylcellulose produced as Tylose TWA/MK–3000, in Germany, ethyl hydroxyethylcellulose produced as Modocoll–600, in Sweden, and the hydroxypropyl methylcellulose, such as Methocel 60 H.G. and Methocel 90 H.G., produced in the United States.

The thermal gelling binders may be used as a dry powder or in liquid form, particularly as an aqueous solution. It is possible to use aqueous solutions up to 18% concentration of the thermal gelling binder, such as methylcellulose ether. The concentrations are actually to be limited only by the solubility of the binders in water. Generally, however, concentrations between 1 and 3% have been found to be workable. The binders can be used in amounts preferably from 1 part per 100 parts of food solids, to about 30 parts per 100 parts by weight of food solids.

When the thermal gelling binder is omitted, the product tends to fall apart when deep fat fried, due to the failure of the inherently weak structure of the vegetable solids to hold a predetermined shape. As the amount of binder used increases, the product becomes increasingly compact and firm, and maintains this firmness during the heat processing.

In general it has been found that when flakes are used, aggregates of the vegetables, such as dehydrated granules or crushed dice, must also be used to prevent the feathering which would inherently occur, due to the deep fat frying. The amounts of the aggregates that should be present are desirably 5 to 60% of the total vegetable product; however, up to 100% aggregates can be used.

When the dry cellulose ether binder, such as Methocel, is used, it was found that rehydration temperatures as low as 45°F. were possible, and that extended mixing of up to 3 min., while not being critical or essential, was found to enhance the hydration, particularly if it otherwise extended beyond the usual mixing time of 30 seconds. The process will be illustrated by an example.

Example: The following dry ingredients were combined: 80 grams potato flakes, crushed through a 20 mesh opening, 20 grams dehydrated dice potatoes, crushed through a 20 mesh opening, 4 grams salt, 4 grams Methocel 65 HG. To this mixture, 200 cc of water at 60°F. were added and mixed thoroughly for one minute at a low speed. The resulting dough-like material was allowed to stand for several minutes to insure complete rehydration at an average temperature of 65°F. of the crushed dehydrated particles, and then extruded at a temperature of approximately 70°F., using a household type cookie press fitted with a three-eighths inch square opening directly into an oil bath containing stabilized cotton-seed oil held at 350°F. After 4 minutes the french fries were a pleasing, uniform, golden brown color, and possessed a firm, well cooked, natural appearing interior.

Part of the product was frozen and stored at 0°F. When prepared by heating in an oven at 400°F. for about 15 minutes, it was judged superior in quality to typical commercial products obtained from raw potatoes and heated in the same manner.

Raw Potato Prepared for French Frying

A process for the manufacture of a dehydrated potato product directly from raw potato without requiring the addition of any binder which is suitable for hot oil cooking or frying has been developed by *C.E. Sech; U.S. Patent 3,615,724; October 26, 1971.*

The process includes the following steps: A slurry of ground raw potato is formed having a moisture content of about 80%; the water content of the slurry is then reduced to at least 50% by application of vacuum at a temperature below the temperature of gelatinization of the potato particles. The mass is then heated or cooked at a temperature sufficiently high to gelatinize the product, the latter then being dried to a moisture content of about 6 to 12%. The critical step of dehydration by vacuum evaporation achieves proper plasticity of the mass while at the same time prevents browning of the product normally occurring due to enzymatic action, or oxidative influence or both. The process is clearly illustrated by reference to Figure 6.6

The process commences by feeding peeled raw potatoes into a rotary cutter or grinder **10**. Here the potatoes are ground or macerated at room temperature to form a slurry which can be readily transported as a fluid. The potato slurry flows to a vessel **12** where a few parts per million of sodium bisulfite or equivalent, are added to reduce enzymatic activity. Sodium bisulfite is here added in an amount of about 0.3% by weight of the potato solids.

From vessel **12**, the slurry flows to a combination colloid mill-pump or homogenizer-pump **14** driven by motor **16** wherein the particle size of the slurry is reduced further, preferably to an average size of about 80 to 100 microns or smaller. A water-cooled heat exchanger **18** is provided to assure that the slurry exiting the grinder is maintained below gelatinization temperature of about 150°F. Such temperature may be reduced to about 100°F., thus giving

FIGURE 6.6: RAW POTATO PREPARED FOR FRENCH FRYING

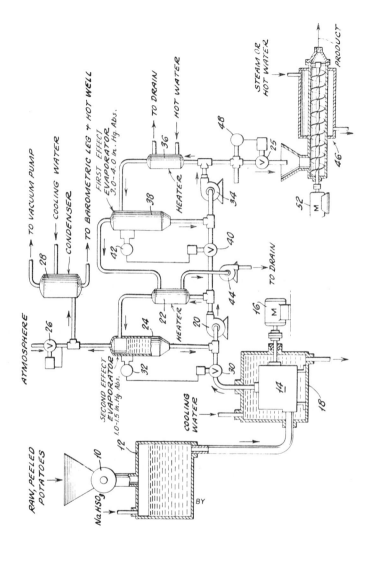

Source: C.E. Sech; U.S. Patent 3,615,724; October 26, 1971

complete assurance on nongelatinization. As shown, this heat exchanger is positioned in contact with the grinder in such fashion as to perform this heat-reduction function. As shown, the mill-pump is completely jacketed, with the cooling fluid completely surrounding it. Without cooling, the friction inherent in the grinding operation could or will heat the slurry above the gelatinization temperature of the starch, thereby making further processing in the form of a slurry impossible, and this because of restriction of flow through the vacuum system.

Although any type of vacuum apparatus may be useful, from the colloid mill-pump **14**, the potato slurry is fed into a series of multiple effect flash evaporator stages. A double effect evaporator is illustrated in Figure 6.6 by way of example only. Other effects can be added or deleted to increase or decrease the economy (pounds of evaporation per unit of heat added) as economics might dictate. The effect receiving prime steam or heat (right hand in this case) is termed the first effect and the one heated with vapor from the first effect is termed the second effect, etc.

Each of these effects comprises a pump, a heater, and a flash chamber. Using the left hand unit, the second effect for illustration, the potato slurry flows from pump **20** into heater **22**. The latter may simply comprise a heated jacket surrounding a flow-through pipe or plate type heat exchanger which has certain advantages for heating viscous liquids or slurries. The addition of heat raises the temperature of the slurry several degrees. The heated slurry then flows to the chamber **24** which is maintained at a subatmospheric pressure, a preferred pressure being 1.0 to 1.5 inches of mercury absolute, and at a preferred temperature of 70° to 80°F. Flashing of water vapor occurs in the flash chamber **24** until the temperature of the slurry from the heater has dropped to a point where its vapor pressure is equal to the pressure in the flash chamber, viz., 1.0 to 1.5 inches of water.

Literature indicates that gelatinization of potato starch may commence in a region between 115° to 122°F. However, it has been found in practice that gelatinization proceeds in a somewhat low rate at temperatures between 150° and 160°F. and does not proceed rapidly, or is not completed, depending in large part upon the type of product feed, until a temperature of 180° to 200°F. is reached. By operating the second effect at subatmospheric pressures, the evaporation of water may thus be accomplished at temperatures low enough to avoid any appreciable conversion of the starch to a gelatin form, i.e., temperatures below 100°F. Maintaining the potato starch in a slurry consistency facilitates flow through the entire process, and such consistency cannot be assured if the temperature utilized is so high as to promote gelatinization.

The vacuum which is applied to the second effect is created by a vacuum pump or steam jet ejector (not illustrated), the pressure being controlled by a vacuum controller **25** and a controlling valve **26** which allows air to bleed into the system at a controlled rate. Between the second effect and the steam jet ejector is a barometric contact condenser **28** which is supplied with cooling water. The condensate formed in the condenser is fed to a hot well through a barometric leg and then to a sewer or returned to a cooling tower.

The flow of the slurry to this second effect is regulated by valve **30**. The latter is governed by a level controller **32** in response to the slurry level in the body **24**. Flow through valve **30** is reduced proportionally when the slurry level rises to a predetermined point, and increased as the level falls, thereby maintaining a substantially uniform level in the evaporator.

The first effect evaporator comprises pump **34**, heater **36**, flash chamber **38**, valve **40** and valve controller **42**, each of which functions in a manner identical to that described above. Heater **36** is supplied with water or other fluid at a temperature of about 150° to 170°F. and evaporator **38** is maintained at 3 to 4 inches of mercury absolute pressure and a temperature of about 115° to 125°F. This temperature is still below that at which little or any significant conversion of the potato starch would occur.

The water vapor liberated in the first effect flash chamber **38** is fed to heater **22** of the second effect, this flow being established by pump **44** or by a barometric leg. By thus utilizing the heat in the vapor from the first effect, the total amount of energy required to dehydrate the slurry is substantially reduced.

In the second effect evaporator, the moisture content of the slurry is preferably reduced in such amount as to have from 45 to 55% retained moisture in the slurry. The effective viscosity of the slurry will increase but not sufficiently, at this juncture, to create a serious problem as far as pumping and handling are concerned; the slurry will continue to flow as a fluid, or possess that level of viscosity permitting free fluid flow.

The slurry from the double effect evaporator is then fed to a combination cooker and extruder 46. A valve 25 and flow responsive controller 48 provide a constant rate of flow input to the cooker and extruder unit. The cooker is heated by steam or hot water supplied at a temperature between 212° and 250°F., which heats the starch to from about 200° to 225°F. Such is sufficiently high to assure complete gelatinization of the potato starch.

The extruder, driven by motor 52, extrudes the product at a pressure of 10 to 75 psig depending upon the shape of the extrusion die and the rate of flow through the die. From the extruder, the gelatinized product may be fed to a drying belt for further dehydration or to a cutting and forming machine to place the potato product in the form of chips, french fries, or the like and then to a drying belt. Although preferred, as an alternative to extrusion, the product can be cooked and then rolled or moulded if desired.

It has been found that extruded material produced by the process is a plastic solid having considerable cohesiveness, or interfacial tenacity. It can be handled quite easily in either a bath or a continuous flow hot air dryer. Alternatively, the cooked-extruded product can be simply air dried.

Upon drying to a moisture content of about 6 to 12%, the partially dehydrated starch product is ready for subsequent frying in oil at a cooking temperature of from about 375° to 400°F. This subsequent frying is greatly facilitated by the prior removal of a substantial portion of the moisture of the potato. In this instance, and as stated, from an initial moisture content in the slurry of about 80% or above, the moisture content thereof is reduced by vacuum evaporation to from about 45 to 55% with a 50% moisture retention being preferred. The following air drying after extrusion removes moisture to the extent above indicated.

Due to a hard, essentially leak proof, hornylike shell on the partially dried product and due to the amount of moisture retained in the product, the heat from the frying causes an internal steam pressure to develop which expands the dehydrated starch into a tasty, crunchy, and appealing food product.

Prior to frying, the dehydrated product can be stored at room temperature in a moisture resistant container for as long as 2 years without any apparent harmful effect. Salt, typically necessary to produce satisfactory puffing in prior art processes, can be added after frying and puffing, thus preventing salt contamination in the main body of cooking oil.

The aspect of vacuum evaporation of this process, and particularly the more economic procedure of multiple effect evaporation, affords a means for efficiently dehydrating raw potato prior to conversion of the potato starch to gelatin form. Naturally, simple or single effect vacuum evaporation is also useful. In any event vacuum of the order of one to four inches of mercury is sufficient to attain the objectives of the process.

Evaporation is accomplished while the starch is in a pumpable, slurry form, and results in a reduced evaporative load. The use of vacuum in the evaporation is to reduce the tendency of water to migrate into the starch granules and thereby decrease the rate at which starch cells swell and burst, thus to release amylose, the primary component causing gelatinization.

Partially Dehydrated French Fry Strips

In a process developed by *C.J. Wilder; U. S. Patent 3,649,305; March 14, 1972; assigned to*

Lamb-Weston, Inc. potatoes are peeled, trimmed, cut into french fry sized strips, washed dehydrated without prior blanching to remove a substantial amount of moisture, blanched in steam, partially fried, and then frozen. Potatoes are held in storage under conventional, normal accepted conditions so as not to accumulate excessive amounts of sugar therein.

They are then peeled, trimmed, sorted, and cut into french fry sized strips of about three-sixteenths inch to nine-sixteenths inch in cross section. The strips are then thoroughly washed to remove the fresh starch remaining on the surface caused by the rupturing of the potato cells in the cutting process. If desired, the sugar content of the potato strips may be adjusted by well-known means at this point in the process. The potato strips are then immersed in a water solution preferably containing 0.75% sodium acid pyrophosphate or 0.1% sodium bisulfite or other commercially acceptable discoloration inhibitor for about five minutes to inhibit discoloration.

The strips are then dehydrated without prior blanching to effect a weight loss of from 10 to 30% depending upon the surface texture desired. Such dehydration is carried out at a temperature of between 150° to 350°F. and preferably takes place by placing the strips in circulating air at a temperature of about 190°F. which may be supplemented with infrared heating. A period of about eight minutes has given desirable results.

During the dehydration, moisture is removed from the strips both internally and from the surfaces thereof. The preferable weight loss is about 20% of the original weight of the strips.

The potato strips are then blanched in steam preferably at atmospheric pressure. The blanching may replace some of the moisture removed by the prior dehydration step, depending on the temperatures of the steam and strips, such that the net weight loss of the strips may remain the same or be reduced to about 16% of their original weight. A desirable length of blanching time is 7½ minutes for one-fourth inch shoestring cuts.

The strips are then partially deep fat fried for from about 30 to 90 seconds in an oil bath at a temperature of from 300° to 400°F. Preferably the strips are so fried for 60 seconds at a temperature of 325°F. During this partial frying step, additional moisture is removed from the strips. The strips are then frozen.

When the frozen product is prepared by the retailer or the final user, it is preferably immersed in a suitable frying oil for about 1½ to about 3 minutes, the oil being at a temperature of from about 325° to 375°F. A preferable time for finish frying is 2¼ minutes at 350°F. The final product resembles french fried potatoes prepared directly from fresh potatoes in quality, color, texture, and most importantly, in flavor and odor.

Agglomerates for Shaping into French Fries

The process described by *M.A. Shatila and R.G. Beck; U.S. Patent 3,622,355; November 23, 1971; assigned to American Potato Company* is directed to agglomerates of potato particles, formed largely of individual potato cells, that are capable of rapid rehydration into a substantially uniform homogeneous dough in the absence of physical agitation (or mixing), even when disposed as a relatively thick mass. The doughlike mass can be shaped into forms for french frying.

The reconstitution can be accomplished merely by subjecting the mass to the desired amount of liquid by allowing the liquid to flow by gravity into and along at least one surface of the mass of agglomerates. The term rapid rehydration is intended to mean rehydration to a uniform, cohesive, but not sticky doughlike mass, within as little as 20 seconds of the introduction of the rehydrating liquid but within as long as 15 minutes. The term relatively thick mass is intended to define a consolidated body of agglomerates having each dimension greater that one-fourth inch, preferably formed by the gravitational flow of the agglomerates in an at least partially enclosed container.

It has been found that by providing an agglomerate of dehydrated potato particles having a unique, unobvious correlation between the properties of bulk density, bulk porosity and rate of moisture absorption, a relatively consolidated mass of these agglomerates can be reconstituted with from 1.5 to 2.5 parts of water per part of agglomerates, in the absence of or at least without significant agitation (or stirring), and within the desired time period. In the preferred aspect, it has now been found that when agglomerates possessing a bulk density of from 0.25 to 0.50 grams per milliliter, a bulk porosity in volume of voids to volume of particles of 0.6 to 0.9, and a moisture absorption rate of at least about 0.12 grams of water per gram of product per second, are reconstituted without mixing, the resulting dehydrated doughlike mass is particularly suitable for shaping into french fry bodies. A minor amount of binder such as guar gum has been found to be essential to the preparation of a suitable dough.

The product of this process can be formed by any method that will produce the desired characteristics for the agglomerate. Included among suitable forming processes are spraying an agglomerating liquid onto a heated fluidized bed of granules or low-pressure forming (extrusion) and subsequent drying of a moist mass of potato particles. A satisfactory mix for extrusion has been found to result from mixing four parts by weight of potato granules with sufficient freshly cooked potatoes to supply one additional part by weight of potato solid matter. The low-pressure extrusion process is preferable.

Starting with dehydrated potato granules, they are first combined with sufficient moisture and binder to provide a damp adhesive mix. The preferred amount of moisture has been found to be from 35 to 55% by total weight of adhesive mix when water at about 60°F. is utilized. With moisture concentrations substantially above 55% and even in the range of 70%, the extruded pieces, when dried, exhibit decreased porosity, and a hard shell-like surface with greatly reduced water absorption power, unless the agglomerating moisture is introduced into the potato solids at an elevated temperature such as 150° to 180°F.

If the granules do not include sufficient free starch, a few percent of binder, i.e., preferably about 1 to 3% by total weight, must also be incorporated with the particles and moisture. Alternatively, the binder can be dispersed as a coating on the agglomerated units or even be incorporated in the rehydration liquid. In any event, the binder is essential to enhance the viscosity of the ultimate dough, thereby improving the quality of the french fries. A suitable mix for extrusion must not have excessive free starch. If ingredients with excessive free starch are used, or if excessive free starch is created during the formation of the dried extruded pieces, a hard nonporous shell is formed which, when reconstituted, forms a lumpy dough.

After ensuring that the mass is of uniform composition, it is formed into units (agglomerates) of agglomerated potato particles. It has been found useful to provide a rotating blade within the extruder between the conventional helical screw and the diehead. This appears to reduce the density of the extruded material. The forming step must be done at sufficient pressure to produce units that can be dried and handled without excessive breakage.

However, the pressure must be low enough to produce the desired bulk porosity and bulk density. With the equipment employed, suitable extrusion pressures were found to be in the range of less than 10 pounds per square inch when taken just in front of the die plate. Although a wide range of agglomerate sizes and shapes have been found to be useful, a diehead that produces pellets in the range of one-eighth to three-eighths inch in diameter has been found to be especially effective.

A cutting device mounted on the extruder head has been found to be useful for dividing the damp mass of particles into damp pellets of suitable length as they are forced from the diehead. Variation in the rotational speed of the cutter produces modifications in the length of the pellets. This in turn produces variations in the bulk density of the dried agglomerates or pellets, i.e., the shorter the pellets, the higher the bulk density of the dried agglomerates. A pellet length in the range of one thirty-second to one-eighth inch has been found to be especially suitable for providing an agglomerate displaying a bulk density in the critical range of 0.25 to 0.50 grams/milliliter. Once the wet mix has been subjected to extrusion and cutting, the resulting pellets are dried, preferably in a fluid bed dryer having an air-circulation rate of about 150 to 400 feet/minute. The moisture content of the agglomerates is reduced to about 7% by total weight, and the agglomerates screened to remove oversized and undersized material which is recycled to the feed hopper. The preferred product is obtained if the particles pass an 8 mesh sieve but are retained on a 20 mesh sieve, although a broader range is also useful. The agglomerates are thereafter reconstituted into a doughlike mass, shaped in the conventional french fry form and deep fat fried.

Example 1: A mixture of 96 parts by weight potato granules, two parts guar gum, and two parts fine salt were added to a fluidized dryer. Sufficient air heated to 160° to 170°F. was forced through the mix to fluidize the mix. Water was sprayed on the surface of the fluidizing bed at a pressure of 35 psig. Small spherical agglomerates were formed and dried quickly to 6 to 8% moisture. The dried product was screened to pass a 6 mesh screen but be retained by a 26 mesh screen.

The material passing the 26 mesh screen was utilized by blending with additional mix to bring back to the original weight and repeating the process. The −6+26 agglomerates had a density of about 0.40 g./ml. When these spherical agglomerates were covered by 2.2 parts by weight water, a uniform dough was formed without mixing. The formed pieces were fried for 1½ minutes at about 370°F. The fried product was judged excellent in appearance, flavor, and texture.

Example 2: Commercial potato flakes were ground to pass a 20 mesh screen. Ninety-six parts by weight were uniformly mixed with 2 parts guar gum and 2 parts salt. This mix was fluidized with air at 140°F., while water was sprayed on the fluidizing surface. The agglomerates which formed were then dried to 7% moisture. The dried product screened over a 10 mesh screen had a density of about 0.35 grams/milliliter. This product was rehydrated by adding 2.3 parts by weight of water to each part by weight product. A uniform dough was formed without mixing. The dough was cut into strips and fried at about 370°F. for 2 minutes. The finished fried product was judged excellent in appearance, flavor and texture.

French Fries Made from Dehydrated Potatoes

A process for making french fries is described by *A.L. Liepa; U.S. Patent 3,396,036; August 6, 1968; assigned to The Procter & Gamble Company.* The process prepares a potato based dough composition comprising from about 21 to 46% by weight potato solids, 1% to 15% by weight milk solids, and about 53 to 73% by weight water, the combined milk solids-potato solids content being at least about 27% by weight of the mixture.

In addition, minor amounts of salt and other flavoring agents and emulsifying substances can be included in the compositions. This mixture is suitable for the formation of strips or sheets, which, when sectioned into appropriate sizes and deep fat fried, result in a desirable fried potato product.

The potato solids used can be dehydrated granules (such as those prepared by fluidized bed dehydration techniques), "flakes" (such as those prepared by drum drying techniques) or fresh potato particles. The potato solids should consist essentially of intact potato cells if flavor and proper dough consistency are to be maintained; for this reason substances such as potato flours and similar potato-derived powders are generally unsuitable for this process. Potato granules are especially durable during mixing and are especially preferred. Since fresh potatoes contain a relatively large amount of water, it is usually necessary to include dried potato granules or flakes in the mix in sufficient quantity to maintain the minimum 27% milk solids-potato solids content of the process.

Fat and nonfat-containing milk solids or whole milk can be used as long as the minimum total of 27% by weight potato and milk solids is maintained in the compositions. Since whole milk contains a relatively large amount of water (approximately 87%), it is preferable to use dehydrated potato granules or flakes with whole milk in order to maintain the necessary minimum 27% milk solids-potato solids content. An example will illustrate the process.

Example: A dry mixture was prepared by admixing the following ingredients in a conventional Hobart mixer for two minutes at 60 rpm: 27.5 parts potato granules (water content 7.5%), 8.6 parts dry milk solids (water content 4%). To this mixture was added 63.9 parts water and a thick dough was prepared in the mixer by mixing for 2½ minutes at 60 rpm. The composition of this dough was about 25.5% potato solids, about 8.3% milk solids, and about 66.2% water. The dough was divided into two portions. One portion was placed in a hand operated piston-type extruder and extruded through a die plate as strips of approximately the same size as conventional potato strips cut for french frying. The other portion was rolled into a sheet about five-sixteenths inch in thickness, using a conventional rolling pin, and was then sliced into strips.

Sections of strips from both portions were deep fried for two minutes at approximately 350°F. in preheated Frymax (a commercial liquid shortening product). The fried strips from both portions were golden brown in color, showed no puffing or any signs of disintegration, and had a crisp, pleasant eating quality.

Apparatus for Extruding Doughlike Dehydrated Potato Mix into French Fry Form

During the past decade there have been many improvements made in the method of preparing and handling food products such as in the dehydration of potatoes into flakes, powder and buds, which when reconstituted in liquid, result in a recognizable product such as mashed, hash brown, or other forms of potatoes. Currently, some major processors of potatoes have developed a potato mix which when reconstituted, forms a solid doughlike substance. The dough may then be extruded or otherwise formed into a shape commonly referred to as french fries or shoe-string potatoes.

Due to the chemical nature of the powdered mix when mixed with water, the liquefied mixture sets extremely firmly in a matter of seconds thereby preventing agitation of the mixture and extrusion of the mixture into either french fries or shoe-string shaped potatoes. In order to overcome this defect, chilled water has generally been used to prolong the mixing cycle. The machines produced to do this mechanically agitate the powdered mix and the chilled water. Although these machines have served the purpose, they have not proved entirely satisfactory under all conditions of service since the time element involved before hardening is still too short and furthermore, while the mix is in a liquid state, it is extremely viscous and sticky, and adheres to most surfaces of the machine. Thus, cleaning of the machine and servicing of the machine due to the inherent properties of the liquified mix have become extremely complicated.

R.L. Brunsing and J.P. Brunsing; U.S. Patent 3,658,301; April 25, 1972 have developed a method and apparatus for manufacturing a doughlike substance from dehydrated potato mixes which contemplates the mixing of the liquid and powdered food mix in predetermined proportions such that the powdered mix is sprinkled into a thin stream of flowing liquid of the proper thickness and this flowing liquid continues to run for a short period after the powdered food mix has been exhausted thereby acting as a cleansing agent and preventing early solidification.

Broadly, the method comprises placing a premeasured amount of dry powdered food mix in a chamber and placing a premeasured amount of water in a second chamber which is separated from the first chamber. Second, releasing the liquid in a continual flow into a receptacle so as to form a thin film of liquid having a thickness sufficient to absorb the powdered potato mix along the sides of the receptacle and substantially simultaneously releasing the powdered mix so as to allow the powder to fall upon the thin film of water, mixing together by the action of gravity and being conveyed downwardly into a collecting of forming container. Thirdly, allowing the remaining liquid to flow after the powder mix has been exhausted so as to act as a cleansing means for the receptacle. Finally, collecting the liquid mixture and allowing it to solidify forming a firm doughlike substance.

FIGURE 6.7: APPARATUS FOR EXTRUSION OF DEHYDRATED POTATO DOUGH

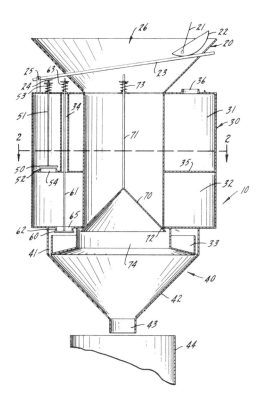

Source: R.L. Brunsing and J.P. Brunsing; U.S. Patent 3,658,301; April 25, 1972

There is shown in Figure 6.7 an apparatus **10** for mixing a predetermined amount of dry powdered potato mix with a predetermined amount of liquid so as to form a liquid mixture which can solidify into a firm doughlike substance. The apparatus comprises a funnel-shaped mixing chamber **20** encircled by a toroidally shaped water storage chamber **30** having two sections, **31** and **32**. Water is introduced into the tank through aperture **36**. A predetermined amount of water from water storage chamber section **31** is released into chamber section **32** and then into an intermediate receptacle **33**.

The overflow from intermediate receptacle **33** flows into a channel **41** of a funnel-shaped mixing receptacle **40**. Dry mix chamber **20** contains a dry mix potato powder having a standard composition, which is commercially obtainable from numerous manufacturers.

This powder is allowed to fall into the liquid stream through valve opening **74**. The liquid mixture of the powder and water then flows into a forming receptacle **44** and solidifies forming a firm doughlike substance which is ready for further processing into any desired shape. The consistency of the dough is such as to allow it to be extruded in the shape of what is commonly known as french fried or shoe-string potatoes. This apparatus, plus alternative apparatus for carrying out the process, are described and illustrated in detail in the patent.

HASH BROWNS

Combination of Potatoes Dehydrated in Two Manners

Hash brown potatoes are conventionally prepared by steaming or boiling whole potatoes, cooling, and thereafter forming potato pieces or strips (commonly referred to as julienne strips) by cutting the cooked potatoes.

This cooking and cutting destroys a number of the cell walls and releases gelatinous starch, which when fried, produces the desirable adherence of the individual strips to each other to obtain the well-known product.

However, when the potato strips are not fried immediately, the presence of this "free" starch is undesirable. The free starch reacts with other unbroken cell walls in the strips, causing their rupture, which in turn destroys the individual nature of the strips. Thus, the eventual hash brown product which is obtained contains lumps or clumps rather than the distinct textured strips desirable in hash browns.

The free starch is additionally undesirable during dehydration of the strips because it causes the individual pieces to stick together, thereby complicating and prolonging the drying process. In addition, the dehydrated strips, from which the free starch has not been removed, do not reconstitute rapidly.

In order that potato pieces, which are to be used for the preparation of hash browns, will maintain their individuality and attractive size and not stick together during dehydration, the strips must be thoroughly washed prior to dehydration to remove substantially all the free starch created during cooking and cutting.

However, such a washed product does not stick together and, therefore, does not produce an attractive hash brown product, because the free starch, essential as the adhesive mechanism between the individual reconstituted potato strips when fried, has been removed.

As a result of these factors, dehydrated potato strips which are commercially available for the preparation of hash browns, do not reconstitute rapidly, do not possess the desired textured characteristics of hash browns prepared from fresh potatoes and generally do not attain an eye-appealing form normally associated with the fried product.

L.J. Frank; U.S. Patent 3,410,702; November 12, 1968; describes a process for producing a combination of dehydrated potatoes which have been prepared by two distinctly different processes, yet reconstitute quickly and when fried, produce a desirable hash brown product. The two potato components of the product are formed and prepared so as to remain in the desired physical proportions during packaging and shipping. The pieces do not stratify or lose this desired proportional arrangement even during long periods of shipping and storage.

This instantly reconstitutable potato product, suitable for the preparation of hash browns, can be prepared by combining, especially in certain desirable proportions, blanched potato pieces from which free starch has been removed prior to dehydration, with other potato pieces which have been fully cooked, riced and dehydrated without washing after the cooking step.

When carrying out the process a quantity of whole, white opaque potatoes are divided into pieces such as julienne-type strips. The pieces are partially cooked (precooked) for a time sufficient to destroy the peroxidase present. This is most effectively accomplished by blanching for from 5 to 10 minutes in water which, preferably contains either sodium chloride, sodium acid pyrophosphate or a combination of the two.

The free starch produced by the cutting and precooking is then thoroughly removed such as, for example, by means of washing to prevent the pieces from sticking together during dehydration. These blanched and washed strips are then subjected to dehydration.

A second portion of either similar or dissimilar whole, white opaque potatoes are conventionally peeled, trimmed and cut into small pieces, for example, diced. These uncooked pieces can be washed, but it is usually not desired. These pieces are then cooked in a conventional manner for at least about 20 minutes.

The fully cooked pieces, now containing free starch, which must not be washed away, are riced in a conventional ricing machine. The ricing process destroys more of the cell structure thereby releasing further amounts of amylose, i.e., free starch. The riced product is dried to a low moisture concentration and relatively short pieces of irregular conformation are obtained.

The dehydrated strips and riced product are then combined. The dried riced potatoes and julienne strips are found to be of roughly comparable shape which enables the two components to be combined without subsequent separation or stratification in the package. The product is further unique in that the individual dehydrated strips and rice are free flowing, i.e., they do not stick together. The combined product has been found to reconstitute rapidly, and forms a coherent mass, which when fried develops the attractive appearance and texture of fresh hash browns.

The term "strips" as used herein includes potato pieces which, prior to dehydration, are about ½" to 1½" long, about ¼" to ½" wide and from about 1/16" to 5/16" thick, and may take various shapes. Preferably a julienne strip is employed. Although the size of the pieces is in no way critical, it is desirable that one dimension be no greater than about ¼" to insure uniform and sufficient drying. The resulting dehydrated strips usually display a minor amount of shrinkage in overall dimensions from that of the raw pieces.

The raw potato portion which is to be treated by fully cooking and ricing is preferably first diced into cube-shaped pieces about ½" on a side. The resulting riced portion, when dehydrated, forms rough, irregular, curled and twisted pieces about ⅛" in diameter and up to about ¾" long. Of course, these dimensions, as well as any others which are set forth in this specification, can be varied by one skilled in the art and will depend on the properties of the actual potato as well as the actual processing procedures.

It has been found that the advantages of this process are obtained when the strips comprise from about 40 to 90% by weight, of the total dehydrated combination and especially when from 60 to 70% by weight, julienne strips are incorporated in the combination.

Both the potato strips and rice pieces are subjected to a sulfite treatment. The sulfite is applied to the strips immediately after washing, for example, as a spray of a solution of sodium sulfite, sodium bisulfite or sodium metabisulfite or combinations thereof. With respect to the portion which will be riced, it is most effective to incorporate the salts during cooking.

The dehydration is accomplished in any of the multiplicity of conventional dehydrating equipment. For example, cabinet, tunnel and conveyor dryers and the like are all effective. The dehydrating step has little effect on the size of the blanched strips but the riced potato breaks up into relatively small pieces of generally uniform size approximating a large rice hull or slightly larger, during ricing and dehydration. To further illustrate the process, the following example is provided.

Example: Ten pounds of whole, white opaque potatoes were cut into julienne strips about ¼" x ⅛" x 1" and blanched in an aqueous 4% sodium chloride solution at 198°F. for seven minutes. The strips were then thoroughly washed in a 2% aqueous dextrose solution and dehydrated to about a 6% moisture content.

An additional 5 lbs. of whole, white opaque potatoes were diced into ⅜" cubes and heated for about 20 minutes at 198°F., until completely cooked. The cooked diced potatoes were not washed as washing would have caused the individual pieces to dissolve into a starchy paste. The diced potatoes were then riced and the resulting product extruded as spaghetti-like strings about ¼" in diameter, breaking irregularly into about ½" lengths. The pieces were dehydrated to about a 7% moisture content and the resulting dehydrated portions were of irregular shape and less than about ½" long.

The dehydrated blanched julienne-type strips and dehydrated cooked riced potato, transported by separate conventional blending equipment, were thoroughly combined. After a lengthy storage period, the product was added to a sauce pan containing 1.5 cups of boiling tap water per 100 grams of dehydrated product. The water was seasoned with salt and pepper. The sauce pan was removed from the heat source and allowed to stand for 3 min. The reconstituted combination was easily spread into thin irregular layers and fried on an oiled skillet. The resulting hash brown cake was appetizing in appearance and texture, closely resembling hash browns prepared from the fresh tuber.

Large Potato Pieces for Hash Browns

R.G. Beck, L.H. Parks and M.A. Shatila; U.S. Patent 3,634,105; January 11, 1972 have developed a process for producing large potato pieces which can be rehydrated and prepared as pan fried potatoes or hash brown potatoes in a fraction of the time required for conventionally dried pieces of the same size. When using raw potatoes as starting material, this process comprises the normal preparation step of producing a cooked potato or potato piece. From this one produces a debris-free slurry of about 17% solids.

The solids content of the slurry is then increased to about 22 to 25% by partial dewatering, addition of dehydrated instant potato, or both. Additives such as antioxidants sulfite salts, calcium salts, sugar, starch or other binders are added as required to control stability, color, and texture. This thickened slurry or mix is now comparable in consistency to a very dry mashed potato. By eliminating precooking and cooling of the potato prior to cooking and by adjusting the slurrying procedure at least a portion of the required adhesive character is obtained from the controlled rupture of a small fraction of the potato cells.

Alternatively, the debris-free potato mix is made by reconstituting debris-free dehydrated potato products such as potato granules or potato flakes, which in commercial form already contain the desired additives. Any combination of potato solids whether from a freshly cooked source or from a dehydrated source can be used in the process by preparing a mix of the proper moisture and containing the proper additives. The following example will illustrate one process for producing hash browns where potato granules are used as the sole source of potato solids.

Example: 330 grams of commercial potato granules are mixed uniformly with 30 grams of raw corn starch. This mix is then added uniformly to 1,170 ml. of water heated to 185°F. in a Hobart mixer operated at low speed for 2 to 3 minutes to form a homogeneous dough of about 22% solids. The dough is then extruded in strips about one-eighth inch by one-fourth inch in cross section and of any length onto a screen. The extruded dough is then steamed for about 10 minutes on the screen. The heat-treated strips are then dried, still on the screen at about 180° to 200°F. for about 3 to 4 hours to reduce the moisture content to about 7½%.

When the end product desired is an instant one-eighth inch potato slice suitable for pan frying, pieces of the desired slice size are formed from the hot sheeted mix. These pieces are then predried out of contact with one another to a moisture content of about 65% by subjecting the individual pieces to air at about 300°F. for about 5 to 6 minutes. Any other set of conditions which gives this moisture reduction without damage would be acceptable. The purpose of this step is to predry the surfaces of the pieces so they do not adhere to each other in final drying.

The final drying step immediately follows before diffusion of moisture to the surface of the pieces can create any stickiness. The final drying step is normally conducted with the pieces in several layers on a continuous perforated belt dryer using air at about 200°F. until the moisture content is reduced to about 6 to 7%.

All raw starches tested were effective adhesive agents, but raw corn starch resulted in finished products with superior rehydration and handling attributes. Effective starch gel formation in the dough requires a minimum heat treatment temperature of about 190°F. In tests where dough was heated to only 170°F., no beneficial result was found. The temperature at which the dough is dried is critical. If excessive drying temperatures are employed, the outer layer of the extruded pieces dries too quickly causing case hardening which results in puffed pieces which slough undesirably when reconstituted.

The starch added for adhesion must be in the raw or ungelatinized state when mixed into the dough. A cooked or gelatinized starch ingredient does not form the proper gel distribution. The most desired quantity of raw starch ranges from 6 to 20%. With less than 6% the reconstituted product sloughs excessively. With more than 20% the end product has less potato flavor, less desirable browning, and has an undesirable tough or rubbery texture. Tests with corn starch and commercial potato granules have shown that about 9% starch and 91% granules is optimum. Although a solids content of the dough of about 23 to 25% is preferred, a range from 17 to 50% gives useful results.

The products have great flexibility in use. The following conditions of rehydration all give properly reconstituted hash brown pieces which can be pan fried in about 4 minutes at about 275°F.: Cold water soaking for 15 to 30 minutes, 140°F. water soaking for 5 to 10 minutes and 200°F. water soaking for 3 to 5 minutes. The excellent rehydration characteristics allow simultaneous rehydration and frying in a single pan. An excellent one pan recipe for hash browns is as follows. 125 grams of the dehydrated hash browns are mixed with salt, fat, and 1 to 1¼ cups of water. When the mix is heated, an excellent hash brown potato with desirable piece identity can be finished in about 10 to 14 minutes.

Potato Pieces Coated with Starch Before Dehydration

The process described by W.J. Englar and D.C. Dew; U.S. Patent 3,635,729; January 18, 1972; assigned to Western Farmers Association relates to the preparation of a dehydrated potato product formed by a combination of partially cooked potatoes with a starch material derived from another source. The potato pieces which are coated with starch are then dehydrated to form the product of this process.

The process is accomplished by following a series of steps which may be generally described as follows: fresh raw white potatoes are peeled, trimmed, cut into strips, shreds, slivers, pieces or the like and blanched or precooked. The potatoes are preferably cut into a size, which may be referred to as a strip, which prior to dehydration has dimensions of approximately ½ to 1½ inches in length, about ¼ to ½ inch in width and from about one-sixteenth inch to one-eighth inch thick, or of various shapes having at least one dimension of substantially less than ½ inch so that uniform and adequate drying may be accomplished by ordinary techniques. The pieces are blanched or precooked for a time sufficient to destroy the peroxidase present. A balancing time of from 5 to 10 minutes is usually required in water which contains a salt such as sodium chloride, sodium acid pyrophosphate or a combination thereof at a temperature of 180°F. up to the boiling temperature.

The resulting precooked strips or potato pieces are then partially dehydrated by use of known apparatus such as moving belt dehydrators and tray dehydrators, or by use of fluidized bed techniques. The precooked and partially dehydrated potato pieces or strips are then coated with a starch solution or dusted with powdered starch to coat the outer surface of potato strips or pieces. A starch solution having a concentration of 5 to 25% starch is used. This solution contains water-soluble or pregelled starches which require no cooking in preparation, or raw starch from the peeling and washing stages of potato-processing industries may be used.

To prepare a solution from raw, dried starch, the powdered starch is supended by agitation in cold water. The agitation is maintained as the water temperature is raised to cook and gel the starch. The starch solution is sprayed on the dried potato strips which are turned and mixed during this spraying operation in such a manner that all strips are uniformly coated. An adequate amount of starch to make the strips adhere during rehydration, browning and eating is applied in the above manner. Alternately, powdered starch may be applied to the potato strips in an amount adequate to cause adherence of the strips during rehydration. Normally between 4 and 10% of starch on a dry weight basis of the weight of the dehydrated potato product is sufficient to bind the rehydrated potatoes.

However, the amount of starch added to the potato pieces will vary according to the viscosity of the starch solution; that is, a starch solution which has a low viscosity will be added to the potato strips being treated in larger quantities than starch solution which has a high viscosity.

When the potato pieces have been treated with the starch solution, they are subjected to drying and further dehydration using a fluidized bed dryer or other method to keep the product from sticking and fusing together.

To increase the color intensity and eye appeal of the cooked hash brown product, dextrose is applied along with the starch solution. Other spices and treating agents may be used to enhance the appearance, flavor and shelf life of the dried product. The dehydrated, starch-coated product is then packaged in convenient packages for use by the consumer.

In preparing the dehydrated, starch-coated strips for consumption, the product is placed in boiling water which rehydrates the potatoes and gels the starch to bind strips in the manner of fresh, hashed potatoes. The rehydrated product is then browned in an oil grill or skillet and served.

The dehydrated and starch coated hash brown potato product does not require costly refrigerated transportation or storage and is unaffected by temperature fluctuations normally encountered in storage of foodstuffs. The dehydrated product is therefore less expensive to transport and store than fresh or frozen hash brown potatoes.

The reconstitution ratio of the dehydrated, starch-coated hash brown potato product is great enough to make the cost per serving competitive with the fresh or frozen product. When prepared in the manner specified above, 1 pound of the dehydrated product of the process will yield, when rehydrated, approximately 3 pounds of reconstituted hash browns thus offsetting to a large degree the higher cost per pound of the dehydrated product. The process will be more readily understood from the following example.

Example: Fresh raw white potatoes were peeled and shredded into shapes typically used in preparation of hash brown potatoes. The shreds were blanched for 10 minutes in water held at a temperature of 200°F. and containing 1 pound of sodium chloride for each 10 gallons of water. After 10 minutes precook, the potato shreds were removed and dehydrated on a moving belt dehydrator, to a moisture content of 20 to 30%.

A starch solution having a concentration of 25% was prepared by adding raw starch to cold water and cooking the starch to the gelled stage. This starch solution was sprayed onto the potato product which was turned and mixed during the spraying operation to uniformly coat the strips. The starch was applied to the precooked and partially dehydrated potato strips in an amount such that the product contained approximately 8% by weight of the starch on a dry weight basis.

The potato pieces were then introduced into a fluidized bed dryer and dried and dehydrated to a total moisture content of approximately 7.5%. The fluidized bed dryer kept the product from sticking and fusing together. The product was then packaged for sale to the consumer.

The product was prepared for cooking and eating by the following steps: the dehydrated, starch-coated strips were placed in boiling water to dehydrate the potato and gel the starch to bind the strips in the manner of fresh hash brown potatoes. An adequate amount of water was used to return the reconstituted potatoes to the approximate moisture content of fresh potatoes. In each case, the product was rehydrated in less than 5 minutes, and was in that condition suitable for cooking and consumption. The rehydrated product was then browned in the oiled grill or skillet, and served. The taste, appearance, texture and aroma were found to be equivalent to freshly prepared hash browns.

A.W. Tschirgi; U.S. Patent 3,650,776; March 21, 1972 describes a similar process for coating potato pieces designed for producing hash browns before they are hydrated. In this process, the liquid system for coating the potato pieces is made by comminuting and adding to water dehydrated potatoes such as dices, slices, or other potato pieces. The mixture of dehydrated potato pieces and water is then heated above the ambient temperature for a sufficient period of time to permit the gelation of the potato starch in the mixture. As the preparation of the liquid system evolves, the resulting system will carry potato proteins, sugars, fiber and other potato constituents in several phases, e.g., colloidal, sol, suspension and solution.

In a preferred embodiment, a dried milk product, such as sweet dried whey, is added to the liquid system prior to the coating step. The addition of a dried milk product, which consists principally of caseinate, gives the rehydrated potato when fried a richer golden brown coloration and permits a wider range of browning possibilities according to the personal preference of the consumer.

It is also possible to add flavor additives, harmonizers, and other agents which may be desirable to enhance the subtle flavor of the fried product, in addition to providing a pleasing and appetite stimulating aroma. Such additives include salt, pepper, monosodium glutamate, onion, and meat flavorings, such as ham, bacon, or chicken, and many others.

The flavoring agents, seasonings, and other related substances are mixed with the liquid system prior to coating the fresh potato pieces with the liquid. The seasoning adheres to the potato pieces with the liquid during the dehydration step and also through subsequent rehydration and frying to provide a flavorful hash brown potato.

The coating of potato pieces with the liquid system enables the pieces to replace any of the potato protein, starch, or cellular materials which may have been lost through the various steps for preparing the fresh potato pieces for dehydration. The potato product is thereby restored to a near natural potato state or greater, and achieves an enhanced potato flavor. It has been found that the coated dehydrated potato pieces can actually be made to gain approximately 10% or up to as much as 25% by weight in potato constituents through the coating action. Upon rehydration, the coated potato pieces acquire a volume approximately double that of rehydrated hash brown potatoes processed according to known methods.

It has been found that any ratio of comminuted dehydrated potato pieces to water may be used, keeping in mind that larger amounts of water dilute the liquid system and provide a thinner coating on the potato pieces. A satisfactory liquid is formed by employing a ratio of approximately one part dehydrated potato to ten parts of water.

The mixture of dehydrated comminuted potatoes and water should be heated until gelation of the potato starch occurs. The preferred temperature at which gelation occurs most readily in a minimum amount of time is approximately 185° to 190°F. The minimum time required under these temperatures is approximately five minutes. Under the requirements of a continuous flow operation, the mixture can be held for longer periods of time at these temperatures, and there is no appreciable change in sugar content or dextrinization.

The coating of the potato pieces by the liquid system is accomplished in any manner sufficient to coat the pieces completely. It has been found that a ratio of one volume of the liquid system to four volumes of potato pieces produces a satisfactory coating.

The coated pieces are dehydrated by known means, such as heating on trays, and can be packaged for distribution. Rehydration and frying are accomplished by adding water to the potato product and permitting the potatoes to absorb water and swell in size. The potato pieces can then be formed into patties, if desired, and fried to a golden brown color with a pleasing texture and flavor. The following example illustrates the process.

Example: Water weighing 500 lbs. was mixed with 51 lbs. dehydrated comminuted potatoes and heated to 185°F. for 10 minutes, during which time the potato starch present in the mixture gelled forming the potato liquid system. Dried sweet whey (Krafen) weighing 39 lbs., 33 lbs. sodium chloride, 3.3 lbs. monosodium glutamate, 2 lbs. onion powder and 4 ounces rosemary were added to the liquid system and thoroughly mixed.

The liquid system containing the seasonings was continuously poured over fresh potato pieces being constantly fed into the top of a loosely packed vessel. The pieces of potato became coated and dropped through an orifice in the bottom of the vessel onto a belt which carried them into a dehydrating apparatus. Dehydration is effected in customary manner and may be carried to any desired extent, for example from 6 to 75% moisture content. Preferably the moisture content is reduced to within a range of 6 to 35%.

Extruded Mash Dried at Controlled Humidity

A process for producing dehydrated potato pieces for making hash brown potatoes has been developed by *P.G. Miller and F.C. Griffith; U.S. Patent 3,725,087; April 3, 1973; assigned to Rogers Brothers Company.* The process is carried out using cooked fresh potatoes, or dehydrated products such as flakes, crushed dice or mixtures thereof to form a mash in which the solids content is 21 to 23% preferably 21.9 to 22.5% solids. The solids content within these ranges is desirable although it is not critical.

The mash is preferably mixed in a large blender, suitably sized to hold the moist mash. Provision must also be made for heating the mash, preferably by steam, so that the mash temperature is raised to 160° to 179°F., preferably 165° to 175°F. and ideally 168° to 172°F. Desirably, although not necessarily, the mash temperature should be raised by the use of steam or the like heating means, prior to the addition of any of the optional ingredients such as sugar or starch or the like.

After the mash has been raised to the proper temperature, it is ready for shaping. The usual shaping procedure includes extrusion, but may include any other form of shaping if desired. The extrusion may be in the form of sheets which may be subsequently cut into any form. Preferably the shaping takes the form of extrusion into ribbons, for instance, those that may be one-eighth by seven thirty-seconds of an inch in cross-section.

In order to provide some shape retention, it is advisable, although not essential to pass the mash, subsequent to extrusion, through a predrying step to lower the moisture content from the 77 to 79% moisture to preferably a moisture content of 64 to 68%. The predrying is preferably brought about at a temperature between 250° to 500°F. for 2 to 15 minutes, most ideally the temperature of the predrying is 430° to 500°F. for a period between 3 and 5 minutes.

It has been found possible to achieve the final drying in a single step where the temperature of the drying medium, air, may be at a temperature of 90° to 210°F., although preferably 170° to 210°F., and ideally 190° to 210°F., provided that the humidity for at least 25% of the total drying time is not less than 20% and that for the remaining 75% of the final drying period the relative humidity of the drying medium should not be below 10%.

The time for the final drying is found to be generally between ½ to 6 hours although this time is not critical, it being only important to obtain the desired moisture content of below 8%, preferably 5.5 to 7.5% with the maintenance of the humidity as stated above. The maximum relative humidity is not critical but generally should be an upper maximum relative humidity of 98% with usually, but not necessarily, a 90% upper range being found acceptable.

Example: A mash of 1,000 pounds of cooked fresh potatoes or dehydrated granules or crushed dice reconstituted to a solids content of approximately 22% was mixed in a suitable blender. To this mash is added 1.0% sugar and 7.5% by weight cornstarch. The wet mash is heated with steam to produce a mash temperature of 170°F., whereupon it is extruded into continuous ribbons one-eighth by seven thirty-seconds of an inch and passed into a predryer with countercurrent air at 500°F. having a relative humidity between 20 to 30%. The moisture of the product coming out of the predryer ranges from 65 to 68%. The ribbons are thereafter cut to lengths of approximately one inch and passed through a final dryer to be dried in moving air at a temperature of 210°F. for approximately 45 minutes to obtain a moisture content of about 7%. The relative humidity of the air in the final dryer is 20%.

For consumption the product of the foregoing example may be added to boiling salted water in an amount of about 1½ cups to ¾ cup of water, to which ½ teaspoon of salt and 2 tablespoons of butter have been added. After the pieces have been added to the water, the mixture should be permitted to stand away from the heat for several minutes, generally 5 minutes.

Thereafter the excess liquid is poured off and the pieces are placed in a skillet greased with butter, the potato pieces are grilled at medium heat for approximately 4 minutes on each side to produce excellent quality hash brown potatoes.

MISCELLANEOUS PRODUCTS

Potato Puff

A process developed by *F.P. Griffiths, P.W. Kilpatrick, W.O. Harrington, C.E. Hendel, and R.L. Olson; U.S. Patent 2,705,679; April 5, 1955; assigned to the U.S. Secretary of Agriculture* for the dehydration of potatoes involves subjecting diced potatoes to a hot air stream of sufficient velocity that the potatoes are supported and constantly agitated by the air stream. The conditions during the hot air treatment are so regulated that the potato pieces are each puffed or expanded so as to leave the centers hollow. At the same time the potatoes are cooked, partially dehydrated, and the surfaces of the pieces are browned to an attractive degree. The products so produced are then further subjected to heat, either while suspended in the air stream or in a conventional dehydrator, to reduce the moisture content to a degree where the products are self-preserving.

During this dehydration the physical dimensions of the potato pieces do not change to any significant extent. The products so produced are very light in weight, being hollow in the center with the sides puffed out much as are the sides of a pillow. Their surfaces are smooth and glazed and their color is tan to golden brown depending on the degree of heat treatment. The texture of the products is crisp and brittle so that they can be eaten directly.

While the potato pieces produced in accordance with the practice of the Griffiths process may be well suited for consumption as snack-type product, the pieces have been found to have relatively poor rehydration properties, and take at least 15 to 30 minutes to recook.

It has been found by *I.L. Adler and A. Apostolina; U.S. Patent 3,338,724; August 29, 1967; assigned to General Foods Corporation* that a quick cooking or quick rehydrating potato product may be produced which is not greatly expanded to give a light product with a hollow center. Instead, it is formed with a shell portion and a core portion, the core portion, consisting of a matrix of potato meat having a plurality of voids spaced from each other and distributed in the matrix. The product has a moisture content not greater than 10%. Preferably, it has a density of 0.15 to 0.20 g./cc. Separation between the shell and core is slight.

To produce products in which voids are distributed throughout a matrix of potato meat in the core portion of the piece, potato pieces of a proper size are immersed in a solution of about from 0.1 to 2.0% sodium chloride in water. The immersion takes place at a temperature range of about 170° to 212°F. for a period of 3 to 15 minutes. The potato pieces are subsequently dried to a moisture content not greater than 10% in a stream of air heated to a temperature of about 310° to 390°F. In order to attain moisture content of less than 10%, it has usually been found necessary to dry the pieces in the air stream for 5 to 20 minutes.

The composition of the aqueous liquid in which the potato pieces are immersed, the drying conditions and the size of the potato pieces, are all important factors in the attainment of a product having a plurality of voids in its core portion, and but slight shell-core separation. It has, for example, been found that if a cubic form of potato piece is used, the maximum dimension of the cube will be ½ inch side.

The amount of sodium chloride in the aqueous solution in which the potato piece of proper size is to be immersed is likewise important. This concentration has been found to vary from 0.1 to 2% sodium chloride in water. Likewise, the time of the so-called blanching treatment and the temperature at which the blanch is effected are critical. It has been found that the range of the temperature of the blanching medium, i.e., sodium chloride in water, may extend from about 170° to 212°F. The time during which the blanching may be effected is about 3 to 15 minutes. Strict control of the sodium chloride concentration, temperature of the blanch medium, and blanch time will result in the desired product.

The dehydration step of the blanched potato pieces is generally known to those skilled in the art, an important feature of this dehydration step being the maintenance of the temperature of the dehydrated medium, e.g., air, at 310° to 390°F. It has been found desirable to dehydrate the pieces to a moisture content of 10% although further dehydration may then be required. To effect such dehydration various techniques may be utilized.

Example: Raw potatoes were peeled and then diced into cubes having a uniform dimension of three-eighths of an inch. These cubes were then immersed in a 0.5% sodium chloride-water solution for 4 minutes, the temperature of the solution being maintained at 180°F. The blanched potato pieces were removed from the solution and fed into a fluid bed dryer in the form of an inverted cone and having an air temperature of 350°F.

The potato pieces were maintained in the form of a bed by the velocity of the heated air and were permitted to remain in the dryer for a period of about 17 minutes after which their moisture content was less than 10%. The core portions of the dehydrated pieces were found to have voids distributed throughout a matrix of potato meat in substantially random arrangement. The entire potato piece had a density of about 0.17 g./cc; and was capable of rehydration in less than 5 minutes.

The dehydrated potato products of this process have been found suitable for rehydration as french fries or hash brown potatoes, and can easily be recooked within 5 minutes. While they are not generally ready to eat immediately after puffing, they can be briefly heated in a suitable oil to give a tasty product suitable for use as a snack.

Whole Potatoes or Large Potato Chunks

While it has been attempted to produce dehydrated potato pieces of various sizes, these attempts on the whole have been unsuccessful with the exception of vitreous, thin sliced, small diced, or granulated products having limited recipe uses and poor rehydration qualities. On the other hand, substantially larger pieces of potatoes, such as whole potatoes, potato chunks, and large slices of potatoes have not been successfully produced due to the unique and difficult dehydration problems involved. Especially is this true in the case where the processed potatoes are to be dehydrated by drying in air, which is the cheapest manner of dehydration, and the most feasible from a commercial standpoint.

G.A. Katucki, A.D. D'Ercole, W.J. Howley and D.A. Alia; U.S. Patent 3,359,123; Dec. 19, 1967; assigned to General Foods Corporation have found that unpeeled, whole potatoes may have their starch cells gelatinized to a swollen, but substantially unruptured state, by cooking under atmospheric conditions which avoid additional moisture pick-up by the potato as well as a deterioration or breakdown in the cellular structure of the potato and render the potato more amenable to freeze-thawing procedures. The whole, unpeeled, and partially gelatinized potatoes are then subjected to a freezing step wherein the potatoes are cooled to their freezing point to initiate freezing of the potatoes and maintained at this freezing point until the entire cellular structure of the potato is frozen into an expanded form which will become porous, and sponge-like in texture upon thawing from its frozen state. The frozen potatoes are then peeled, thawed and subjected to a dehydration step which reduces the moisture content of the potatoes to a stable level without disturbing the cellular structure or sponge-like texture of the potato product.

Large dehydrated pieces of potato such as whole potatoes and potato chunks as well as small slices may be rapidly rehydrated and reconstituted to a boiled potato texture and taste in water which is at boiling temperature or below boiling temperature, typically 180° to 212°F., in a matter of about 10 minutes.

The reconstituted product has a uniform boiled potato-like texture and flavor and is suitable for use in a wide variety of potato recipes. If the particular recipe requires further heating of the potato, the rehydration time can be reduced to a matter of minutes, say 3 to 7 min., and this rehydration may be carried out in cold or warm water.

In practicing this process raw unpeeled potatoes are washed, graded according to their size and then subjected to the first critical step, namely the cooking under limited moisture conditions. The controlled cook or partial gelatinization is conducted under heating conditions which avoid any pick-up of moisture during gelatinization. Essentially, cooking is carried out at elevated temperatures under noncondensing atmospheric moisture conditions which maintain the potato at a sufficiently high temperature to gelatinize the potato while avoiding any substantial condensation of water on the potatoes being cooked. For most operations the relative moisture of the heating air should therefore be below 100% or the saturation point of the heated air.

The unpeeled potatoes are heated preferably under substantially dry conditions, typically below 30% relative humidity, with a draft of air having a temperature of less than 200°F., say 175°F. During the heating cycle the internal potato temperature should never exceed 180°F. and should be preferably within the range of 155° to 175°F. during the partial gelatinization process. All the temperatures recited herein are dry bulb temperatures wherein the relative humidity is below 100%.

The controlled cooking and partial gelatinization should be performed under conditions which avoid moisture pick-up by the potato, and preferably under conditions where the potato loses a minor proportion of its original moisture content. Therefore, the potatoes are cooked in an unpeeled state, that is, with their skins or peel portions substantially unremoved. However, several of the outermost layers of the potato epidermis may be removed incident to a gentle abrasion during preliminary washing.

The optimum potato temperature during cooking is in the neighborhood of 170°F. During cooking the desired potato temperature should be retained for a sufficient period to effect a partially gelatinized state wherein the starch molecules assume a swollen but unruptured state, such as exists in a cooking period of 30 to 60 minutes, typically 45 minutes.

This cooking operation is further characterized by a gradual heat transfer accomplished by means of a controlled heat source which avoids condensation of water on the potato surface during the cooking. The heating means may take the form of a dry air oven, dry air forced-draft heating on a belt dryer, a vacuum-steam chamber, and other similar devices. Whatever the means employed, the cooking operation will essentially involve elevating the internal potato temperature to a requisite level, say 165° to 175°F., whereat partial gelatinization may proceed at a controlled rate. The gelatinization process then proceeds for about 30 to 60 minutes without exceeding the aforestated maximum internal potato temperature of 180°F. The cooking operation should avoid the excessive cooking effects created by the more conventional cooking operations involving the heating of potatoes in an excess of moisture, such as by immersing the whole potatoes in boiling water, or by cooking the whole potatoes with excess steam which condenses into free water on the potato surface.

The preferred method for carrying out this cooking operation involves subjecting the cold, washed, graded and substantially unpeeled potatoes to a forced draft of hot air, having a typical relative humidity of less than 40%, say 10 to 20%, in two separate stages.

In the first stage, air at a temperature in the neighborhood of 190°F. or slightly higher (depending upon the rate of circulation of the air, the capacity of the oven, the size of the oven, and the charge of the potato in the oven) is circulated about the potatoes for a period of time sufficient to elevate the temperature of the potatoes to the aforestated maximum internal potato temperature range of not more than 180°F. and preferably 160° to 170°F. Thereafter, the second phase of the potato cooking operation proceeds for a period of 30 to 60 minutes while keeping the temperature relatively constant, the internal potato temperature never exceeding 180°F.

After this cooking operation the whole potatoes with the skin portions thereon are subjected to a sponge-forming freeze operation which greatly increases the porosity of the potato. Freezing is accomplished by precooling the potatoes to 30°F. and then maintaining the potatoes at their freezing or crystallization plateau wherein the internal potato temperature is at 30°F. for a sufficient period of time to form a sponge-like porous product.

The sponge-like texture of the final potato product can be varied by changing the residence time at which the potatoes are kept at the freezing plateau. This period will vary from several minutes to several hours depending on the final texture desired in the end product and the size of the potato piece. For small slices of potato a holding period at the plateau of 15 to 60 minutes is suitable, for transverse potato slices or chunks of potatoes of five-eighths of an inch thickness 60 to 100 minutes is suitable, while for whole potatoes having a length of 3 to 6 inches one to eight hours is usually required.

Freezing may be carried out by a variety of methods; even immersion freezing, that is, immersing the potato in a refrigerated liquid or brine having a temperature of about -30°F. to +30°F. may be used with advantage.

After freezing the potato is subjected to a peeling operation. Peeling may be carried out by any conventional means, such as by abrasion, or brush peeling of the frozen potato. Brush peeling easily removes all of the frozen peel.

The peeled potato may now be subjected to a cutting operation while still in a frozen state or may be thawed prior to cutting. Thawing is preferably carried out by immersing the potato in room temperature water. The cut potatoes may be relatively large in size, or small in size. The pieces may be sliced, diced or cut into any desired form.

However, it is a particular feature of this process that potatoes may be cut into large pieces which may be effectively dehydrated and then reconstituted to a texture, flavor and quality similar to freshly cooked potatoes. The preliminary soak prior to cutting may of course include various additives such as antioxidants, texturizing and/or flavoring ingredients.

The potatoes are then preferably subjected to a prolonged soaking step in an aqueous liquid. The soaking step serves to improve the rehydration characteristics of the final product. This soaking results in only a moderate increase in the moisture content of the potato (say in the neighborhood of 5%). As a result of this soaking, residual ice crystals melt leaving a sponge-like texture of potato mass containing expanded potato tissue with accompanying channels for both the egress and ingress of moisture whereby dehydration as well as rehydration are facilitated. Incident to such thawing or soaking step will be the practice of employing a preservative or antioxidant treating agent, such as a sodium meta bisulfite or other soluble SO_2-containing salt solutions, inhibiting browning and development of off-flavors.

Ultimately the sponge-like potato product is subjected to dehydration under such temperature conditions as to avoid any extensive further gelatinization of the potato pieces, deterioration of the sponge-like cellular structure, or any undue charring of the product. Preferably, dehydration will be carried out in a forced draft hot air chamber which removes the moisture content of the product and carries it down to a stable moisture of below 10%.

The method of drying should be such that the structural identity of the potato is preserved. In the case of drying in air, it has been found that air temperatures of less than 190°F. should be employed. For whole potatoes drying air temperature of 170°F. for about 10 hours is suitable, for large potato pieces or chunks of potatoes a drying air temperature of 120°F. for about 14 hours is suitable.

The potato slices of five-eighths of an inch thickness may be reconstituted to an edible consistency in about 10 minutes in water at a temperature of 180° to 212°F. The reconstituted potatoes had a taste, texture, mouthfeel and other qualities similar to potato pieces prepared by the conventional household cooking method and were suitable for use in all recipes wherein boiled potato pieces are used. Potato pieces to be used in a recipe calling for further heat treatment, such as a potato casserole, were rehydrated for only 3 minutes in cold water prior to the final cooking operation while still giving a desirable potato texture.

Potato Crumbs with Baked Potato Flavor and Texture

It was the object of *A.I. Nelson, J.N. McGill and M.P. Steinberg; U.S. Patent 3,063,849; November 13, 1962; assigned to International Minerals & Chemical Corporation* to provide a dehydrated cooked potato product which may be reconstituted without mechanical admixing of the product with water to provide a baked potato. The process comprises:

   (1) Precooking potato slabs at a temperature of about 145° to 170°F. for a time period of from 15 to 45 minutes in water-containing calcium ions as the sole essential mineral ingredient;

   (2) Cooling the precooked potato slabs to a temperature below about about 60°F. with cold water essentially free of minerals other than calcium and maintaining the potato slabs at a temperature below about 60°F. for at least about 10 minutes;

   (3) Cooking the cooled precooked potato slabs with steam for a time period of from about 10 to 20 minutes;

   (4) Mashing the hot cooked potatoes;

   (5) Rapidly drying the mashed potatoes as a thin film on a heated surface to a moisture content of not more than about 5%; and

   (6) Crumbling the dried potato to provide a potato crumb not greater than about one-eighth of an inch in cross-section.

This process is carried out in the substantial absence of iron contamination. The dehydrated cooked potato material of this process is in the form of crumbs which may be mechanically admixed with hot water or milk to provide a mashed potato product, or which may be reconsititued to provide a product having the flavor and texture of a baked potato simply by adding milk and heating for a short period of time.

The utilization of demineralized water in the process insures a product having superior flavor. Iron apparently reacts with tannin or phenol products to produce complexes which tend to darken the product and impart a bitter taste thereto. In order to avoid this undesirable effect, the water employed throughout the process, including the final drying of the sliced raw potato, the precooking, the intermediate cooling and the final drying should be substantially free of iron.

Demineralized water most appropriately may be employed, but the demineralized water for the precook should contain added calcium ions to maintain the desired body in the final product. The use of totally demineralized water in the precook tends to provide a pasty product. Generally, the water will contain at least 10 ppm of calcium. Calcium gluconate advantageously can be added to the water in amounts preferably 200 ppm to provide about 35 ppm of calcium.

Example 1: Idaho russet potatoes were washed, peeled, trimmed, subjected to a second washing and finally were sliced in ⅝ inch slabs. The potatoes were subjected to a precooking for 30 minutes at 165°F. The amount of water employed was about equal to the weight of the potatoes. Following the precooking, the potatoes were cooled at 45°F. with cold water. The potatoes were maintained at 45°F. for about one hour and then were subjected to steam cooking at 210°F. for about 16 minutes. The potatoes, while hot, were mashed and dried on a double drum dryer to a moisture content of slightly less than 5%. Finally, the precooked potato flakes were crumbled through a standard ⅛ inch wire screen.

The above process was conducted utilizing stainless steel equipment and demineralized water containing an added 200 ppm of calcium gluconate. When reconstituted with boiling water or with a mixture of water and milk, the potato crumb product was characterized by good texture, flavor, and color.

Example 2: Small aluminum pouches of about 1½ by 3½ inches by 2 inches high were formed and about 14 grams of each of the dried crumb products of Example 1 were added to a pouch. The potato crumb product further was seasoned with about 0.5 gram of salt and about 0.1 gram of monosodium glutamate. 70 cc of cold milk was added to each of the pouches without stirring, the top of the pouch was folded to close the package and the package was placed in an oven preheated to 450°F. for 10 minutes. The product was hot and it exhibited the texture, flavor, and color of a baked potato.

Toasted Particles to Impart Baked Potato Flavor

A process for toasting potato particles before dehydration has been developed by *E.V. Kwiat and D.W. Andreas; U.S. Patent 3,495,994; February 17, 1970; assigned to The Pillsbury Company.* The toasted potato particles impart to the resultant whipped product a baked potato flavor.

Reconstituted and whipped products having a baked potato flavor are provided when the total dry potato solids of the dehydrated potato product contain from 5 to 30 weight percent toasted potatoes. The preferred range of toasted potatoes for imparting baked potato flavor and proper texture is about 5 to 15 weight percent.

Dehydrated, toasted white potatoes in contradistinction to untoasted, dehydrated white potatoes possess a characteristic color ranging from tan to dark brown. Such toasted potatoes are obtained from heat treatment of white, fleshed potatoes at elevated temperatures. For a particular level of texture, baked flavor, taste, etc., the prerequisite amount of toasted potatoes depends largely upon the degree of toasting imparted to the toasted potato product. Toasted potatoes ranging from a light to dark (e.g., heavy) toast may be employed.

The degree of toasting may be determined by a differential colorimeter or other analytical methods and apparatus. Suitable toasted potato products may be prepared by employing conventional potato flake producing drum dryers. In drum drying, previously cooked and mashed potato products are dried as a thin layer. However, for toasting, the drum drying operation is allowed to continue for a longer time interval and preferably at a higher temperature to provide the brown color development therein. Thereafter, the potato layer is scraped from the drum in the form of a toasted potato flake.

Another method of providing a toasted potato product adaptable herein is to toast uncooked, dehydrated potato cubes. In such a method, previously peeled potatoes are blanched in the conventional manner. The peeled, blanched potatoes are then cut into relatively small cubes (e.g., ⅜" by ⅜" by ¼") and subsequently dried to about 5% moisture level at a temperature generally ranging from about 100° to 180°F. The resultant dried potatoes are then toasted in a suitable oven at a temperature generally ranging from about 300° to 350°F. for a period of time sufficient to provide substantial toasting throughout the entire cube (e.g., 8 minutes). Toasting of the aforementioned cubes is completed upon the attainment of a cube surface temperature ranging from about 240° to 250°F. Excessive charring of the toasted product should be avoided (e.g., 260°F. cube surface temperature will provide an undesirable, charred, hard, brittle product). Excessive charring adversely effects both the flavor and texture characterisics thereof.

In order to prevent grittiness and nonuniform texture, it is preferred that at least 98% of the toasted product employed be less than 420 microns, 75% less than 300 microns with a major portion less than 200 microns.

The dehydrated potato products are prepared by toasting white fleshed potatoes (e.g., potatoes which in nature contain less than about 500 international units of vitamin A per pound), forming a potato mash by mixing toasted potato with untoasted potato in an aqueous medium to uniformly disperse the toasted potato therein and drying the resulting potato mash to provide a dehydrated potato product containing less than 10% moisture.

It has been discovered that by uniformly distributing the toasted potatoes along with the untoasted, cooked potato solids in an aqueous medium in the potato mash, processing and product uniformity is greatly enhanced thereby. By employing the toasted potato in the mash, subsequent dehydration thereof is facilitated (e.g., drum sticking during drying is greatly reduced).

By providing a potato mash comprising water, toasted potato product and untoasted potatoes, the toasted potato product is substantially and uniformly distributed throughout the mash. By preparing a potato mash and subsequently dehydrating the mash, the resultant dehydrated potato exhibits an improved resistance to flavor changes which generally occur when the product is stored over prolonged periods of time.

Since the toasted potato particles are firmly bonded in a matrix of untoasted dehydrated solid, product uniformity during packaging and shipping is maintained. Suitable untoasted potatoes include peeled raw potatoes, cooked mashed potatoes, blanched mashed and unmashed potatoes, dehydrated mashed potatoes such as potato flakes and granules, mixtures thereof and the like.

In general, depending upon the properties of the starting materials, a suitable mash comprises about 15 to 55% potato solids and about 45 to 85% water. Advantageously, the mash consists essentially of from about 20 to 30% by weight potato solids and from about 70 to 80 weight percent water. It should be apparent herein that the term water as used herein includes aqueous equivalents such as milk which is primarily comprised of water.

The potato mash may be prepared by various methods. An appropriate amount of toasted potato product can, for example, be added to unmashed, untoasted potatoes such as raw peeled potatoes and/or blanched and/or cooked potatoes as well as the untoasted mash thereof. The mashed product of such whole potatoes generally contains sufficient water to provide uniform distribution of the toasted product therein. When untoasted, dehydrated potatoes such as potato flakes and granules are employed, sufficient water should be added to insure uniform distribution of the toasted potato therein and bonding or adherence of the toasted portion to the untoasted portion (e.g., an untoasted, dehydrated potato matrix).

The potatoes made by this process can be prepared as flakes or granules by methods which have been described in previous patents. The following example will illustrate the process.

Example: (A) Preparation of the Toasted Potato Product — Field run potatoes (Kennebecs and Pontiacs) were lye peeled in a conventional manner, sliced into ⅜" thick slices and rinsed to remove free starch. One hundred pounds of the sliced potatoes were then sliced into cubes of about ⅜" by ⅜" by ¼", blanched 10 minutes in 180°F. antibrowning solution, and placed in an air dryer provided with a heated air flow maintained at 140°F. and dried therein to a moisture content of 5%. The dried cubes were then placed into a rotary tumble type roaster and toasted at a temperature of 340°F. for 8 minutes. Surface temperature (as measured by a thermocouple) of the product upon completion of the roasting was 245°F.

The toasted potato cubes were ground to provide a pulverized, toasted product having the following characteristics: 99.2% less than 420 microns; 73.6% less than 297 microns; and 53.6% less than 210 microns.

(B) Preparation of Dehydrated Potato Flakes Containing Toasted, Dehydrated Potatoes — Field run potatoes (Kennebecs and Pontiacs) were lye peeled in a conventional manner, sliced into ½" thick slices and rinsed to remove the free starch. In conventional cooking equipment, potatoes were cooked at a temperature of 170°F. for about 30 minutes and then steam cooked for an additional 30 minutes. The cooked potatoes were then transferred to a conventional potato ricer, mashed and then conducted through a screw conveyor 20 feet long and 6 inches in diameter. Ten parts by weight toasted potato product from the above (A) per 90 parts by weight dry untoasted potato solid was metered at a uniform rate into the screw conveyor. Sufficient glyceryl monostearate to provide 0.3% by weight in the dehydrated product was also introduced into the screw conveyor.

The conveyed potato product (i.e., the mash) had a 20% by weight solids content. Dehydration of the potatoes was accomplished by a typical commercial, hollow, single drum dryer with a roller diameter of 5 feet and length of 16 feet which was provided with a doctor blade and 4 feed rollers positioned parallel to the axis of a hollow dryer roller at a clearance of 0.50 inch. The hollow dryer roller surface temperature was maintained at 330°F. with a 5 to 8 second contact time. The resultant dried product had a thickness of 0.009 inch and a moisture content of 6 weight percent. Uniform distribution of the toasted potato product and mashed, cooked potatoes was accomplished by means of a screw conveyor and mastication of the mashed product on the drum dryer.

(C) Reconstitution and Whipping of the Resultant Product — Several portions of the dehydrated products thus prepared were reconstituted and whipped. One and one-quarter cups water, ½ teaspoon salt and 2 tablespoons butter were placed in a saucepan and heated to boiling, after which the heating was discontinued and ⅓ cup milk was added. Seventy grams of dehydrated potato product flakes were then added and stirred gently. The potato flakes after becoming soft and moist (30 seconds after addition) were whipped with a standard household fork.

The reconstituted and whipped potato products containing the combination of toasted and untoasted potatoes provided a stiff but yet fluffy whipped potato product without pastiness and heaviness. The whipped product had an improved appearance and texture and a flavor very similar to that of baked potato. Whiteness of the product was not substantially altered by the addition of the toasted potato.

It was also observed during the processing of the dehydrated product that the addition of the toasted potato flakes enhanced the drum drying thereof in that there was less tendency of the product to stick to the drum dryer. Storage tests at 72° and 100°F. indicated that the product was stable against flavor degradation even after 100 days of storage. Similar tests on potato flakes without the toasted potato therein indicated flavor degradation.

Potato Patty Mix

As is the case with many other foodstuffs, potato patties were first prepared as a means for disposing of potatoes which were left over after a meal. However, the flavor and certain aspects of the eating quality of potato patties were so well liked that they are deliberately prepared for use as the potato dish in a meal. However, it is inconvenient to peel potatoes, boil and mash them (usually with the addition of milk and butter), and then to use the mashed potatoes as a base for potato patties.

C.J. Bates and A.A. Andre; U.S. Patent 3,089,773; May 14, 1963; assigned to The Procter &

Gamble Company describe a process for preparing a culinary mix from which a light, fluffy potato patty product can be prepared with a minimum of effort.

The composition which enables one to prepare a potato patty product by the operation of merely adding water or other appropriate liquids, stirring into a batter, and ladling onto a heated surface comprises a flour component in which the characteristics are balanced to obtain the proper eating quality, a dried potato component properly balanced to obtain the necessary batter consistency, and a leavening system.

A combination of a hard wheat flour and a soft wheat flour must be used in the correct proportions. The ratio of soft wheat flour to hard wheat flour may be varied in a range from about 8 to 1 to 1 to 2, while still maintaining satisfactory product eating quality. The preferred ratio of soft wheat flour to hard wheat flour is from about 4:1 to about 1:1, with an optimum apparent at about 2:1.

All dried potatoes are rehydratable in hot water, that is, water at a temperature of 150°F. up to the boiling point. All dried potatoes are not, however, rehydratable in cold water. To obtain proper batter consistency a mixture of dried potato which does not rehydrate in cold water (hereinafter referred to as noncold water-rehydratable dried potato) and dried potato which will rehydrate in cold water (hereinafter referred to as cold water-rehydratable dried potato) is required.

To obtain the necessary batter consistency the ratio of noncold water-rehydratable dried potato to cold water-rehydratable dried potato should be in the range from 1:7 to 1:1, and preferably between about 1:6 and about 1:3. The amount of dried potato component present may be varied from about 25 to 65%, preferably about 30 to 45%.

The leavening system of this process should produce some gas as the batter is mixed to facilitate blending of the ingredients and to achieve some flow properties in the batter. The leavening system should also provide gas during the cooking operation to maintain a light, open structure. For this reason it is preferred to use a leavening system which combines features of both the fast-acting and slow-acting leavening systems. Such a combined leavening system is well-known in the art as a double-acting leavening system because it acts both in the batter and during the cooking operation. From about 2 to 4.5% by weight of a leavening system has been found to give the desired results.

It has been found helpful to add sucrose as a flavoring ingredient at a level of from 1½ to 4½% by weight of the mix composition. Sucrose also aids in producing a golden brown color in the fried patty. Salt at a level of from about 2 to 5% may be used for flavoring. For a low-sodium content product the salt should, of course, be omitted.

While salt and sucrose are generally used basic flavoring ingredients, other ingredients may be used to impart a positive flavor to the patties or to accentuate and highlight some of the flavors already present. For instance, dried onion may be present in the amount of from about 0.3 to 1.5% by weight of the mix. The use of about ½% by weight of dried onion has been found to produce a flavor which is well-liked by a large proportion of persons.

Example:

	Percent by Weight
Straight-grade soft wheat flour (average protein, 8.8%)	19.90
All-purpose hard winter wheat flour (average protein, 10.7%)	8.00
Cold water-rehydratable dried potato flakes	48.00
Noncold water-rehydratable dried potato granules	12.00
Sucrose	3.54
Fast-acting monocalcium phosphate-monohydrate	1.13
Slow-acting coated anhydrons monocalcium phosphate	1.13
Sodium bicarbonate	1.80
Salt	3.00
Dried onion	1.50

The flour, potato granules, sucrose, phosphates, soda, salt, and onion were blended in an upright planetary mixer. After the ingredients were thoroughly blended, they were placed in a ribbon mixer, the potato flakes were added, and blending was continued just long enough to incorporate the flakes homogeneously through the batch.

Sixty-eight grams of the mix was commingled with 140 grams of water and stirred until the ingredients were evenly blended to form a batter. One-quarter cup of batter was ladled onto a greased griddle, the griddle being maintained at medium heat (380°F.). The patty was cooked for about 1½ minutes until golden brown and then turned over and cooked for a like period to produce a golden brown color on the other side. The patty was light and fluffy, with a tender eating quality and pleasant potato flavor.

It will be noted that the flakes used in the above example were added last, after the other ingredients were blended. This is to minimize fracture of the flakes with the consequent release of starch from ruptured cells. As is known to those skilled in the art, the fracturing of potato flakes will tend to cause poor performance due to the aforementioned release of starch and the resultant sticky and gummy character of the rehydrated material.

In preparing the mix for cooking, water is added in an amount equal to between about 1.9 and 2.4 times the weight of the mix. The liquid used in constituting the batter may be milk or egg and milk. A richer patty product is produced if milk and an egg are used in preparation of the batter.

MISCELLANEOUS PROCESSES

Dehydration by Liquid Ammonia

M. Anderson; U.S. Patent 3,049,430; August 14, 1962 discovered foodstuffs of substantially solid consistency may be dehydrated by extraction with liquid ammonia. The process is operable at low temperatures and avoids irreversible thermal and hydrolytic degradation. While this process may be performed at temperatures lower than the boiling point of liquid ammonia (-33.4°C.), it is not at all necessary that the foodstuff be frozen prior to extraction, and the well-known possibility of damage to certain foodstuffs by freezing may thus be minimized or avoided.

The process is especially of value in the dehydration of potatoes. Raw potatoes are advisably treated in a comminuted condition, that is, shredded, sliced, chopped, etc. Cooked potatoes may be comminuted, mashed, or otherwise subdivided prior to dehydration. Dehydrated mashed potatoes prepared according to the process may be obtained in the form of a fine, free-flowing powder which may be instantly dehydrated upon the addition of water. The following examples are illustrative of the process.

Example 1: One hundred grams of peeled potatoes of approximately 78% water content are thinly sliced (approximately 1 mm. thickness) and immediately immersed in 1,000 grams of liquid ammonia. The mixture is stirred under a nitrogen atmosphere and is maintained below -50°C. by means of a Dry Ice-acetone bath. When analysis shows approximately 7% water dissolved in the ammonia, the potatoes are removed and the ammonia distilled off at the ambient temperature. Final traces are removed in a stream of inert gas and then under reduced pressure with gentle warming.

Example 2: To 100 grams of whole white potatoes, cooked by steaming and then mashed and cooled, are added with agitation 1,000 grams of liquid ammonia. The slurry is stirred for 30 minutes, the ammonia being maintained at its boiling point without external cooling. The potatoes are filtered by suction under an inert atmosphere, broken up into a powder, and dried. The details of this process may vary according to the particular foodstuff and the apparatus available.

Foam-Mat Process

A foam-mat process is described by *A.I. Morgan, Jr., L.F. Ginnette and R.P. Graham; U.S. Patent 3,031,313; April 24, 1962; assigned to U.S. Secretary of Agriculture* which is particularly adapted for the production of dehydrated potatoes. This product on mixing with hot water (or other hot edible liquid) instantly forms mashed potatoes of a desirable mealy texture like those prepared from freshly cooked potatoes.

Since the basic foam-mat process has been described in Chapter 1, a description will not be repeated here. In the case of potatoes, they are cooked and mashed, and the resultant pulp is diluted, where necessary, with water to the proper consistency and is then converted into a stable foam by incorporating therewith a minor proportion of a foam-stabilizing agent and a substantial volume of air or other gas. The foam so produced in the form of a relatively thin layer, or small extruded pieces, is then exposed at normal (atmospheric) pressure to a current of a hot gaseous medium until it is dehydrated.

The foam consists of a body of the pulp throughout which is interspersed a multitude of gas bubbles. The presence of the bubbles gives the foam a volume substantially greater than that of the pulp, per se.

During the dehydration step, the mass of foam retains this expanded volume with the result that the final product is a brittle, sponge-like, porous mass consisting of a matrix of solid fruit or vegetable material in which is interspersed a multitude of voids. This porous mass can be readily crushed without damage to individual cells to form a product in the form of porous particles. These particles on admixture with water or other edible liquid form a reconstituted product free from lumps, grit or other nonrehydrated particles. The fact that the pulp is applied to the dehydration in the form of a foam and that the volume thereof is essentially maintained during dehydration are keys to the formation of the easily rehydrated porous product. Moreover, by such means the dehydration takes place rapidly and efficiently because moisture can diffuse readily out of the expanded mass.

Example: Potatoes were peeled and quartered. They were then blanched in steam and air at 190°F. for one hour. They were pressed through a 10 mesh screen and mixed with half their weight of water. This resulted in a thick paste. To 300 parts of this paste, one part of glyceryl monostearate was added in the form of a 20% alcoholic solution. This mixture was whipped at room temperature for five minutes in a double eggbeater mixer. The result was a stiff foam of 0.45 g./ml. density.

The foam was spread on a tray in a layer ⅛ inch thick. This layer was dried in a 160°F. air stream for 60 minutes. The resulting dry mat was pressed through a 10 mesh screen. This product was capable of rapid rehydration in hot water to form good quality mashed potatoes.

Potato Nuggets by Foam-Mat Process

A process is described by *J.H. Rainwater, R.G. Beck and L.H. Parks; U.S. Patent 3,407,080; October 22, 1968; assigned to American Potato Company* in which a potato slurry is foamed, the foam is extruded, the extruded foam pieces are coated with dry potato solids, and the coated pieces are dried to form a porous nugget of potato. The process may be more clearly understood by the following example.

A damage-free potato slurry of about 15 to 18% solids is first produced. Such a slurry may be produced directly from cooked potatoes. This slurry is homogenized and aerated to form a stable foam in a commercial machine such as an Oakes mixer using a monoglyceride stabilizer such as Myverol 1807 at a concentration of about 2% based on potato solids. The foam density should be about 0.60 g./ml. This foam is then extruded through a 0.010 in. thick plate with 3/32 inch orifice holes into a fluidized bed of dehydrated potato granules maintained at about 140° to 160°F. The granules may be the well-known product of the add-back process now in large scale commercial use and consist of almost entirely undamaged potato cells, dried to about 10% moisture or less and existing in the form of single detached cells or a minor fraction of clumps of a small number of cells. The granules average about 75 to 150 microns in size and coat the foam pieces which form as the foam drops onto the granule bed. After the pieces are formed and partially dried in the fluidized bed, they may be finish dried in the fluid bed or separated by screening from the fluid bed material and finish dried separately. Finish drying may be conducted with 175° to 225°F. air. The finished low-density nuggets consist of about 25 to 35% solids from the foam and 75 to 65% solids from the granule bed.

In appearance the nugget is relatively fragile, porous, and has a rough exterior formed by the visible potato cells and agglomerates. The center has one or more large voids depending on how much the walls of the particle have collapsed during drying.

The finished nugget of the above conditions is mostly −3+16 mesh with a density of about 0.25 g./ml. and is more easily used by the housewife in preparing mashed potatoes than are standard potato granules which have a density of about 0.9 g./ml.

When potato granules are rehydrated, it is necessary to add them while stirring into several volumes of liquid. Continuous stirring is required because the granules must be thoroughly wetted before appreciable swelling occurs. If this is not carefully conducted, dry balls of granules can be present in the mesh. With this nugget form of instant mashed potato, however, the proper amount of hot liquid just covers the particles and the large voids allow all of the cells to be wet rapidly. The result is that the proper amount of water may be easily determined and the product needs little or no stirring or whipping to re-hydrate to an excellent mashed potato.

Although the potato slurry for foaming is preferably made from freshly cooked potatoes, it could be made satisfactorily from any dried potato product from which a slurry of damage-free potato cells can be produced. The physical characteristics of the dried potato can be controlled within wide limits by controlling the solids content and density of the foam, by selecting an appropriate orifice size and shape and the orifice plate thickness of the extruder, and by regulating the temperature of the fluidized potato granule bed.

Crystal-Like Particles

One type of instant mashed potato consists of granules which are obtained by spraying hot mashed potatoes through a fine nozzle into a heated chamber. A second type of instant mashed potato is in the form of flakes, produced by dehydrating cooked mashed potatoes on a heated surface in a film substantially of unicellular thickness. The product so obtained consists of flakes of substantially unicellular thickness.

E.A.M. Asselbergs, H.A. Hamilton and P. Saidak; U.S. Patent 3,260,607; July 12, 1966; assigned to Canadian Patents and Development Limited, Canada have developed a process for providing an instant mashed potato having a physical structure differing from the granular and flake products of the prior art.

The process comprises forming a perforated layer of mashed potato and drying the layer on a heated surface to form crystal-like particles having an average thickness of from about 3 to 4 potato cell thicknesses. The potatoes are cut into pieces of such size that their cooking can be effected by exposure to an aqueous cooking medium at a temperature in the vicinity of the boiling point of water for a period of less than fifteen minutes, the cooked potatoes are mashed and formed into a continuous but perforated layer and dried on a heated surface to form crystal-like particles having an average thickness of from about 3 to 4 potato cell thicknesses.

Emphasis has been laid in the prior art on the use of a prolonged cooking period at rela-tively low temperatures. By contrast, for obtaining this product a relatively short cooking period is preferably used. Advantageously this is not more than 15 minutes and is prefer-ably about 7 to 10 minutes. The aqueous cooking medium which is used can be steam or water, the latter being preferred. A temperature of from 210° to 212°F. is generally used.

In order to ensure that cooking of the potatoes is effected in less than the prescribed preferable maximum cooking time of 15 minutes, it is necessary to cut the potatoes into pieces of relatively small cross-section. Conveniently these pieces can be in the form of french fry strips which generally have a cross-section of about ⅜ inch by ⅜ inch.

Advantageously the cooked pieces of potato are subjected to a fluffing procedure prior to being mashed. This involves causing circulation of air around the pieces of potato to disperse the moisture-laden microatmosphere surrounding each piece of potato and facilitate removal of moisture from the surface of the potato. The fluffing operation can conveniently be carried out by spreading the pieces of potato on a perforated screen and blowing air over them or the pieces of potato may be tumbled in a rotating wire basket.

The formation of the layer of mashed potato can be effected by forcing the mashed potato through any slot of suitable width. However, the preferred method of producing the layer is to pass the mashed potato through the nip between two roller members. Advantageously, the roller members take the form of rotatable heated drums, a layer of mashed potato then being formed on each of the drums, and drying of the layers of mashed potato is actually effected on the drum.

Experiments with several varieties of Sebago and Katahdin potatoes have shown that two perforated layers of mashed potato are formed (one on each drum) when mashed potato is passed through the nip between two rotating drums having a clearance between the drums of from about 0.003 to 0.004 inch and that an excellent instant mashed potato product is obtained by drying these layers.

The perforated layer of mashed potato which is formed and dried according to the process has an appearance resembling that of fine lace work and readily breaks up to give crystal-like particles of instant potato which resemble nothing so much as snow crystals. The product is of course not truly crystalline and the adjective crystal-like must be interpreted as referring to the visual appearance of the instant mashed potato rather than to its physical constitution. Because of its open texture the crystal-like product rehydrates readily and with a minimum of stirring. Thus the reconstituted product is uniformly moist and free from lumps which have absorbed less liquid than the remainder of the product. A grainy appearance is also avoided.

Example: Freshly dug potatoes of the Sebago variety grown in Prince Edward Island, were peeled in an abrasion peeler and trimmed to remove defects such as eyes and spots. They were then sliced into french fry portions. Cooking of the potatoes was then carried out in boiling water in a steam jacketed kettle of stainless steel with a weight ratio of water to potatoes of 1 to 1. The cooking time was 6.5 min. After draining off the water the potato strips were spread out on a perforated screen and air at room temperature blown over them. This had the effect of forcibly removing the moisture-saturated air surrounding the potato strips thereby effecting surface drying and fluffing of the potato strips to give a mealy appearance.

The potatoes were then fed without delay, and while still hot, to a conventional mashing machine, commonly called a ricer, comprising a hopper into which the potatoes were dropped and a conical screw conveyor which served to withdraw the potatoes from the base of the hopper and forced them through the perforations in a perforated conical screen closely surrounding the screw conveyor. In the formation of a commercial product suitable additives could be introduced at this point for ensuring a desired texture, flavor, color, stability and odor of the final product.

The mashed potato falling through the perforated conical screen of the mashing machine was fed to a double drum dryer, the rotating drums of which had a length of 7.75 inches and a diameter of 6 inches. Steam under 30 lbs. pressure was supplied to the interior of each of the drums. The clearance between the drums was between about 0.003 and 0.004 inch and the drums rotated at a speed of about 4 revolutions per minute. A continuous but perforated layer of mashed potato formed on the surface of each drum. In appearance the layer resembled a lace work or felted fibrous structure having a random orientation of the fibers. Knife scraper devices were used for removing the layers from the drums each of the layers having been in contact with the associated drum for a period of about ten seconds prior to its removal from the drum.

The layers of dried mashed potato thereby obtained were readily disintegrated by slight mechanical force into a mass of crystal-like particles resembling snow crystals. This product contained about 8% moisture. The moisture content could however be reduced below this value by operating in an environment the humidity of which was more carefully controlled than that used in this example.

The product was reconstituted by admixture with a suitable amount of water or milk or a milk-water mixture. Preferably the liquid employed for reconstitution is at a temperature somewhat above room temperature, for example about 180°F. The texture and flavor of the reconstituted product were found to be excellent.

Prevention of Browning

One disadvantage which has characterized certain dehydrated potato products is their uncertain and comparatively short shelf life. The shelf life of dehydrated mashed potatoes is measured by the amount of darkening or browning which is present in the product. A highly desirable product would be white after an extended period of time, while the undesirable product will become partially or completely brown upon storage. Browning is indicative of deterioration and is accompanied by undesirable odor and flavor changes in the dehydrated product.

R.C. Reeves and F. Hollis, Jr.; U.S. Patent 3,136,643; June 9, 1964; assigned to General Foods Corporation have developed a process which separates completely from the potato products cells which contain a majority of browning precursors (soluble reducing sugars and proteins) without substantial damage to the cell walls so that such potato cells can be employed in the production of dehydrated mashed potatoes.

This can be accomplished by classifying raw or cooked potatoes into a core fraction and peel fraction. By classifying is meant the division of potatoes into a core fraction containing the bulk of the core portion and a peel fraction containing the peel, eyes, rot and some of the core portion.

The peel fraction which is cooked before or after classification is mashed to produce discrete free potato cells, aggregates thereof, and particles of peel, eyes and rot. Water is then added to the mashed peel fraction to form a slurry and the slurry is then screened to remove the peel, eyes and rot.

The slurry of now clean potato cells and small aggregates thereof is dewatered thereby removing free starch, undesirable soluble reducing sugars and proteins. The dewatered slurry is combined with the core which was separately mashed, and dried in granular form or riced to form filaments and then dried.

The potatoes may be sliced and leached prior to cooking and classification. It has been noted that when potato slices are leached prior to cooking and classification considerable improvement in the storage stability of the dehydrated mashed potatoes is obtained. This is attributed to the leaching out of limited amounts of soluble reducing sugars and proteins from the potato slices. The leaching, although insufficient to remove the bulk of the soluble reducing sugars and proteins (browning precursors) from the potato portion most proximate to the peel (peel fraction), is sufficient to remove the bulk of the browning precursors from the core portion which naturally contains less browning precursors than the peel fraction.

Although classification can be effected before or after cooking, it is desirable to classify the potato slices subsequent to cooking so that the slices are cooked in their skins which results in a dehydrated mashed potato product with a highly preferable strong potato flavor. The potato flavor appears to develop in the peel portion of the potato during cooking as

potatoes peeled prior to cooking do not develop the flavor. If the potatoes are not classified until after cooking then a good part of the flavor which develops is retained in the
core fraction which is not subjected to the same process conditions as the peel fraction.

It is desirable in the course of producing a slurry to employ at least one part by weight of
water per part by weight of potato. The use of lesser amounts of water had a detrimental
effect upon the screening operation. The use of excessive amounts of water in slurrying,
e.g., above 10 parts by weight of water per part by weight of potato, is detrimental in that
an excess of potato flavor and/or solids is lost when the potatoes are dewatered.

It has also been noted that the potatoes prepared by earlier disclosed processes all tend to
deteriorate rapidly on storage and have a relatively short shelf life. However, this process
by exposing the largest possible potato cell surface area of the peel fraction (which contains
the bulk of the soluble reducing sugars and proteins in the potato) to the slurry water removes the maximum amount of reducing sugars and proteins from the cells and thereby
increases the shelf life of the potato product considerably and decreases the probability of
hay flavor developing.

Recovery of Mashed Potato Solids from Waste in the Dehydration Process

M.J. Willard; U.S. Patent 3,535,128; October 20, 1970 describes an apparatus for reclaiming
edible potato solids discarded as waste in the dehydration of mashed potatoes and for dehydrating such reclaimed potato solids. The apparatus is particularly well adapted for use
with a dehydrating device which produces dehydrated mashed potatoes in flake form.

The commercial process for drum-drying mashed potatoes to produce flakes has advantages
which make competition with other forms of dehydrating mashed potatoes difficult. This
process uses one or more spreader rolls whose first function is to deposit succeeding layers
of mashed potatoes onto a hot drum to increase the thickness of the dried flakes that will
later be removed from the drum. Thin flakes have a light bulk density, are expensive to
package, and are therefore less desirable.

The spreader rolls are disposed around the periphery of the drum with their axes parallel
to the axis of the drum to perform a second function. During the time the mashed potatoes
stay on the spreader rolls, defective material such as skin particles, rot, corky tissue, etc.
are progressively concentrated and transferred from an upper spreader roll toward the lower
spreader rolls. In the past the material on the last roll was periodically removed by an
operator with a scraper and was then discarded. Thus, the rolls serve as a separation device
for removing undesirable portions of the potato from the mash prior to dehydration.

It has become apparent that potato flakes produced according to this method are at a
competitive disadvantage compared to dehydrated mashed potatoes produced by other
methods. The main reasons are that the above described method entailed high labor costs
caused by the fact that an operator must be present to control the spreader rolls and, as
is frequently the case, must shovel the mash from the lower spreader rolls to the upper
spreader rolls because the mash tends to drop from one roll to the next.

Another and even more important problem is that the mashed potatoes held on the spreader
rolls are extensively exposed to the heat from the drying drum. As the mash is transferred
from the first spreader roll toward the last, an undesirable change in flavor and color of
the mashed potatoes occurs, which is accentuated as the exposure time is increased. During
storage of the dried product a change in flavor becomes increasingly apparent and makes
the product less palatable. Portions of the mash remain on the spreader rolls for up to
two hours before being deposited on the drying drum or before being finally discarded
together with the waste.

Lastly, the removal of the accumulated waste material from the last spreader roll results
in a loss of usable mashed potato solids. When the drum-dryer briefly described above is
operated in a normal manner the loss of edible and potentially recoverable mashed potato

solids averages about 5% of the weight of potato solids fed into the dehydrating device. Holding the mash on the spreader rolls longer to effect a more complete removal of potato solids results in greater amounts of defective material being transferred to the drying drum. These waste materials detract from the quality of the finished product and require expensive hand-picking for their removal. Simultaneously, if the mash is held longer on the spreader rolls to reduce losses of edible potato solids, the flavor changes become more apparent, resulting in a shorter shelf life of the finished product and in a generally lesser quality.

The apparatus comprises a collection means which receives the part of the mashed potatoes that has been discarded by the dehydration apparatus together with the waste. The discarded material is mixed with water in the collection means to form a slurry. The waste, such as rot and peels, is separated from the slurry. The potato solids left suspended in the slurry are dehydrated according to any one of several known methods including recirculating the solids through the dehydrating apparatus from which they were discarded. The following example illustrates the process.

Example: Idaho russet potatoes of a processor grade were converted into mashed potatoes ready for processing by first peeling them in steam at a pressure of 90 lbs./sq. in., washing the peeled potatoes to remove the peels, trimming the defective portions of the potato, and slicing the peeled and trimmed potatoes into slices of ½ inch thickness. The potatoes were thereafter washed to remove free starch and precooked at a temperature of 160°F. for twenty minutes. They were cooled in tap water for approximately twenty minutes to modify the starch structure, whereby the internal temperature of the potatoes fell to 70°F.

The precooked and cooled potatoes were finish-cooked for 30 minutes in steam at a temperature of 204°F. and, immediately following the final cooking, were mashed through smooth one-half inch diameter cylindrical rods spaced three-eighths inch apart. At this point, chemical additives were added and they were present in the final dehydrated mashed potatoes in the following amounts: 0.5% monoglyceride emulsifier (Myverol 18–07); 0.1% sodium acid pyrophosphate; 250 ppm sulfur dioxide (from sodium bisulfite); and 50 ppm BHA.

This mash was combined with filtered and recovered potato mix in the quantities shown below, and was drum-dried at a steam pressure of 125 lbs./sq. in. (350°F.) and at a drum speed of 3 rpm to obtain potato flakes with a final moisture content of 6%.

Automatic roll scrapers were timed to discharge 20% of the mash fed to the drum into the collection trough where the discharged mash was combined with water and mixed by a rotating screw that simultaneously conveyed the mixture out of the trough. The added water was a mixture of ½ fresh water and ½ filtrate water recovered at the filter. The slurry obtained in the collection trough contained 12.5% potato solids. It was then passed through a cylindrical sieve provided with 0.045 inch openings in which a standard paddle rotated at 50 rpm.

The paddle blades were set one-eighth inch away from the inside surface of the screen. The potato waste materials were screened and discarded and the slurry, now containing water and potato solids, was conveyed to a continuous rotary vacuum filter operating at a vacuum of 15 inches mercury to yield a potato solids mix of a thickness of about one-quarter of an inch and a solids content of 18%.

A screw conveyed the potato solids mix to where it was discharged onto and mixed with the potato mash flowing toward the drum-dryer. One-half of the filtrate water was discarded, the remainder was mixed with an equal amount of fresh water and used for subsequent dilution of the discarded potato mash in the collection trough. The overall recovery of potato solids amounted to 98% of the potato solids fed to the drum-dryer.

The quality of the finished potato flakes obtained on the single drum-dryer was superior in every respect to flakes made from the same raw material by standard procedures. The quality differential was especially apparent after the obtained mashed potato flakes had been stored for 9 months at room temperature. Flakes produced according to the conventional flaking method showed distinct oxidative rancidity type flavors. The taste of potato flakes produced according to this process after they had been stored for 9 months was practically indistinguishable from the taste of freshly produced potato flakes. Complete drawings and a detailed description of the apparatus are contained in the patent.

Additions to Improve Flavor

A flavor-enhancing agent derived from a plant of the Cruiciferae family is added to the potatoes prior to cooking to provide an enhanced natural potato flavor in the process developed by *A.L. Liepa; U.S. Patent 3,594,187; July 20, 1971; assigned to The Procter & Gamble Company.*

A wide variety of potato products having improved flavor obtained by the addition of a flavor-enhancing agent selected from plants of the Cruciferae family can be made in accordance with this process by cooking a mixture of dehydrated potatoes and water. Examples of such products are: mashed potatoes prepared by adding water or milk to dehydrated potatoes to form a mixture of the desired consistency, stirring or whipping the mixture and then heating; potato soup prepared by cooking a liquid mixture of dehydrated potatoes and water; and french fries, pancakes, or potato chips prepared by baking or frying a dough based on dehydrated potatoes and water.

Preferred products which are flavor improved by this process are fried potato products prepared from a dough based on dehydrated potatoes and water, i.e., a dough prepared by partially rehydrating dehydrated potatoes.

Preferred doughs utilized in this process are prepared from dehydrated potatoes which have a portion of their cells ruptured and thus contain free starch as indicated by a preferred iodine index of from 0.03 to about 6.

The dehydrated cooked potatoes (hereinafter dehydrated potatoes) used in the process can be either in flake, granular, or powdered form (potato flour). Dehydrated potato flakes typically have a moisture content of about 7% by weight and have their potato cells substantially intact with a minimum of free starch.

Dehydrated potatoes in granular form have a moisture content of about 6% by weight and are composed of substantially unicellular potato particles which have their cell walls intact and which are capable of passing through about a No. 60 to No. 80 sieve.

Potato flour is made by drying cooked mashed potatoes to a moisture level of about 6% by weight and grinding the dry product to a given particle size, generally from 70 to 180 microns. Unlike the dehydrated potato flakes and granules described above, however, potato flour is composed of substantially 100% ruptured potato cells.

Although the flavor of products abovedescribed is generally satisfactory, it is frequently considerably inferior to the same type of product prepared in the conventional manner using raw or cooked potatoes. However, improved flavor can be obtained by adding to the dehydrated potatoes a flavor-enhancing agent selected from plants of the Cruciferae family. This family of plants is sometimes referred to be the mustard family; preferred members are mustard, horseradish, rutabaga, and radish.

Of the plants of the Cruciferae family, dried, powdered horseradish root and mustard oil represent the most preferred flavor-enhancing agents for use herein. The flavor-enhancing agents are preferably added to the dehydrated potatoes in the amount of from 0.05 to 2% by weight of the potatoes, most preferably from 0.1 to 1% by weight, either in solid form, e.g., as a dry powder, in wet form, e.g., as ground or macerated roots, or in oil form.

When added at a preferred level of from 0.05 to 2% by weight, the inherent sharp flavor of the material is not noticeable and the effect is to enhance the natural flavor of the products to provide potato products having improved potato flavor. The flavor-enhancing agents can be added to the dehydrated potatoes after the dehydration process or at the time the potatoes are fully or partially rehydrated, or they can be added to the potatoes during the dehydration process at the stage wherein the potatoes are in the form of a wet mash prior to the drying step. The contribution of these flavor-enhancing agents to the flavor of the potato products involves the addition of the earthy character of raw, uncooked potatoes, which is normally lost in the dehydration process and as a result weakens the flavor of the resulting dehydrated potatoes and products prepared therefrom.

Example 1: 308 grams of dehydrated cooked potato flakes having an average reducing sugar content of about 0.8% are pulverized in a hammermill to provide pulverized flakes with ruptured cells and with the particles having a maximum size capable of passing through a No. 50 sieve. The pulverized flakes have an iodine index of about 0.04. The flakes contain 8.3% water, 1.1% lipid, and a total of about 0.2% of sodium acid pyrophosphate, sodium bisulfite, BHA, and BHT.

1.3 grams of lipid in the form of commercially available monoglycerides, diglycerides, and triglycerides are added to 192 grams of boiling water in a suitable vessel. The monoglycerides, diglycerides, and triglycerides are prepared by superglycerinating soybean oil to obtain a mixture of monoglycerides, diglycerides, and triglycerides comprising about 40% monoglycerides, about 40% diglycerides, and about 20% triglycerides, and having an iodine value of 65. The lipid is permitted to melt in the boiling water and 1.0 gram of dried, powdered horseradish is added to the water, after which the mixture is agitated by hand stirring for about 1 minute to completely disperse the added materials. The water solution thus contains 0.67% lipid by weight and 0.52% horseradish by weight.

The pulverized flakes are intimately intermixed with the above prepared boiling water solution in a Hobart Model C–100 vertical, planetary, paddle mixer by slowly adding the water solution to the pulverized dehydrated potatoes to provide a dough having a total moisture content of 43.1% which includes both the added water and the water present in the flakes. The combination is intimately blended at a mixer speed of 69 rpm for 4 minutes to completely rehydrate the potatoes and form a workable dough.

The total lipid content of the dough is 0.94% and the amount of horseradish added is 0.20%, based on the total weight of the dough. The amount of horseradish based on the weight of of potatoes is 0.3%.

The dough is at a temperature of about 115°F. and is passed between the rolls of a two-roll mill to provide a coherent, easy-to-handle dough sheet which has a thickness of 0.015 in. The dough is immediately cut into substantially elliptical pieces which have a major diameter of 3.0 inches and a minor diameter of 2.0 inches and the pieces are then deep-fat fried for 15 seconds in a cottonseed-based frying oil which is maintained at a temperature of 350°F. The resulting fried product is a tasty, crisp, chip-type food product which is then salted uniformly over one surface to provide a chip having 2% salt.

The taste, texture, color, eating quality and appearance of the potato chip product prepared according to this example closely resembles that of conventional potato chips made by frying sliced raw potatoes. When compared with a control product prepared by the same process and including the same ingredients as indicated above but without the addition of horseradish, the example chip with the added horseradish has a better potato chip flavor than the control chip.

Example 2: 120 grams of fresh whole milk, 15 grams of butter, 2.3 grams of salt, and 0.3 gram of dried powdered horseradish were placed in a mixing bowl and 360 grams of boiling water were added. The mixture was hand stirred for 1 minute to completely disperse the ingredients in the water. 93 grams of dehydrated potato flakes were added to the mixture and intimately and gently mixed therewith in a Hobart mixer for one minute at low speed until the potato flakes had become uniformly rehydrated to provide mashed potatoes of a uniform consistency. The mixture was at a temperature of 140°F.

The mashed potatoes so prepared were compared with hot mashed potatoes prepared by mashing 465 grams of freshly boiled U.S. No. 1 Idaho potatoes and adding thereto 120 g. of fresh whole milk, 15 grams of butter, and 2.3 grams of salt and intimately admixing the combination to provide a uniform consistency. When the mashed potatoes prepared by rehydrating dehydrated potato flakes together with added horseradish was compared with the mashed potatoes prepared from freshly boiled potatoes, a taste panel of four members judged the former product that included horseradish to be at least equal in flavor to the latter product (freshly prepared mashed potatoes).

Mashed potatoes prepared from dehydrated potatoes and without flavor additives have a lower flavor level than those mashed potatoes prepared from freshly boiled potatoes and thus the added horseradish improves the flavor of such dehydrated potato products.

COMPANY INDEX

The company names listed below are given exactly as they appear
in the patents, despite name changes, mergers and acquisitions
which have, at times, resulted in the revision of a company name.

INVENTOR INDEX

U.S. PATENT NUMBER INDEX

3,318,708 - 147	3,410,702 - 254	3,574,643 - 231
3,323,923 - 43	3,417,483 - 227	3,577,649 - 15
3,335,015 - 23	3,418,142 - 230	3,594,187 - 277
3,337,349 - 175	3,425,848 - 48	3,597,231 - 124
3,338,724 - 261	3,425,849 - 238	3,607,316 - 167
3,340,068 - 150	3,428,463 - 94	3,615,724 - 245
3,340,071 - 107	3,431,119 - 57	3,622,355 - 249
3,343,970 - 224	3,447,934 - 230	3,628,967 - 55
3,352,693 - 99	3,457,088 - 206	3,634,103 - 101
3,353,969 - 187	3,458,325 - 207	3,634,105 - 256
3,355,304 - 205	3,459,562 - 202	3,635,729 - 257
3,359,123 - 262	3,463,643 - 241	3,649,305 - 248
3,365,298 - 80	3,484,253 - 62	3,650,776 - 259
3,365,309 - 140	3,490,355 - 4	3,658,301 - 252
3,366,497 - 94	3,493,400 - 163	3,692,546 - 127
3,378,380 - 165	3,494,050 - 15	3,694,236 - 178
3,379,538 - 44	3,495,994 - 266	3,705,814 - 143
3,384,496 - 113	3,500,552 - 6	3,718,485 - 19
3,386,838 - 119	3,506,447 - 134	3,723,133 - 78
3,388,998 - 154	3,510,313 - 151	3,725,087 - 260
3,391,009 - 96	3,511,671 - 50	3,728,131 - 122
3,394,012 - 184	3,516,836 - 131	3,728,134 - 168
3.396,036 - 251	3,535,127 - 115	3,738,848 - 173
3,399,062 - 244	3,535,128 - 275	3,743,513 - 105
3,404,012 - 186	3,565,636 - 203	3,758,313 - 106
3,407,080 - 271	3,567,469 - 53	3,752,677 - 179
3,408,209 - 60	3,573,937 - 120	3,761,282 - 170
3,408,210 - 18	3,573,938 - 46	3,764,716 - 208
3,409,999 - 3		

NOTICE

Nothing contained in this Review shall be construed to constitute a permission or recommendation to practice any invention covered by any patent without a license from the patent owners. Further, neither the author nor the publisher assumes any liability with respect to the use of, or for damages resulting from the use of, any information, apparatus, method or process described in this Review.

DESICCANTS AND HUMECTANTS 1973

by R. W. James

Chemical Technology Review No. 7

Such chemicals are routinely used in a wide variety of commercial applications to reduce or increase the moisture content of gases, liquids, and solids. Their mode of action is based on the following principles:

1. Chemical reactions resulting in absorption or release of water vapor.
2. Physical absorption and desorption (the process of removing water by superior attraction from a solid on which it is adsorbed accidentally).
3. Regeneration by heating.

Not all drying agents can be regenerated, but most of the newer substances are amenable to this practice. Silica gels have the highest capacity among the solids, taking up to 40% water, which is released on heating. Molecular sieves are effective even at high temperatures, yet can be regenerated in an oven. Activated alumina can be made very porous and is extremely effective for drying gases.

This book describes 121 processes from the patent literature for all types of drying agents. Several hundred different compounds and their applications are discussed. A partial and condensed table of contents follows. Numbers in ( ) indicate the number of processes per topic.

ISBN 0-8155-0483-7   313 pages

SOFT DRINK MANUFACTURE 1973

by M. T. Gillies

Food Technology Review No. 8

In the manufacture of many of the so-called soft drink beverages, the effectiveness of carbonation is a main factor in determining the quality and consumer-acceptance of the final beverage. A prolonged visible and sparkling effervescence is sought after to produce the "soda" taste which is considered desirable in such a drink. Rapidly fizzing products, which upon opening of the container, suffer an almost immediate loss of available carbon dioxide, quickly disappear from this highly competitive soft drink market.

Carbonation techniques must necessarily be changed, when other factors, such as sweeteners, flavors, or other ingredient formulations have to be changed.

The market for carbonated beverages has increased at a dramatic rate over the past few years. It is estimated that the U.S. market alone is on the order of four billion dollars annually. While carbonation of beverages at the point of consumption, as at soda fountains, or from dry mixtures in the home, has been practiced for years, the bulk of the market is served almost exclusively by beverages which are precarbonated and then packaged in bottles or cans. However, an increase of sales in the dry beverage mix field has occurred lately, so these are also included in this book.

The soft drink beverage industry has in reality two separate technologies in that it is made up of both carbonated and noncarbonated drinks. Since both facets of the industry command an enormous consumer market, the problems inherent in each phase and the means for their resolution are covered in considerable detail in this book.

About 130 distinctly different processes are described. A partial and condensed table of contents follows. Numbers in ( ) indicate the number of processes per topic. Chapter headings are given, followed by examples of important subtitles.

1. APPARATUS (12)
Improved Carbonation Systems
Continuously Operated Carbonation
   Apparatus
Using Solid Carbon Dioxide
Using Electromagnetic Energy
Carbonation by Supersonic Waves
Quick Dissolution of CO_2 by Liquid
CO_2-Saturated Liquid Mixed with
   Pressurized, Gaseous CO_2
Drink Dispensing Machines

Improved Carbonator
Improved Reliability of Foam
   Production
Root Beer and Birch Beer
Preparing Beverages of Various Flavors
Automated Dispensers

2. CARBONATED BEVERAGES (21)
Dry Mixes
Stabilization by Gums
Producing Protracted Effervescence
Sustained Carbonation by Controlling
   Ingredient Particle Size
Coating Acids and Carbonates with
   Dextrin
Use of Carbonic Anhydride
Adsorption of CO_2 by Starches
Improvement of Cola Flavor
Addition of Polyvinylpyrrolidone
Concentrated Beverage Bases
Foam Improvements
Better Carbonation Retention
Carbonation by Yeasts

3. ACIDULANTS (27)
Adipic and Fumaric Acids
Use of Sorbitan and Sulfosuccinate
   Esters
Use of Chelating Agents
Dispersing Acids in Dextrose Malts
Low Calorie Agglomerates
Phenylphosphoric Acid
Other Acidulants

4. CITRUS DRINKS — ADDITIVES & PROCESSES (38)
Sweetening Agents
Ammoniated Glycyrrhizin Synergist
Use of Maltol
Aspartic Acid + Dipeptide Lower
   Alkyl Esters
Coloring Agents
Water-Dispersible Carotenoids
Clouding Agents
Use of Brominated Oils
Fats and Water-Soluble Gums
Pulping Agents
Flavors and Clarifiers
Elimination of Ringing

5. SLUSH DRINKS (10)
Carbonating in the Cup

6. VARIOUS OTHER BEVERAGES (12)
Fruit and Milk Drinks
Soybean Drinks
Use of Licorice
Colors for Cola Drinks
Supersaturated CO_2 Solutions

7. CONTAINERS AND PACKAGING (7)

ISBN 0-8155-0507-8   336 pages

FEEDS FOR LIVESTOCK, POULTRY AND PETS 1973

by M. H. Gutcho

Food Technology Review No. 6

Food fads, health claims and high prices notwithstanding, meat remains Western man's most preferred source of high quality protein.

The trend in modern livestock production is toward rearing animals with a rapid rate of growth. Feed efficiency in poultry and cattle has been increased 50%, and 40% in swine, with a corresponding weight gain, over what was considered normal 15 years ago.

Since estrogens and antibiotics are no longer accepted components of feeds, new accessory factors are of great importance. Many processes in this book are concerned with just such growth-promoting compositions, making the natural nutrients easier to digest and to assimilate.

Among the fastest selling animal feeds in a gaining market are those for pets. In the U.S. a pet population of 55 million dogs and cats eats 3 million net tons of food every year. Formulations for dog and cat food are therefore included in this book, as are some for birds, tropical fish, fur bearers, and laboratory animals.

The book describes 396 patent-based processes and formulations. A partial and condensed table of contents follows. Numbers in ( ) indicate the number of processes per topic.

ISBN 0-8155-0496

389 pages

EDIBLE COATINGS
AND SOLUBLE PACKAGING 1973

by R. Daniels

Food Technology Review No. 3

Edible and soluble packaging can be used to
— retard moisture transmission
— serve as a barrier to oxygen or other gases
— prevent loss of volatile components and flavor
— resist transfer of oils and grease
— serve as a barrier to bacterial invasion
— provide protection from mechanical injury — or at least minimize crumbling and breakage
— isolate reactive ingredients from each other.

The processes in this book are grouped according to the food products to which the edible or soluble coatings are applied. The subjects of edible and soluble capsules for medicines and of casings for sausages are so vast that they are treated in separate volumes.

118 patent-based processes. A partial and condensed table of contents follows. Numbers in ( ) indicate the number of patents per topic. Chapter headings are given, followed by examples of important subtitles.

ISBN 0-8155-0475-6

360 pages